MECHANISMS OF ENVIRONMENTAL MUTAGENESIS-CARCINOGENESIS

MECHANISMS OF ENVIRONMENTAL MUTAGENESIS-CARCINOGENESIS

Edited by

A. Kappas

Institute of Biology
National Research Center "Democritus"
Athens, Greece

SPRINGER SCIENCE+BUSINESS MEDIA, LLC

Library of Congress Cataloging-in-Publication Data

European Environmental Mutagen Society. Meeting (19th : 1989 :
  Rhodes, Greece)
    Mechanisms of environmental mutagenesis-carcinogenesis / edited by
  A. Kappas.
        p.   cm.
    "Proceedings of the 19th Annual Meeting of the European
  Environmental Mutagen Society ... held October 21-26, 1989, in
  Rhodes, Greece"--T.p. verso.
    Includes bibliographical references and index.
    ISBN 978-1-4613-6698-0   ISBN 978-1-4615-3808-0 (eBook)
    DOI 10.1007/978-1-4615-3808-0
    1. Genetic toxicology--Congresses. 2. Mutagenesis--Congresses.
  3. Carcinogenesis--Congresses. I. Kappas, A. II. Title.
    [DNLM: 1. Carcinogens, Environmental--pharmacology--congresses.
  2. Cell Division--genetics--congresses. 3. Chromosome
  Abnormalities--chemically induced--congresses. 4. Cytogenetics-
  -congresses. 5. Monitoring, Physiologic--congresses. 6. Mutagens-
  -pharmacology--congresses.   QZ 202 E89m 1989]
  RA1224.3.E93 1989
  616'.042--dc20
  DNLM/DLC
  for Library of Congress   90-14296
                                                                 CIP

Proceedings of the 19th Annual Meeting of the European Environmental
Mutagen Society, on Environmental Mutagens-Carcinogens,
held October 21-26, 1989, in Rhodes, Greece

ISBN 978-1-4613-6698-0

© 1990 Springer Science+Business Media New York
Originally published by Plenum Press, New York in 1990

PREFACE

The 19th annual meeting of the European Environmental Mutagen Society was held in Rhodes, Greece, from October 21st to 26th, 1989.

The programme was chosen to explore what is currently known about the mechanisms of mutagenesis and carcinogenesis, induced by environmental agents, and the questions regarding the relationship of these two processes. Recent findings, techniques and methodologies in the area of biomonitoring of humans exposed to environmental mutagens-carcinogens were presented and considerable attention was also paid to the aspects and issues of collaborative environmental policy. Researchers from all over the world contributed to the programme of the meeting with posters and oral presentations, providing a variety of new data and interesting scientific approaches.

A number of outstanding scientists were invited to present the results of their work. It is only their presentations which are included in this book, covering the following topics: Mutations and carcinogenesis; mechanisms of chemically-induced genetic effects on molecular, chromosomal and cell division level; adaptability and repair mechanisms; chemical carcinogenesis and oncogenes; structure and metabolism of mutagens-carcinogens; biomonitoring and epidemiology of humans exposed to environmental mutagens-carcinogens.

For the sake of evaluating and controlling the mutagenic and carcinogenic potential of our environment it is indispensable to understand the mechanisms and processes by which chemicals act on the genetic material, causing either hereditary disease or cancer. The publication of these proceedings will hopefully contribute to this task.

A. Kappas

ACKNOWLEDGEMENTS

Financial support for the Meeting was provided by the following organizations:

CIBA-GEIGY LTD, Basle, Switzerland
COMMISSION OF THE EUROPEAN COMMUNITIES, Brussels, Belgium
CREDIT BANK, Greece
HELLAFARM S.A., Athens, Greece
HELLENIC CANCER SOCIETY, Athens, Greece
ICI CENTRAL TOXICOLOGY LABORATORY, United Kingdom
MINISTRY OF INDUSTRY, ENERGY AND TECHNOLOGY, Greece
NATIONAL RESEARCH CENTER "DEMOCRITUS", Athens, Greece
OLYMPIC AIRWAYS, Greece
SANDOZ LTD, Basle, Switzerland
SHELL CHEMICALS HELLAS LTD, Greece
THE MAYOR OF THE CITY OF RHODES, Greece
WELCOME FOUNDATION LTD, Greece

CONTENTS

MUTATIONS AND CARCINOGENICITY

MECHANISMS OF CHEMICALLY-INDUCED GENETIC EFFECTS ON MOLECULAR, CHROMOSOMAL AND CELL DIVISION LEVEL

ADAPTABILITY AND REPAIR MECHANISMS

MUTATIONS AND CARCINOGENICITY

MUTATION SPECTRUM IN CARCINOGENICITY

Claes Ramel

Institute of Genetic and Cellular Toxicology
Wallenberglaboratory
University of Stockholm
S-106 91 Stockholm, Sweden

ABSTRACT

The correlation between mutagenicity and carcinogenicity has been a major approach in genetic toxicology. Although the critical importance of changes in DNA in cancer induction is beyond any reasonable doubts, the actual relationship between mutagenic and carcinogenic properties of chemicals is more complex than previously conceived. A primary reason for this complexity is the multistep nature of cancer induction, which implies both genetic and non genetic events. A wealth of data shows that more than one mutational event is generally required for tumour formation. In fact the progression stage of carcinogenicity implies a cascade of mutational events, indicating a stress induced instability of the genetic machinery, ending up in a variety of genetic lesions. The endpoints involved in cancer induction do not only include conventional mutations like point mutations and chromosomal aberrations, but also genetic changes, which are rarely taken into consideration in short-term assays for carcinogenicity. The genetic endpoints, involved or suspected to be involved in cancer induction, comprise insertion mutations, recombination events, gene amplification, methylation of 5-cytosin, mitochondrial mutations and different indirect mutagenic effects. The present paper focuses on these "unconventional" genetic endpoints and attempts to give an overview of their possible role and their mechanism of action in carcinogenicity. It is emphasized that the testing strategy for carcinogenicity has to take into account these genetic endpoints as well as the rapidly growing knowledge of the molecular mechanism behind neoplastic changes.

INTRODUCTION

I feel deeply honoured and grateful to have been elected for this prestigious award by the European Environmental Mutagen Society. I have chosen as a title of my presentation

---

*The author of this article was awarded the 1989 EEMS Award.

Mechanisms of Environmental Mutagenesis-Carcinogenesis, Edited by
A. Kappas, Plenum Press, New York, 1990

"Mutation spectrum in carcinogenicity" for several reasons.
Justified or not, it is a fact that genetic toxicology has
focused the attention on the relationship between mutagenicity
and carcinogenicity in order to identify environmental
chemicals, which may imply a carcinogenic hazard. The present
concept of the role of mutational changes in the development
of cancer furthermore is a suitable subject for some retros-
pective look on the development of this area of basic and
applied research and practical applications. I think that
this development constitutes an important chapter in the
history of natural sciences. But the connection between
mutation and cancer is and will also be in the future a
crucial area from both practical and theoretical viewpoints.
Relevant questions in this connection concern the present and
future position and direction of environmental mutagenesis and
genetic toxicology. The use of short-term tests for mutagenicity has been under particular scrutiny during the last few
years and it has to adapt to the rapid increase in the
knowledge of the mechanism of cancer induction in order to
survive.

     I will therefore take the opportunity not only to discuss
the actual scientific questions related to the mechanism of
carcinogenicity but also to look backward to the historical
background and forward to the future of environmental
mutagenesis with special consideration to cancer induction.

     Some important events in the development of genetic
toxicology and the relationship between chemical mutagenesis
and carcinogenesis is illustrated in Fig. 1. The genetic
hazards from chemicals did not become any generally recognized
issue until the 1960ies in spite of the fact that the ability
of chemicals to induce mutations has been demonstrated two
decades before by Auerbach, Robson and others. Previously the
issue of induced mutations and genetic hazard had been brought
up with Muller's discovery 1927 of radiation induced mutations
and almost all discussions of genetic hazards were confined to
this source of mutations. The atom bomb 1945 and the
subsequent use of nuclear power emphasized that approach to
the hazards of mutations.

     When chemical mutagenesis was brought into the discussion
of human genetic hazards the concern was mainly directed
towards effects on germ cells and hereditary hazards. This
was in fact the situation when the American and European
Environmental Mutagen Societies were founded 1969 and 1970
respectively.

     Chemical carcinogenesis dates back much further; it
usually is referred to the discovery of scrotum cancer among
chimney sweepers by Sir Percival Pott (1775). Other milestones
in that field were the recognition of multistep carcinogenicity by Berenblum (Berenblum, 1941; Berenblum and Shubik, 1947)
and the metabolic activation of chemical carcinogens in the
1960ies by the Miller's (see Miller and Miller, 1971).

     Gradually data accumulated, which pointed to a connection
between chemical mutagenicity and carcinogenicity and the two
areas became united in the beginning of the 1970ies. It was
manifested by the introduction of the term "genotoxic", which
Druckrey suggested at a conference in Stockholm 1972 (Ramel,

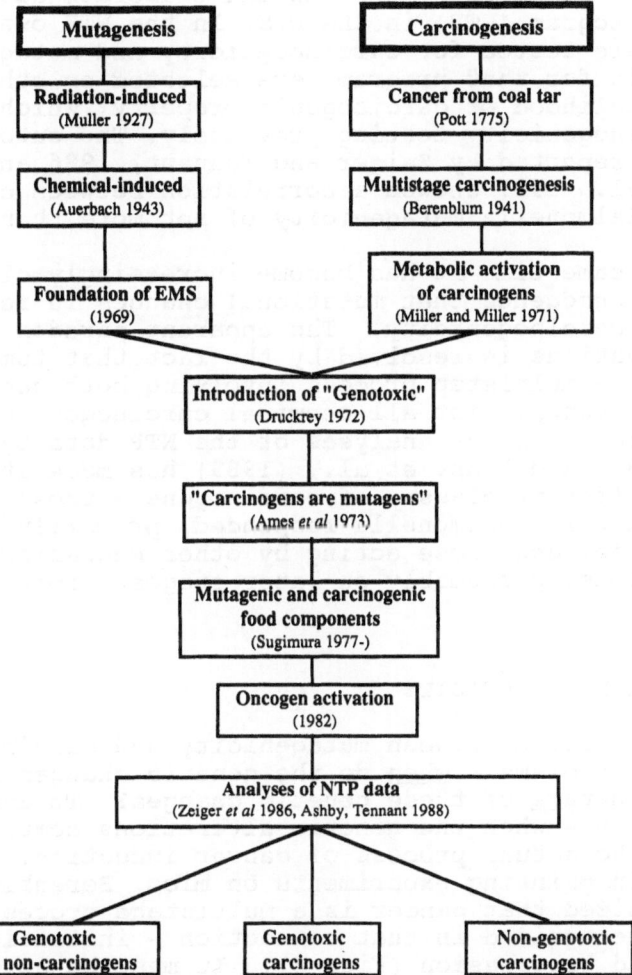

Fig. 1. An overview of the development of chemical mutagenesis and carcinogenesis.

1973). The introduction of the Salmonella microsomal test by
Ames 1973 in his paper with the provocative title "Carcinogens
are mutagens" (Ames et al., 1973) was another step in the same
direction. The subsequent research on chemical mutagenesis by
Ames, Sugimura and others indicated a very high correlation
between short term tests for mutagenicity and animal carcinogenicity. The hope that simple tests on bacteria would
provide a reliable method to identify carcinogenic chemicals
in our environment came to an end, however. In 1979 Rinkus and
Legator pointed out that some categories of chemical carcinogens did not exhibit any correlation with mutagenicity and this
was further emphasized by the new and elaborate National
Toxicology Program (NTP) in the U.S. In the NTP over 300 new
chemicals were tested for carcinogenicity and mutagenicity.
The chemicals for that program were selected on other criteria
than the likelihood of carcinogenic property, which had guided
animal carcinogenicity testing previously. The outcome of the
NTP work as reported by Zeiger and Tennant, 1986 and by
Tennant et al., 1987 showed a correlation between carcinogenicity and Salmonella mutagenicity of not more than about 50%.

At the same time it has become increasingly clear through
research on oncogenes that mutational changes in fact are
critical in carcinogenicity. The apparent paradox between
these observations is resolved by the fact that tumour
induction is a multistep process involving both genetic and
non genetic events. Not all chemical carcinogens act on
genetic steps. Further analyses of the NTP data by Ashby and
Tennant (1988) and Ashby et al., (1989) has made it clear that
there are different classes of carcinogens - those acting as
mutagens, to which Salmonella responded, primarily electroph-
ilic compounds, and those acting by other mechanisms than
point mutations, presumably on later stages, promotion and
progression.

MULTISTAGE CARCINOGENICITY

The connection between mutagenicity and carcinogenicity
implies two problems - when do the genetic changes occur and
what is the nature of these genetic changes? To answer the
first question - when the genetic alterations occur - we have
to look at the actual process of cancer induction. On the
basis of skin painting experiments on mice, Berenblum (op.
cit.) recognized that cancer is a multistage process. Three
stages are recognized in that connection - initiation,
promotion and progression (Fig. 2). It must be stressed,
however, that the course of events in cancer induction is not
rigidly determined. There are no precise dividing lines
between the stages and evidently there are wide variations in
the processes, depending on tissues, inducing agents, differentiation and so forth. It is, however, a useful model and
appropriate for our purpose to discuss the variety of genetic
and other mechanisms in cancer induction.

When it comes to the initiation of cancer a wealth of
data strongly indicates that it is an irreversible process
based on mutations. In accordance with the mutational nature
of initiation no threshold can be expected.

The promotion stage is more complex and the data indi-

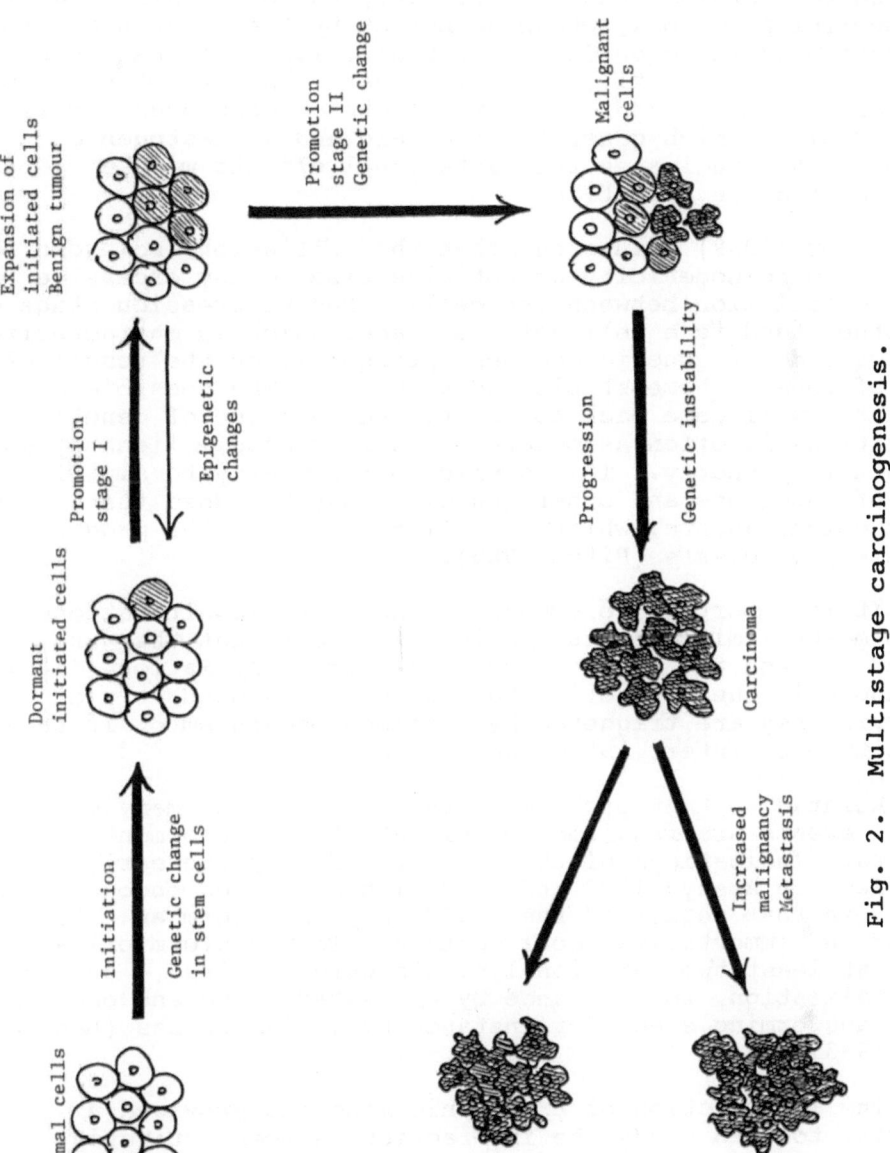

Fig. 2. Multistage carcinogenesis.

cates that more than one step is involved. In skin carcinogenesis the formation of benign tumours is a reversible and
evidently epigenetic process, which is caused by promoting
agents like phorbol esters, while the further development to
malignant tumours responds to mutagenic agents (Hennings et
al., 1983). At the molecular level the classical promoting
agents such as phorbol esters act on cell proliferation by
binding to protein kinase C. However, there are also examples
of powerful promotors, which do not ct by that mechanism, for
instance okadaic acid (Suganuma et al., 1988). It has been
pointed out by Cerutti (1985) and others that promoting agents
also are characterized by their effects on membranes and lipid
peroxidation, which generates radicals and a clastogenic
factor. This indicates that mutational effects may be
involved also in promotion.

Pitot (1989) points out that the initiation and promotion
steps in carcinogenicity do not give rise to any excessive
genetic variation between the cells. The progression stage on
the other hand is a well defined stage, which is characterized
by a cascade of genetic changes, presumably as the result of a
loss of genetic homeostasis and stability. The possible cause
of that I will come back to later. An increase of genetic
alterations is often associated with increased malignancy and
metastatic property. This applies particularly to amplification of oncogenes and other genes. It may be added that there
are chemical agents, which specifically act at the progressive
stage - progressers (Pitot 1989).

It is important to emphasize that genetic alterations are
by no means limited to initiation of cancer, but they are
spread through the entire process of tumour formation. It is
not known if these genetic changes occur independently of each
other, if they are triggered by a common mechanism or if they
constitute an interrelated sequence of events.

Related to that problem is the question how many mutational events are required to initiate the development of
tumours. Indications of that was provided by the early
transfection assays 1982 and 1983 with activated oncogenes. It
was shown that activated ras could only transform an established and immortalized cell culture. To transform primary
cells at least two mutational events were required — one for
immortalization, for instance by activated c-myc and one for
the transforming step, for instance by activated ras (Land et
al., 1983).

The introduction of transgenic mice has provided a
powerful tool to study the interaction between activated
oncogenes to cause cancer _in vivo_. The technique of
transgenic animals implies that DNA is injected in fertilized
eggs, which are then reimplantated into females and allowed to
develop. Often a high frequency of the injected eggs incorporate the DNA in the genome and develop a new generation with
this hybrid DNA piece actively transcribed.

The technique has developed rapidly and several research
groups have studied transgenic mice with activated oncogenes.
Leder's group in the U.S. found that transgenic mice with
activated ras or myc oncogenes developed hyperplasia, but
incorporated together they gave rise to carcinomas (Sinn et

al., 1987). The fact that not all cells became carcinogenic even with both those activated oncogenes suggested that a third event was necessary.

There are a number of other observations, which also show that more than one mutational event is necessary for the development of tumours. An overview of some relevant data in that connection is given in Table 1. Several experiments, particularly with transgenic mice, show that activated oncogenes give rise to monoclonal tumours, indicating that the activation of oncogenes is not sufficient for all cells in a tissue to develop tumours. One exception has been reported by Leder's group (Muller et al., 1988) on transgenic mice with activated c-neu, which gave rise to polyclonal adenocarcinoma, which affected all cells in glandular epithelium, indicating that one event in fact might have been sufficient. However with another strain of mice Bouchard et al., (1989) found that activated c-neu was not sufficient to induce adenocarcinoma.

Illustrative examples of genetically multifactorial carcinogenesis is provided by human colorectal carcinomas. The development of carcinoma proceeds via several steps of adenomas. Vogelstein et al., (1988) showed that this process implies a gradual increase of genetic lesions, including activation of ras and deletions in the chromosomes 5, 17 and 18.

It may be added that nuclear interactions between the products of oncogenes and suppressors of oncogenes pave the way for interactions between mutations in oncogenes and antioncogenes. Interesting examples are the cooperation of FOS and JUN oncoproteins to form a transcription factor and the inactivation of the retinoblastoma gene product by binding to the products of specific oncogenes (Green, 1989).

From these observations one can draw the conclusion that more than one mutation is required for the development of tumours - at least for most cancer forms.

MUTAGENIC CHANGES IN CARCINOGENICITY

Before entering the question of the nature of mutagenic changes in carcinogenicity, it should be emphasized that the concept of both spontaneous and induced mutations has changed considerably during the last two decades. The reason for that is that DNA has proved to be far more unstable and dynamic than one had anticipated previously. Even Crick's "central dogma", which stated that the flow of information was DNA-RNA-protein, has turned out to be inadequate. It is now well known that RNA can be transcribed into DNA by means of reverse transcriptase and the question has even arisen whether there may be a control from the protein end. The issue has been brought up by some observations of directed mutagenesis. Cairns et al., (1988) reported mutagenicity studies on bacteria, which indicated directed mutagenesis through the contact with the substrate. Related observations were reported by Hall (1988) on metabolism in bacteria of the beta-glucoside suger salicin, which required two events - one point mutation and one excision of an insertion. No relative advantage of either of these alterations could be found. The

Table 1. Multifactorial Carcinogenesis

Assay system	Genetic alterations oncogenes activations	Induction of cancer	Comments	Author
Transgenic mice	ras myc ras + myc	- - +	Polyclonal adenocarcinoma	Sinn et al. (1987) " "
Transgenic mice	neu	+		Muller et al. (1988)
Transgenic mice	neu myc ras Int-1	(+) - -	Monoclonal adenocarcinoma	Bouchard et al. (1989)
Reconstituted prostate (mouse)	ras myc ras + myc	- - +	Monoclonal induction	Thompson et al. (1989) " "
Mouse skin test	ras	-	Papilloma	Aldaz et al. (1988)
Induction of rat mammary	ras		Activated ras independent of hormone response	Sukumar et al. (1988)
Human colon cancer	Chromosomal deletion (CD) Familiar polyposis (FPC) CD + FPC	- - +		Okamoto et al. (1988) " "
Human colon cancer	ras		High frequency activated ras in premalignant polypes	Forrester et al. (1987)
Human colon cancer	Chromosomal deletions ras		Successive increase of genetic alterations during tumor development	Vogelstein et al. (1988)

chance for the occurrence of both alterations was about 10^{-20}. In spite of that extremely unlikely event double mutants were recorded in bacteria exposed to salicin.

Other illustrations of the dynamics of DNA are the occurrence of restriction enzymes, mobile DNA elements and the various forms of genetic recombination and methylation of cytosin residues in DNA.

As a result of this instability and dynamics of DNA the spectrum of genetic alterations is wider than the once counted with in mutation research twenty years ago. The old concept of DNA as a highly stable entity only subjected to rare stochastic mutational processes has guided much of the practical application of mutation research, such as the use of short-term tests. Practically all the current assay systems are focused on "conventional" genetic alterations, point mutations, chromosomal aberrations and nondisjunction. But there are important additions to be made and I have compiled a list of such genetic alterations (Table 2). Most of them have in fact been demonstrated in connection with cancer induction and others can be suspected to be involved and may even be of critical importance in some cancer forms. The conventional mutations dealt with in standard short-term assays only comprise the upper part of the list.

I will concentrate my discussion on those more unconventional genetic endpoints we rarely deal with in short-term testing. However, before I get into that subject I would like to comment on the mutation spectrum in general, as it is of relevance for short-term testing. Thanks to modern DNA technology it has been possible to classify mutations at the nucleotide level. A remarkable variation in mutagenicity spectrum has then been found, both for spontaneous and induced mutations. Thus three of the most studied mammalian genes -aprt, hprt and TK have a quite different mutational spectrum (Breimer 1988). Both spontaneous and X-ray induced mutations are primarily base substitutions in the aprt gene, while hprt exhibits a much higher frequency of deletions. TK mutations are practically only deletions. It can thus be concluded that mutagenicity studies of single loci can not reflect the variation in response between loci.

TRANSPOSITIONS

The formation of tumours implies a cascade of genetic alterations, the background of which can be suspected to be stress-induced, although the actual mechanism is not known. Many of these genetic changes are of an "unconventional" type.

One category of genetic changes, which should be discussed in this context is "jumping genes" - transpositions and insertion mutations. Various kinds of such mobile elements occur throughout the whole organism world, but the contribution to mutations apparently is very different between organisms. In Drosophila about 50% of spontaneous mutations are due to insertion of mobile DNA elements (Rubin 1983). The advanced genetic analysis with Drosophila has enabled extensive studies of transposing elements. There is a whole series of mobile elements, which can be roughly divided in three groups:

11

Point mutations
- Base substitution
- Frameshift

Deletion

Translocation

Inversion

Sister chromatid exchange

Non-disjunction

Insertion mutation, "jumping gene"
- DNA transposon
- Retrovirus-like transposon. Retrofection

Recombination
- Somatic crossing over
- Gene conversion
- Sister chromatid unequal exchange   - Amplification
- Unequal somatic crossing over   - Amplification

Disproportional replication of DNA   - Amplification

Methylation of cytosin. Changes of gene expression

Mitochondrial mutations

Indirect mutagenesis
- Imbalance of nucleotide pool
- Secondary formation of oxygen radicals
- Inactivation of defence mechanisms against free radicals
- Endogen formation of DNA adducts

  - Copia and copia-like transposons, which are almost
    identical in organization with retrovirus and which
    transpose by means of RNA intermediate and reverse
    transcriptase.

  - P-element, containing a transposase and interacting
    with a cytoplasmic factor.

  - Fold back elements with long inverted repeats at the
    ends. These transposing elements can be very complex,
    containing more than one coding gene.

    Transpositions of mobile elements are affected by both
external and internal factors, which do not interact with DNA
directly. The now classical example of that is Mc Clintock's
recognition of "genomic stress", which leads to her discovery
of jumping genes and the mobilization of insertion elements.

In Drosophila copia and other retroviruslike elements are mobilized by stress situations. An interesting example has been reported by Biemont et al., 1987. They followed the distribution of copia and another similar element, mdg-1, in the salivary glands in 17 highly inbred lines for 70 generations. Most lines showed a stable pattern of distribution of the mobile elements, but in one line a complete reshuffling of copia occurred during one or two generations.

It was suggested that the mobilization of copia was caused by a stress situation created by the inbreeding and that the response may even imply an adaptation to inbreeding. A mobilization of mobile elements would be a most rapid way of acquiring heterozygocity. A burst of genetic instability has also been recorded in plants after inbreeding (Marx, 1984).

A variety of other stress situations can induce mobilization of transposing elements in plants. This applies for instance to viral infection. RNA mosaic virus in maize caused an increased rate of mutations in the bronze locus through mobilization and insertion of transposing elements (Johns et al., 1985). Gershenson reported 1986 that the injection of DNA and RNA virus in Drosophila increased the mutation rate in certain loci 10 000 times. Unfortunately no further analysis of such mutations has been done to my knowledge, but it is tempting to speculate that mobile elements were involved.

It can a priori be suspected that the cellular environment in developing tumours implies a stress situation, which would be favourable for mobilization of insertion elements. However, very little data is available from mammalian cells in general with respect to mobile elements. No indications of insertion mutations by mobile elements have been obtained from mutagenicity studies of standard mammalian loci such as hprt, aprt and TK. It is however quite possible that we do not have sufficiently effective systems to detect insertion mutations in mammals. There are however some suggestive data from neoplastic cells, as summarized in Table 3.

Kirschmeyer et al., (1982) found that transformed C3H 10T1/2 cells contain RNAs with long terminal repeats (LTR), which are characteristic of retroviral type of transposons.

Morse et al., (1988) found in a human breast tumour an insertion in the c-myc protooncogene of LINE 1 - a long repetitive DNA with all the characteristics of a mobile element. There are two other observations of insertion of LINE in c-myc in cancer cells - one in a canine transmissible tumour and one in a rat immunocytoma (see Table 4).

An insertion of the retrovirus-like A-particle has been observed in the protooncogene c-mos (Recahvi et al., 1983).

In conclusion it can be said that insertion of mobile elements do take place also in mammals. Of particular importance in the present context is the fact that this occurs in protooncogenes. This does not prove that insertion mutations play a role in carcinogenicity, but it certainly is suggestive.

Table 3. Observations Indicating Insertion Mutagenesis in Cancer Cells

1. Transformed cells (C3H 10T1/2 contain RNAs with long terminal repeats (LTR) characteristic of retroviral transposons (Kirschmeyer et al., 1982).

2. Activation of c-mos by insertion of endogenous retroviral like elements in mouse myeloma (Rechavi et al., 1983).

3. Integration of LINE at c-myc locus in canine tumours (Katzir et al., 1988).

4. Integration of LINE in c-myc rat immunocytoma (Pear et al., 1985).

5. Integration of LINE in c-myc in human breast tumour (Morse et al., 1988).

6. Integration of Alu in a plasmid, introduced in human lung carcinoma cells (Lin et al., 1988).

RECOMBINOGENIC EFFECTS AND AMPLIFICATION

Genetic recombination, the exchange of genetic material between chromosomes and between DNA molecule is a central property of all organic life. Genetic recombination constitutes a wide variety of processes with different mechanisms, enzymes and endpoints. It seems that recombinogenic events play a much more important role for alterations of the genetic material than one had conceived before. This applies to both spontaneous and induced events. It is from that viewpoint problematic that the standard short-term tests rarely are directed towards recombinogenic effects of chemicals. One test system, which includes somatic recombination is the Drosophila SMART tests - somatic mutation and recombination test. It was clear from the investigation by Vogel and Zijlstra (1987) that chemical mutagens cause far more recombinations than ordinary mutations. Only compounds like ethylnitrosourea (ENU), which give rise to a high frequency of O^6-guanine alkylation, give a substantial proportion of point mutations.

Somatic recombination occurs also in mammals, but the frequency can be expected to be lower than in Drosophila, which has somatic chromosome pairing. One important consequence of somatic recombination in carcinogenicity is the effect of anti-oncogenes like retinoblastoma. Recombination can cause the loss of the normal allele in heterozygotes and render cells homozygous for a defect allele or a deletion. Also there are observations indicating that homozygosity of activated oncogenes is involved in some tumours (Sager 1986).

Another genetic change, which has turned out to be important in carcinogenicity is gene amplification, which also is brought about by genetic recombination. In many tumours gene amplification is responsible for an increased expression of nuclear oncogenes such as c-myc. But also cytoplasmic oncogenes like c-ras are sometimes amplified at least at later stages of carcinogenicity. In general, gene amplification is associated with progression of cancer and amplification of oncogenes is sometimes correlated with malignancy. This is the case with breast - and ovary-cancer and the amplification of the oncogene c-neu. The prognosis of those cancer forms is associated with the degree of amplification, which is of importance for the therapy (see Marx 1989 for an overview).

Spontaneous amplification is a wide spread phenomenon particularly in connection with resistance to chemicals (for reviews see Hamlin et al., 1984 and Schimke, 1984). The best studied case is the amplification of dehydrofolate reductase gene (DHFR), which gives resistance to the anticancer drug methotrexate. Gene amplification can also be induced by chemicals, for instance inhibitors of DNA synthesis. Among amplifying agents can be mentioned arsenic salts (Lee et al., 1988) and intercalating agents (Pall and Hunter, 1987).

At least two basic mechanisms of amplification can be recognized (Fig. 3). One is unequal recombination - both sister chromatid exchange and somatic crossing over. The result is tandem duplications. Deamplification can also occur through intrastrand crossing over (see Fig. 3). The other mechanism is unscheduled replication of DNA - "onion skin" replication. This can result in insertions of the amplified piece at other loci (Wahl et al., 1984). At the cytological level amplification gives rise to homogenously stained regions (HSR) and double minutes, which have been recorded in cancer cells (Cowell, 1982).

The frequency of amplification apparently is greatly elevated in cancer cells and again this probably is another result of the genetic instability of DNA during cancer progression. However, already under normal conditions amplification occurs at a much higher frequency than point mutations. The actual frequency of amplification of DHFR gene has been estimated at 10^{-3}/cell generation (Schimke, 1984).

The initial occurrence of repeated sequences is, as Flavell has stated it "biological dynamite". Such DNA pieces constitute the primary target for further amplification and presumably also other genetic alterations.

One can speculate that repeated sequences are responsible for rapid changes in genetic variation and high mutation frequency. It seems to me that amplification could be one explanation of the high mutation frequency of polygenes that has been reported particularly in Drosophila (for a review see Ramel, 1983). A system based on amplification and deamplification by means of recombination would meet the need of rapid changes in quantitative traits much more efficiently than random point mutations.

Recombination

Between chromosomes

Within chromosomes

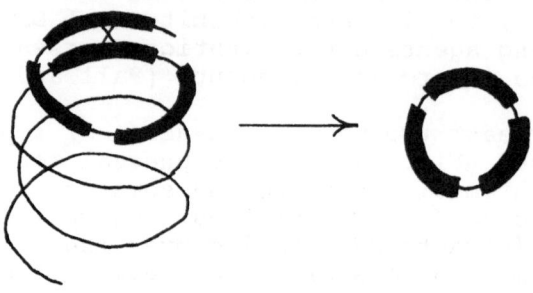

Sister chromatid exchange

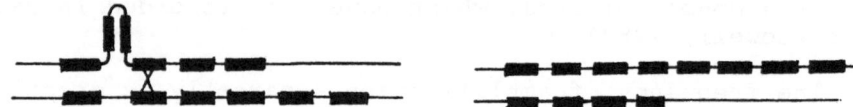

Unscheduled DNA synthesis

"Onion skin" replication

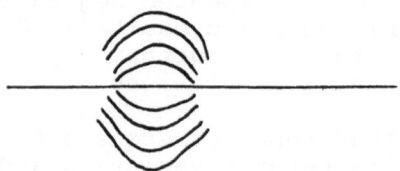

Fig. 3.   Schematic outline of mechanisms of amplification and deamplification.

METHYLATION

Altered gene expression is characteristic of cancer cells, which can be aquired by a number of mechanisms at the level of DNA and the genes - chromosomal rearrangements, amplification of genes, increased number of chromosomes and demethylation of cytosin. All of these mechanisms have been associated with cancer cells.

The methylation of 5-cytosin is a reversible process, which turns off gene transcription. Holliday has emphasized the importance of methylation of DNA in differentiation, ageing and cancer induction (Holliday and Pugh, 1975, Holliday 1979, 1985). According to his model carcinogens may act by interfering with the methylase activity with demethylation and release of gene expression as a result.

The suggestion that methylation of 5-cytosin plays a role in carcinogenesis has been supported by a number of observations of neoplastic cells, which often are hypomethylated, also including oncogenes (Feinberg and Vogelstein, 1983; Goelz et al, 1985; Jones 1986, Rao et al., 1989). The relation between hypomethylation and cancer is however complex and involves for instance tissue specificity, which may explain organotropic effects in cancer (Silva and White 1988).

It is of interest in this context that 5-azacytidine, which is an inhibitor of cytosin methylation, is also a potent carcinogen with a wide range of action (Holliday, 1989). The methylation pattern may furthermore have connections with other events in carcinogenicity. There are thus indications that overexpression of genes may initiate amplification and the insertion of mobile elements in maize occurs in unmethylated sites of DNA (Wessler, 1988). It can be concluded that hypomethylation may very well play an essential role in carcinogenicity, although the evidence at this point is only suggestive.

MITOCHONDRIAL DNA

It should be remembered that DNA is not limited to the nucleus, but it also occurs in mitochondria. It has been suggested that there may be an interaction between mt DNA and nuclear DNA in carcinogenicity. Shay and Werbin (1987) present the following observations, which point to an association between cancer and mitochondria:

- Enhanced aerobic glycolysis in cancer cells
- Reversible nature of some cancer forms
- Suppression of tumorigenicity by normal cytoplasm
- Immortalization of lymphocytes by tumour cell cytoplasm
- Sometimes unique restriction enzyme pattern of mt DNA from neoplastic cells
- Mt DNA is more accessible to mutagens
- Distribution of mixed populations of mitochondria is not necessarily random - sometimes one type is completely lost

The actual mechanisms by mitochondria in carcinogenesis can only be a matter of speculation at this point, but a number of possibilities have been suggested (Shay and Werbin, 1987).

- Mitochondria mutations may provide growth advantage
- Mitochondria may function as suppressors of tumorigenicity
  Mutations diminish the number of mitochondria and release the action of activated oncogenes
- Gross alteration of mt DNA modulates nuclear gene DNA involved in membrane biogenesis
- Exposure to carcinogens cause break down of mitochondria and integration of mt DNA into nuclear DNA

INDIRECT MUTAGENIC EFFECTS

In many cases chemicals do not act directly on chromosomes and DNA but can still have a mutagenic effect by indirect pathways. An imbalance of the nucleotide pool can thus affect a wide variety of genetic events and in particular cause a synergistic effect in combination with chemical mutagens. The imbalance can cause an error in the incorporation of bases at DNA replication and repair. As an example treatment with hydroxyurea or thymidine, which cause an imbalance of the nucleotide pool, strongly increased the effect of chemical mutagens in cultured hamster cells (Jenssen 1986).

The mutagenic effects of free radicals can often be of an indirect nature. The defence mechanisms against oxygen radicals, which under normal conditions prevent interactions with DNA, can be saturated or be put out of order by chemicals. For instance certain carbamates inactivate superoxide dismutase and caused a secondary increase of mutations in Salmonella (Rannug and Rannug, 1984).

A puzzling carcinogenic action is induced by agents which cause a proliferation of peroxisomes - organelles which are involved in fat metabolism. Such a proliferation of peroxisomes in the liver is connected with lipid peroxidation and the generation of oxygen radicals. It is believed that the induction of liver cancer by peroxisome proliferation somehow is related to this production of oxygen radicals, but in spite of attempts by many laboratories, including ours, to reveal the exact mechanism it is still unknown.

Indirect mutagenic effects are often difficult to reveal in standard test systems and to foresee them requires a thorough knowledge of the mechanism of action of such agents.

GENETIC INSTABILITY

The experimental evidence from cancer research during the last ten years indicates that practically all conceivable alterations of the genetic material are or can be suspected to be involved in carcinogenicity. There is indication that the mutation frequency in developing cancer cells is far higher than in normal cells. Both the wide spectrum of mutational changes and the high rate of mutations can probably be brought back to a collapse of the control of the genetic machinery.

This genetic instability is particularly manifested in the induction of recombination, amplification and presumably transpositions. There are reasons to believe that this genetic instability is a late event and associated with progression, as emphasized by Pitot (1989). Although the background of such an instability is not known there are some observations, which at least may be used for some speculation of possible mechanisms.

Transformation of cells implies a series of changes in the cellular physiology and the formation of defect proteins, which are substrates for degradation by ubiquitin. An overload of the system may result in a stress response involving heat shock genes, as suggested by Munro and Pelham (1985). It is possible that this result also affects the proteins involved in DNA repair and DNA synthesis. It is of interest that stress conditions in E. coli results in a reduced fidelity of DNA repair because of a loss of the proof-reading function of polymerase III and a subsequent drastic increase of the mutation frequency (Kunkel, 1988).

Another possibility might be an alteration of the DNA nucleotide pool. Transformation with HSV virus is dependent on ribonucleotide reductase, which also controls the DNA nucleotide pool and thereby also DNA replication and fidelity (Huszar and Bacchetti, 1983). An unbalanced nucleotide pool can cause altered proof-reading or error prone repair. A disturbance of this balance is associated for instance with mutator phenotypes.

Poly(ADP-ribose)polymerase is involved in several cellular processes, including DNA repair and cellular trans-formation. A genetic instability may be brought about by an overload of this system. Nakayasu et al. (1988) reported that inhibition of poly(ADP-ribose)polymerase caused deletion of transfected oncogenes and such inhibition also causes an increase of chemically induced somatic recombination (Ramel unpublished).

When it comes to mobilization of transposons an associa-tion with hypomethylation has been suggested. Wessler (1988) reported that most insertions of the transposable element Ac in maize occurs to undermethylated loci. He suggested that agents, which cause chromosome breakage change the pattern of methylation and activate mobile elements. This activation of mobile elements is amplified, because there is a kind of chain reaction to other elements to transpose. Following excision the elements are likely to transpose to active genes, as there is a preference for integration into unmethylated target sites.

SHORT-TERM TESTING

Now what does this highly complicated process of various genetic alterations mean from the point of view of risk evaluation of chemicals? How many of the mutational steps in carcinogenicity are actually affected by external factors like chemicals? Is the cascade of genetic changes during the later stages of carcinogenicity just an inevitable consequence of earlier events during tumour promotion in combination with cellular proliferation? These are crucial questions for the strategy of testing chemicals for carcinogenicity. The attention in short-term testing has hitherto focused on initiation of cancer, where the data indicates at least a predominant role of those conventional mutations covered by short-term tests. It can however not be excluded that the whole series of events in tumour formation, including increase in malignancy, is induced or affected by chemicals. It is a well known fact that specific chemicals act on the initiation and promotion steps but, lately, agents acting on tumour progression, that is specific progressors, have also been identified as pointed out by Pitot (1989). It is thus very possible that chemicals can affect not only mutation steps during initiation, but also the whole sequence of genetic alterations at later stages. If that is the case it seems unlikely that our standard short-term tests are adequate. They can not be expected to cover all the various endpoints, even at a qualitative level. The procedure has to be adapted to the rapid increase of our knowledge of tumour development. But it must be emphasized also that genetical alterations caused by chemicals in our environment have to be evaluated in their own right as genotoxic.

REFERENCES

Aldaz, C.M., Conti, C.J., Yuspa, S.H. and Slaga, T.J., 1988, Cytogenetic profile of mouse skin tumors induced by the viral Harvey-ras gene, Carcinogenesis, 9:1503.
Ames, B.N., Durston, W.E. and Yamasaki, E., 1973, Carcinogens are mutagens: A simple test system combining liver homogenates for activation and bacteria for detection, Proc. Natl. Acad. Sci., U.S.A., 70:2281.
Ashby, J. and Tennant, R.W., 1988, Chemical structure, Salmonella mutagenicity and extent of carcinogenicity as indicators of genotoxic carcinogenesis among 222 chemicals tested in rodents by the U.S. NCI/NTP, Mutation Res., 204:17.
Ashby, J., Tennant, R.W., Zeiger, E. and Stasiewicz, S., 1989, Classification according to chemical structure, mutagenicity to Salmonella and level of carcinogenicity of a further 42 chemicals tested for carcinogenicity by the U.S. National Toxicology Program, Mutation Res., 223:73.
Berenblum, I., 1941, The mechanism of carcinogenesis, Cancer Res., 1: 807.
Berenblum, I. and Shubik, P., 1947, A new quantitative approach to the study of stages of chemical carci- nogenesis in the mouse's skin, Br. J. Cancer, 1:383.
Biemont, C., Aouar, A. and Arnault,C.,1987, Genome reshuffling of the copia element in an inbred line of Drosophila melanogaster, Nature, 329:742.
Bouchard, L., Lamare, L., Tremblay, P.J. and Julicoeur, P., 1989, Stochastic appearance of mammary tumors in transgenic mice carrying the MMTV/c-neu oncogene, Cell, 57:931.
Breimer, L.H., 1988, Ionizing radiation-induced mutagenesis, Br. J. Cancer, 57:6.
Cairns, J., Overbaugh, J. and Miller, S., 1988, The origin of mutants, Nature, 335:142.

Cerutti, P.A., 1985, Prooxidant states and tumor promotion, Science, 227:375.

Cowell, J.K., 1982, Double minutes and homogenously stained regions: gene amplification in mammalian cells, Ann. Rev. Genet., 16:21.

Feinberg, A.P. and Vogelstein, B., 1983, Hypomethylation distinguishes genes of some human cancers from their normal counterpart, Nature, 301:89.

Forrester, K., Almoguera, C., Han, K., Grizzle, W.E. and Perucho, M., 1987, Detection of high incidence of K-ras oncogenes during colon tumorigenesis, Nature, 327:298.

Gershenson, S.M., 1986, Viruses as mutagenic factors, Mutation Res., 167:203.

Goelz, S.E., Vogelstein, B., Hamilton, S.R. and Feinberg, A.P., 1985, Hypomethylation of DNA from benign and malignant human colon neoplasms, Science, 228:187.

Green, M.R., 1989, When the products of oncogens and anti-oncogens meet, Cell, 56:1.

Hall, B.G., 1988, Adaptive evolution that requires multiple spontaneous mutations. I. Mutations involving on insertion sequence, Genetics, 120:887.

Hamlin, J.L., Milbrandt, J.D., Heintz, N.H. and Azizkhan, J.C., 1984, DNA sequence amplification in mammalian cells, Int. Rev. Cytol., 90:31.

Hennings, H., Shores, R., Wenk, M.L., Spangler, E.F., Tarone, R. and Yuspa, S.H., 1983, Malignant conversion of mouse skin tumours is increased by human initiators and unaffected by tumour promotors, Nature, 304:67.

Holliday, R., 1979, A new theory of carcinogenesis, Br. J. Cancer, 40:513.

Holliday, R., 1985, The significance of DNA methylation, in: "Molecular Biology of Ageing", A.D. Woodhead, A.D. Blackett and A. Hollaender, eds., Plenum, New York.

Holliday, R., 1989, A different kind of inheritance, Scientific American, 40:48.

Holliday, R. and Pugh, J.E., 1975, DNA modification mechanisms and gene activity during the development, Science, 187:226.

Huszar, D. and Bacchetti, S., 1983, Is ribonucleotide reductase the transforming function of herpes simplex virus 2? Nature, 302:76.

Jenssen, D., 1986, Enhanced mutagenicity of low doses of alkylating agent and UV-light by inhibition of ribonucleotide reductase, in: "Genetic Toxicology of Environmental Chemicals", Part A, "Basic Principle and Mechanism of Action", A. Liss, New York.

Johns, M.A., Mottinger, J. and Freeling, M., 1985, A low copy number, copia-like transposons in maize, EMBO J., 4:1093.

Jones, P.A., 1986, DNA methylation and cancer, Cancer Res., 46:461.

Katzir, N., Rechavi, G., Cohen, J. B., Unger, T., Simoni, F., Segal, S., Cohen, D. and Givol, D., 1985, "Retroposon" insertion into the cellular oncogene c-myc in canine transmissible venerial tumor, Proc. Natl. Acad. Sci. U.S.A., 82:1054.

Kirschmeyer, P., Gattonicelli, Dina, D. and Weinstein, I.B., 1982, Carcinogen-transformed and radiation-transformed C3H 10 1/2 cells contain RNAs homologous to the long terminal repeat sequence of a murine leukemia virus, Proc. Natl. Acad. Sci. U.S.A., 74:2773.

Kunkel, T.A., 1988, Exonucleolytic proof reading, <u>Cell</u>, 53:837.

Land, M., Parada, L.F. and Weinberg, R.A., 1983, Tumorigenic conversion of primary embryo fibroblasts requires at least two cooperating oncogenes, <u>Nature</u>, 304:596.

Lee, T.C., Tanaka, N., Lamb, P.V., Gilmer, T.M. and Barrett, C., 1988, Induction of gene amplification by arsenic, <u>Science</u>, 241:79.

Lin, C.S., Goldtwait, D.A. and Samols, D., 1988, Identification of Ala transposition in human lung carcinoma cells, <u>Cell</u>, 54:153.

Marx, J.L., 1984, Instability in plants and the ghost of Lamarck, <u>Science</u>, 224:1415.

Marx, J.L., 1989, Gene signals relapse of breast, ovarian cancers, <u>Science</u>, 244:654.

Miller, E.C. and Miller, J.A., 1971, The mutagenicity of chemical carcinogens: correlations, problems and interpretations, <u>in</u>: "Chemical Mutagens, Principles and Methods for their Detections", A. Hollaender, ed., Plenum Press, New York.

Morse, B., Rothberg, P.G., South, V.J., Spandorfer, J.M. and Astrin, S.M., 1988, Insertional mutagenesis of the <u>myc</u> locus by a LINE-1 sequence in a human breast carcinoma, <u>Nature</u>, 333:87.

Muller, W.J., Sinn, E., Pattengale, P.K., Wallace, R. and Leder, P., 1988, Single-step induction of mammary adenocarcinoma in transgenic mice bearing the activated c-neu oncogene, <u>Cell</u>, 54:105.

Munro, S. and Pelham, H.R., 1985, Use of peptide tagging to detect proteins expressed from cloned genes: deletion mapping functional domains of Drosophila hsp 70, <u>EMBO J.</u>, 3:3087.

Nakayasu, M., Shima, H., Aonuma, S., Nakagama, H., Nagao, M. and Sugimura, T., 1988, Deletion of transfected oncogenes from NIH 3T3 transformants by inhibitors of poly(ADP-ribose)polymerase, <u>Proc. Natl. Acad. Sci. U.S.A.</u>, 85:9066.

Okamoto, M., Sasaki, M., Sugio, K., Sato, C., Iwama, T., Ikeuchi, T., Tonomura, A., Sasazuki, T. and Miyaki, M., 1988, Loss of constitutional heterozygosity in colon carcinoma from patients with familiar polyposis coli, <u>Nature</u>, 331:273.

Pall, M.L. and Hunter, B.J., 1987, Induction of genetic tandem duplication by polycyclic aromatic hydrocarbons and amine carcinogens, <u>Mutation Res.</u>, 182:5.

Pear, W.S., Wahlstrom, G., Nelson, S.F., Axelson, H., Szeler, A., Wiener, F., Bazin, H., Klein, G. and Sumegi, J., 1988, 6;7 chromosomal translocation in spontaneously arising rat immunocytomas: evidence for c-<u>myc</u> breakpoint clustering and correlation between isotypic expression and the c-<u>myc</u> target, <u>Mol. Cell. Biol.</u>, 8:441.

Pitot, H.C., 1989, Progression: The terminal stage in carcinogenesis, <u>Jpn. J. Cancer Res.</u>, 80:599.

Pott, P., 1775, Chirurgical observations, Haukes, Clarke and Collins, London.

Ramel, C. (Ed.), 1973, Evaluation of genetic risks of environmental chemicals. Report of a symposium held at Skokloster, Sweden. Ambio Special Report 3.

Ramel, C., 1983, Polygenic effects and genetic changes affecting quantitative traits, <u>Mutation Res.</u>, 114:107.

Rannung, A. and Rannug, U., 1984, Enzyme inhibition as a
        possible mechanism of the mutagenicity of dithiocar-
        bamic acid derivatives in Salmonella typhimurium, Chem.
        Biol. Interact.,49:329.

Rao, P.M., Antony, A., Rajalakshmi, S. and Sarma, D.S.R.,
        1989, Studies on hypomethylation of liver DNA during
        early stages of chemical carcinogenesis in rat liver,
        Carcinogenesis, 10:933.

Rechavi, G., Givol, D. and Canaani, E., 1983, Activation of a
        cellular oncogene by DNA rearrangement: possible
        involvement of an IS-like element, Nature, 300:607.

Rinkus, S.J. and Legator, M.S., 1979, Chemical characteriza-
        tion of 465 known or suspected carcinogens and their
        correlation with mutagenic activity in the Salmonella
        typhimurium system, Cancer Res., 39:3289.

Rubin, G.R., 1983, Dispersed repetitive DNAs in Drosophila, in
        "Mobile Genetic Elements", J.A. Shapiro, ed., Academic
        Press, New York.

Sager, R., 1986, Genetic suppression of tumor formation: A
        new frontier in cancer research, Cancer Res., 46:1573.

Schimke, R.T., 1984, Gene amplification, drug resistance and
        cancer, Cancer Res., 44:1735.

Shay, J.W. and Werbin, H., 1987, Are mitochondrial DNA
        mutations involved in the carcinogenic process? Mutation
        Res., 186:149.

Silva, A.J. and White, R., 1988, Inheritance of allelic
        blueprints for methylation patterns, Cell, 54:145.

Sinn, E., Muller, W., Pattengale, P.K., Tepler, I., Wallace,
        R. and Leder, P., 1987, Coexpression of MMTV/v-Ha-ras
        and MMTV/c-myc genes in transgenic mice: Synergistic
        action of oncogenes in vivo, Cell, 49:465.

Suganuma, M., Fujiki, H., Suguri, H., Yoshizawa, S., Hirota,
        M., Nakayasu, M., Ojika, M., Wakamatsu, K., Yamada, K.
        and Sugimura, T., 1988, Okadaic acid: An additional
        non-phorbol-12-tetradecanoate-13-acetate-type tumor
        promotor, Proc. Natl. Acad. Sci. U.S.A., 85:1768.

Sugimura, T., Nagao, M., Kawachi, T., Honda,M., Yahagi, T.,
        Seino, Y., Matsushima, T., Shirai, A., Sawamura, M.,
        Sato, S., Matsumoto, H. and Matsukura, N., 1977,
        Mutagen-carcinogens in food with special reference to
        highly mutagenic pyrolytic products in broiled foods,
        in: "Origin of Human Cancer", H. Hiatt, J.D. Watson and
        J.A. Winsten, eds., Cold Spring Harbor Conference on
        Cell Proliferation, Vol. 4, Cold Spring Harbor,
        New York.

Sukumar, S., Carney, W.P. and Barbacid, M., 1988, Independent
        molecular pathways in initiation and loss of hormone
        responsiveness of breast carcinoma, Science, 240:524.

Tennant, R.W., Margdin, B.H., Shelby, M.D., Zeiger, E.,
        Haseman, J.K., Spalding, J., Caspary, W., Resnick, M.,
        Stasiewicz, S., Anderson, B. and Minor, R., 1987,
        Prediction of chemical carcinogenicity in rodents from
        in vitro genetic toxicity assays, Science, 236:933.

Thompson, T.C., Southgate, J., Kitchener, G. and Land, H.,
        1989, Multistage carcinogenesis induced by ras and myc
        oncogenes in a reconstituted organ, Cell, 56:917.

Vogelstein, B., Fearon, E.R., Hamilton, S.R., Kern, S.E.,
        Preisinger, A.C., Leppert, M., Nakamura, Y., White, R.,
        Smits, A.M.M. and Bos, J.L., 1988, Genetic alterations
        during colorectal-tumor development, New England J. of
        Medicine, 319:525.

Vogel, E.W. and Zijlstra, J.A., 1987, Mechanistic and methodological aspects of chemically induced somatic mutation and recombination in <u>Drosophila melanogaster</u>, <u>Mutation Res.</u>, 182:243.

Wahl, G.M., de Saint Vincent, B.R. and de Rose, M.L., 1984, Effects of chromosomal position on amplification of transfected genes in animal cells, <u>Nature</u>, 307:516.

Wessler, S.R., 1988, Phenotypic diversity mediated by the maize transposable element <u>Ac</u> and <u>Spm</u>, <u>Science</u>, 242:399.

Zeiger, E. and Tennant, R.W., 1986, Mutagenesis, clastogenesis, carcinogenesis: Expectations, correlations and relations, <u>in</u>: "Genetic Toxicology of Environmental Chemicals", Part B, "Genetic Effects and Applied Mutagenesis", C. Ramel, B. Lambert and J. Magnusson, eds., Alan Liss, New York.

MECHANISMS OF CHEMICALLY-INDUCED GENETIC EFFECTS ON MOLECULAR, CHROMOSOMAL AND CELL DIVISION LEVEL

ULTRAVIOLET LIGHT MUTAGENESIS IN BACTERIA: THE POSSIBLE ROLE

OF A DNA POLYMERASE III COMPLEX LACKING PROOFREADING

EXONUCLEASE

B. A. Bridges

MRC Cell Mutation Unit, University of Sussex
Falmer, Brighton, BN1 9RR, Great Britain

Exactly ten years ago I addressed this Society on the topic of ultraviolet light (UV) mutagenesis in bacteria and the burden of my talk was that the appearance of a newly induced mutation reflected a failure of the normal error-correcting mechanisms (Bridges, 1980). The three error-correcting mechanisms discussed then were DNA polymerase base selection, exonucleolytic proofreading, and mismatch correction. Ten years later the same topics appear, the jigsaw is nearer completion, but still lacks important pieces of information.

Before discussing the roles of DNA polymerase and the proteins associated with them it is necessary to refer to a model which we have proposed which divides the process of UV mutagenesis into two steps (Bridges and Woodgate 1984, 1985a,b). The first step is heavily influenced by a multifunctional protein called RecA and involves the incorporation of a wrong base opposite a photoproduct in the template strand. The second step, which requires the products of the umuD and C genes, is postulated to be the reinitiation of chain elongation using the misincorporated base as a priming end. We believe we can visualize the first step (misincorporation) in the absence of the second (bypass) by means of a trick called delayed photoreversal mutagenesis. We take a strain which is not UV mutable, for example because it cannot make UmuC protein. After exposure to UV bases are wrongly incorporated but the second step cannot occur. If after a period of time the bacteria are exposed to light, pyrimidine dimers are split and no longer act as a block to continued DNA synthesis. Bases which have been wrongly incorporated opposite pyrimidine dimers may be "fixed" as the chain elongates and give rise to mutant clones. In our experience it is bases opposite cytosine-containing dimers that show this effect (Bockrath et al., 1987; Bockrath, 1989).

Escherichia coli possesses three well-established DNA polymerases. Pols I, II and II. All three are altered in nature or quantity after exposure of bacteria to DNA damaging agents that are capable of inducing the series of genes controlled by the LexA repressor (the SOS response). Pol I is the main repair polymerase in E. coli and also has a minor but important role in chromosomal replication. Its activity is increased in SOS-induced cells (Goodman et al., 1988) and an altered form (Pol I*) is present which is capable of internalizing mismatched primer termini during synthesis (Lackey et al., 1982, 1985; S. Martin-Moe and S. Linn, pers. comm.). In principle, therefore, Pol. I* has properties that might be expected of an error-prone polymerase. Studies with a deletion of its gene (pol A) have clearly shown, however, that UV mutagenesis goes on quite normally in the complete absence of Pol I (Bates et al., 1989). Therefore, although it is possible that Pol I* may contribute to UV mutagenesis when it is present, it is not essential for the major UmuD, C-dependent pathway and its overall contribution is likely to be small.

The situation with Pol II is less clear. There is no known role for Pol II but as it is seven-fold SOS-inducible and has little difficulty in polymerizing past abasic sites (Goodman et al., 1988) it could also be important as a mutagenic polymerase. Cells with a mutation in the gene specifying Pol II (pol B-100) appear to be normally mutable by UV (M. Goodman, pers. comm., and author's unpublished results), which tends to argue against an essential role for Pol II. It would be wise to reserve judgement, however, until deletion or insertion mutations in pol B have been studied.

The normal replicative enzyme in E.coli, "DNA polymerase III", has been isolated in a number of forms. The holoenzyme (Pol III HE) is about 900 kDa in size and is made up of a catalytic polymerase core of three subunits: α(132kDa); ε(27kDa) and θ(10kDa); and seven auxiliary subunits which are responsible for processivity (Maki et al., 1988): τ(71kDa), γ(52kDa), β(37kDa), δ(35kDa), δ'(33kDa), χ(15kDa) and ψ(12kDa). The pure α subunit has been identified as the polymerase (Maki and Kornberg, 1985), the ε subunit as the 3'-5' exonuclease (Scheuermann and Echols, 1984), and a complex of the two as the proofreading entity (Maki and Kornberg, 1987). In all known forms of "DNA polymerase III" the core of α, θ and ε is believed to be retained.

Pol III activity is induced to a small extent (about 2-fold) in SOS induced cells (Bonner et al., 1988) but there is as yet no evidence from in vitro studies that suggests that Pol III is particularly error-prone, indeed the contrary seems to hold. Pol III is like many other polymerases in generally stopping synthesis immediately before a lesion in the template strand (Moore and Strauss, 1979; Moore et al., 1981). About 30% of the time, at least under certain in vitro conditions, it continues to polymerize right through the region containing the template lesion (Livneh, 1986). There are now several lines of evidence which establish that the α subunit of Pol III is essential for UV mutagenesis (Bridges et al., 1976; Hagansee et al., 1987; Bridges and Bates, 1989; Sharif and

Bridges, 1989). In the first of those papers the suggestion
was made that Pol III could interact with an SOS-inducible
cofactor such that its normal high specificity would be
overridden and bases would be inserted, possibly at random,
opposite a UV photoproduct. Arising out of studies with Pol
I, Villani et al. (1978) suggested that repeated exonucleolytic proofreading might be involved in the halting of replication at sites of DNA damage and that inhibition of the
nuclease function might be responsible for SOS-induced
translesion synthesis. Following the demonstration by Fersht
and Knill-Jones (1983) that proofreading could be inhibited by
RecA protein in vitro it seemed possible that RecA protein
could be the postulated inducible cofactor and some evidence
consistent with this has been presented (Lu et al., 1986;
Bridges and Woodgate, 1985). The evidence is far from
compelling, however, and in vitro work has failed to
demonstrate any increase in the ability to insert bases
opposite photoproducts under conditions where proofreading is
inhibited (Bridges et al., 1988; Schwartz et al., 1988).
Nevertheless, even if inhibition of proofreading alone is not
sufficient to allow Pol III to synthesize past a photoproduct,
if any form of "DNA polymerase III" is involved in inserting
incorrect bases opposite sites of damage where base pairing
specificity is lost, it would seem necessary that the proofreading function be non-functional.

A further discussion of the involvement of DNA polymerases can be found elsewhere (Bridges et al., 1989).

THE EPSILON PROOFREADING EXONUCLEASE

The ϵ subunit is the product of a gene known as dnaQ or
mutD and two mutant alleles are fairly well studied. The
first to be studied in the context of UV mutagenesis was mutD5
which confers a dominant mutator phenotype. The mutation is
known to be due to two base substitution in the same gene
resulting in two amino acid changes, $_{73}$Leu(TTG) $\rightarrow$ Trp(TGG)
and $_{164}$Ala(GCA) $\rightarrow$ Val(GTA) (Takano et al., 1986). Because of
its dominance it must be assumed that the MutD5 protein binds
to the holoenzyme at least as well as the wild-type protein
but results in a complex that is defective in proofreading.
Woodgate et al. (1987) reported that introduction of mutD5
into umuC or recA430 strains that are not UV mutable did not
restore UV mutability so that it could be concluded that
blocking proofreading was not sufficient to allow mutagenesis
to occur, a conclusion also supported by in vitro data
(Schwartz et al., 1988). Mutagenesis in umuC mutD5 bacteria
was also normal following exposure to UV plus delayed photoreversal, and in umu$^+$mutD5 following UV alone. In other words
the presence of a defective proofreading activity, although it
had a dramatic effect on the spontaneous mutation rate, was
irrelevant to UV mutagenesis. Of the possibilities discussed
by Woodgate et al. the following cannot so far be discounted.

(1)   The ϵ subunit is excluded from the DNA polymerase III
       holoenzyme when it acts in an error-prone repair
       capacity (i.e. the polymerase exists as an Epsilon Minus
       Complex (EMC).

(2)   Proofreading is normally inhibited at photoproducts

because RecA protein also binds there and is activated. Proofreading would be automatically inhibited and the mutD5 defect would be without discernible effect.

(3)   It is possible that although activated RecA protein inhibits proofreading, this is not sufficient _in vitro_ to cause misincorporation or bypass and some other factor is required which is rate limiting in both mutD[+] and mutD5 strains.

The fact that UV followed by delayed photoreversal is mutagenic to strains carrying a deletion through recA (Bridges, 1988) tends to argue against the latter two possibilities, at least as far as the misincorporation step is concerned. Since 1986, however, a great deal more has been learned about mutD/dnaQ which needs to be considered. The mutator activity of mutD5 has been shown to consist of two effects. On minimal medium the effect is (as expected) largely due to impaired proofreading but on rich medium and/or in the presence of thymidine mismatch correction is also impaired (Schaaper, 1988).

A second, very different mutation in the gene specifying the epsilon subunit is dnaQ49. This mutation is a single base substitution causing one amino acid change, $_{96}$Val(GTG) -> Gly(GGG) (Takano et al., 1986). In dnaQ49 strains the mutator effect increases with temperature and the bacteria are temperature-sensitive for growth at 43.5° on low salt agar. Becasue the mutation is recessive and because the α subunit (the polymerase itself) is known to be temperature sensitive _in vitro_ when it is not complexed with ε, Takano et al. 1986 concluded that DnaQ49 protein was defective in binding to the α subunit. DnaQ49 bacteria therefore constitute a model system in which there may be presumed to be an abnormally large amount of Pol III EMC. As with mutD5, however, dnaQ49 seems to confer two distinct mutator activities (as originally proposed by Piechocki et al., 1986). Around 30°, the mutator effect appears to be largely due to defective proofreading but as the temperature is increased, mismatch correction becomes progressively more defective until at 37° it dominates the picture (Isbell and Fowler, 1989).

The effects of dnaQ49 on spontaneous mutagenesis have been extensively studied by Foster and Sullivan (1988) and some of their results are relevant to experiments to be described below on UV mutagenesis. In particular they found that the presence of activated RecA protein in a cell results in a 2-3fold increase in the dnaQ49 spontaneous mutator effect. There was also a smaller effect of UmuC protein.

THE EPSILON MINUS COMPLEX (EMC) HYPOTHESIS

The suggestion that the state of the DNA polymerase III holoenzyme found in dnaQ49 cells might be the same as the which mediates SOS mutagenesis appears in the work of Piechocki et al., (1986) and Woodgate et al. (1987). They specifically mention the possibility of a Pol III complex lacking epsilon being involved in SOS mutagenesis. There are several lines of evidence that bear upon this question and these will be discussed in detail.

Let us suppose that a form of DNA polymerase III holoenzyme can exist that lacks the epsilon subunit (Pol III EMC), then one may suppose that Pol III EMC and Pol III HE are in equilibrium and that the level of Pol III EMC may be influenced by the level of free epsilon. By overproducing epsilon one would expect that any Pol III EMC in a cell would tend to be sequestered into the normal holoenzyme form. This in turn would result in an inhibition of UV mutagenesis if Pol III EMC had a vital role to play. Such a result has been reported by Jonczyk et al. (1988) who used a strong heterologous promoter linked to dnaQ[+]. However, as these authors acknowledge, other interpretations of this result are possible, the simplest of which is that the excess epsilon can act independently of Pol III and prevent misincorporation by "overediting". This alternative interpretation is made more attractive by the observation of Perrino and Loeb (1989) that epsilon can act independently in a proofreading capacity in conjuction with mammalian DNA polymerase α.

The finding of Jonczyk et al. has been confirmed and extended by Foster et al. (1989) who have found that overproduction of wild-type epsilon blocks both UV and MMS mutagenesis. The effect was abolished either by a mutation in dnaQ or one in rnh, the gene for RNaseH which shares a common promoter region with dnaQ but is transcribed in the opposite direction. Foster et al. also found the inhibition of UV mutagenesis was partially suppressed by excess UmuD, C proteins, a result which led them to suggest that excess epsilon might bind to UmuDC proteins and deplete the number of copies available for UV mutagenesis. This interpretation, however, is complicated by the factor that increased levels of umuDC in any case exert a mild mutator effect. Moreover, the antimutagenic effect of excess epsilon was also exerted against some classes of mutations induced by MNNG independently of umuDC.

More recently, Ciesla, Jonczyk and Fyalkowska (personal communication) have shown that overproduction of epsilon also blocks (a) mutagenesis by UV plus delayed photoreversal, a umuDC independent process which is believed to reflect the first step in UV mutagenesis, and (b) untargeted mutagenesis where a mutation arises at a site close to, but not actually at the position of a photoproduct. Overall, although the effect of an excess of epsilon are consistent with Pol III EMC being involved in UV mutagenesis, they can equally well be explained in terms of excess epsilon acting independently to reverse the misincorporation step.

A different prediction of the hypothesis that Pol III EMC is required for UV mutagenesis is that dnaQ49 bacteria might be hypermutable by UV (in constrast to mutD5 bacteria) because they may have an excess of Pol III EMC. We have examined this prediction and found approximately three times as many His[+]mutants in a dnaQ49 strain as in dnaQ[+] controls (Table 1). We also found that the extent of UV mutagenesis was similar between 32° and 43.5° despite the increase in spontaneous mutator activity temperature due to progressive blockage of mismatch correction (Isbell and Fowler, 1989).

Before drawing the simple conclusion from these data, however, we must consider that Foster and Sullivan (1988) have shown that the dnaQ49 spontaneous mutator effect is

Table 1. UV-induced His[+] mutation frequencies per 10°
surviving bacteria in the dnaQ49 strain CM1158pSE117
compared with two dnaQ[+] isogenic strains.

UV dose	Strain		
(J m^{-2})	TK603	TK610pSE117	CM1158pSE117
0.5	1.33±0.68	3.04±0.42	4.41±3.42
1.0	3.80±0.78	5.16±0.72	16.1±9.32
1.5	5.56±1.14	7.32±1.38	17.6±6.32
2.0	8.52±1.61	9.71±2.77	32.7±9.50

TK603 and TK610 have been described elsewhere (Bridges and
Woodgate, 1985). CM1158 is a derivative of TK610 prepared by
Dr. R. Woodgate in which dnaQ49 was introduced by conjugation
with TAM21. In TK603 the umuDC operon is carried chromosom-
ally, in the other two strains, it is carried on a plasmid
made by Dr. S. Elledge and kindly supplied by Dr. M. Blanco.
The experiments were carried out three times at 32° using
published methodology (Bridges and Woodgate, 1985). Surviving
fractions were greater than 0.5 at all doses.

increased by activated RecA protein and UmuDC proteins. Since
all these proteins will be induced with increasing UV dose,
the increase in mutation frequency observed in the dnaQ49
strain might be due to an increase in spontaneous mutation
occurring on the plate rather than genuine UV mutagenesis. We
have therefore examined UV mutagenesis in a dnaQ49 umuC
strain. In such a strain genuine UV mutagenesis should be
absent due to the umuC mutation, but RecA protein will still
be activated and could lead to an increase in the spontaneous
mutator effect giving a spurious impression of UV mutagenesis.
We found, however, that there was no evidence for any UV
mutagenesis in the umuC dnaQ49 strain CM1158 (data not shown).
From these data we conclude that the excess of UV mutagenesis
in the dnaQ49 strain is real and not an artifact due to
stimulation of the spontaneous mutator effect.

The result is readily reconcilable only with the first
possibility mentioned above, namely that Pol III EMC is
invovled in the UV induction of His[+] mutations and is
effectively rate limiting; in dnaQ49 bacteria there is more
Pol III EMC and therefore more UV mutagenesis. Unfortunately,
such a conclusion does not emerge when the induction of
mutation to rifampicin resistance is examined in the same
strains. Preliminary experiments show little excess of UV
mutagenesis in the dnaQ49 strain compared with the dnaQ[+]
strain. This may be related to the observation (Foster et
al., 1989) that the suppression of SOS mutagenesis by overproduction of DnaQ protein is also much less pronounced with
rifampicin resistant mutations than with mutations to Arg[+],
which like His[+] mutations in these strains are largely ochre

suppressors. Why should DnaQ49 protein affect the induction of His$^+$ but apparently not rifampicin resistant mutations? It is too early to say. More work needs to be done. At present, therefore, it must be concluded that evidence for an epsilon-minus complex of DNA polymerase III being involved in UV mutagenesis is far from convincing. Yet it remains an attractive idea worthy of further exploration.

ACKNOWLEDGEMENTS

I thank Ms Vanessa Mogre-Taebi for expert technical assistance.

REFERENCES

Bates, H., Randall, S.K., Rayssiguier, C., Bridges, B.A., Goodman, M.F., and Radman, M., 1989, Spontaneous UV-induced mutations in Escherichia coli K-12 strains with altered or absent DNA polymerase I, J. Bacteriol., 171, 2480.

Bockrath, R., 1989, Streptomycin-resistant and dependent mutants of E.coli: possible indicators of two important types of DNA alteration by UV mutagenesis, Mutagenesis, 4: 78.

Bockrath, R.C., Ruiz-Rubio, M., and Bridges, B.A., 1987, Specificity of mutation by ultraviolet light and delayed photoreversal in umuC defective Escherichia coli K-12: a targeting intermediate at pyrimidine dimers, J. Bacteriol., 169:1410.

Bonner, C.A., Randall, S.K., Rayssiguier, C., Radman, M., Eritja, R., Kaplan, B.E., McEntee, K. and Goodman, M.F., 1988, Purification and characterization of an inducible Escherichia coli DNA polymerase capable of insertion and bypass at abasic lesions in DNA, J. Biol. Chem., 263:18946.

Bridges, B.A., 1980, Ultraviolet light mutagenesis in bacteria: a result of the failure of normal error-correcting mechanisms? in: "Progress in Environmental Mutagenesis", M. Alacevic, ed., Elsevier, Amsterdam.

Bridges, B.A., 1988, Mutagenic DNA repair in Escherichia coli XVI. Mutagenesis by ultraviolet light plus delayed photoreversal in recA strains, Mutation Res., 198:343.

Bridges, B.A., and Bates, H., 1989, Mutagenic DNA repair in Escherichia coli XVIII. Involvement of DNA polymerase III alpha subunit (DnaE protein) in mutagenesis after exposure to ultraviolet light, Mutagenesis, in press.

Bridges, B.A., Bates, H, and Sharif, H., 1989, Polymerases and UV mutagenesis in Escherichia coli, Genome, in press.

Bridges, B.A., Kelly, C., Hubscher, U. and Sedgwick, S.G., 1988, Possible roles of RecA protein and DNA polymerase III holenzyme in UV mutagenesis in Escherichia coli, in: "DNA Replication and Mutagenesis", R.E. Moses and W.C. Summers, eds., American Society for Microbiology.

Bridges, B.A., Mottershead, R.P., and Sedgwick, S.G., 1976, Mutagenic repair in Escherichia coli. III. Requirement for a function of DNA polymerase III in ultraviolet light mutagenesis, Molec. Gen. Genet., 144:53.

Bridges, B.A., and Woodgate,R., 1984, Mutagenic repair in
    Escherichia coli X. The umuC gene product may be
    required for replication past pyrimidine dimers but not
    for the coding error in UV-mutagenesis, Mol. Gen.
    Genet., 196:364.
Bridges, B.A., and Woodgate, R., 1985, The two-step model of
    bacterial UV mutagenesis, Mutation Res., 150:133.
Bridges, B.A., and Woodgate, R., 1985, Mutagenic repair in
    Escherichia coli: products of the recA gene and of the
    umuD and umuC genes act at different steps in
    UV-induced mutagenesis, Proc. Nat. Acad. Sci. USA,
    82:4193.
Fersht, A.R., and Knill-Jones, J.W., 1983, Contribution of
    3'-5' exonuclease activity of DNA polymerase III
    holoenzyme from Escherichia coli to specificity, J.
    Mol. Biol., 165:669.
Foster, P.L., Sullivan, A.D., and Franklin, S.B., 1989,
    Presence of the dnaQ-rnh divergent transcriptional unit
    on a multicopy plasmid inhibits induced mutagenesis in
    Escherichia coli, J. Bacteriol., 171:3144.
Goodman, M.F., Petruska, J., Boosalis, M.S., Bonner, C.,
    Randall, S.F., Sowers, L.C. and Mendelman, L., 1988,
    Molecular mechanisms of DNA synthesis fidelity and
    isolation of a possible SOS induced polymerase. In:
    "DNA Replication and Mutagenesis", M.E. Moses and W.C.
    Summers, eds., American Society for Microbiology.
Hagensee, M.E., Timme, T.L., Bryan, S.K., and Moses, R.E.,
    1987, DNA polymerase III of Escherichia coli is
    required for UV and ethyl methanesulphonate mutagene-
    sis, Proc. Natl. Acad. Sci. USA, 84:4195.
Isbell, R.J., and Fowler, R.G., 1989, Temperature-dependent
    mutational specificity of an Escherichia coli mutator,
    dnaQ49, defective in 3'-5' exonuclease (proofreading)
    activity, Mutation Res., 213:149.
Jonczyk, P., Fijalkowska, I., and Ciesla, Z., 1988, Overpro-
    duction of the ε-subunit of DNA polymerase III counter-
    acts the SOS mutagenic response of escherichia coli,
    Proc. Natl. Acad. Sci. USA, 85:9124.
Lackey, D., Krauss, S.W., and Linn, S., 1982, Isolation of an
    altered form of DNA polymerase I from Escherichia coli
    cells induced for recA/lexA functions, Proc. Natl.
    Acad. Sci. USA, 79:330.
Lackey, D., Krauss, S.W., and Linn, S., 1985, Characterisation
    of DNA polymerase I*, a form of DNA polymerase I found
    in Escherichia coli expressing SOS functions, J. Biol.
    Chem., 260:3178.
Livneh, Z., 1986, DNA polymerase III holoenzyme of Escherichia
    coli: Evidence for bypass of pyrimidine photodimers,
    Proc. Natl. Acad. Sci. USA, 83:4599.
Lu, C., Scheuermann, R.H., and Echols, H., 1986, Capacity of
    RecA protein to bind preferentially to UV lesions and
    inhibit the editing subunit of DNA polymerase III: a
    possible mechanism for SOS-induced targeted mutagene-
    sis, Proc. Natl. Acad. Sci. USA, 83:619.
Maki, H., and Kornberg, A., 1985, The polymerase subunit of
    DNA polymerase III of E. coli II. Purification of the
    alpha subunit, devoid of nuclease activity, J. Biol.
    Chem., 260:12987.

Maki, H., and Kornberg, A., 1987, Proofreading by DNA polymerase III of _Escherichia coli_ depends on cooperative
interaction of the polymerase and exonuclease subunits,
Proc. Natl. Acad. Sci. USA, 84:4389.

Maki, H., Maki, S., and Kornberg, A., 1988, DNA polymerase III
holoenzyme of _Escherichia coli_. IV The holoenzyme is
an asymmetric dimer with twin active sites, _J. Biol.
Chem._, 263:6570.

Moore, P.D., Bose, K.K., Rabkin, S.D., and Strauss, B.S.,
1981, Sites of termination of _in vitro_ DNA synthesis on
ultraviolet- and N-acetylaminofluorenetreated 0X174
templates by prokaryotic and eukaryotic DNA polymerases, _Proc. Natl. Acad. Sci. USA_, 78:110.

Moore, P.D., and Strauss, B.S., 1979, Sites of inhibition of
in vitro DNA synthesis in carcinogen and UV treated
0X174 DNA, _Nature_, 278:664.

Perrino, F.W., and Loeb, L.A., 1989, Proofreading by the ε
subunit of _Escherichia coli_ DNA polymerase III
increases the fidelity of calf thymus DNA polymerase,
Proc. Natl. Acad. Sci. USA, 86: 3085.

Piechocki, R., Kupper, D., Quinones, A., and Langhammer, R.,
1986, Mutational specificity of a proof-reading
defective _Escherichia coli dnaQ49_ mutator, _Mol. Gen.
Genet._, 202:162.

Schaaper, R.M., 1988, Mechanisms of mutagenesis in the
Escherichia coli mutator _mutD5_: role of DNA mismatch
repair, _Proc. Natl. Sci. USA_, 85:8126.

Scheuermann, R., and Echols. H., 1984, A separate editing
exonuclease for DNA replication: The epsilon subunit
of _Escherichia coli_ DNA polymerase III holoenzyme,
Proc. Natl. Acad. Sci. USA, 81:7747.

Schwartz, H., Shavitt, O., and Livneh, Z., 1988, The role of
exonucleolytic processing and polymerase-DNA association in bypass of lesions during replication _in vitro_.
Significance for SOS targeted mutagenesis, _J. Biol.
Chem._, 263:18277.

Sharif, F., and Bridges, B.A., 1989, Mutagenic DNA repair in
Escherichia coli XVII. Effect of temperature sensitive
DnaE proteins on the induction of streptomycin resistant mutations by ultraviolet light, _Mutagenesis,_ in
press.

Takano, K., Nakabeppu, Y., Maki, H., Horiuchi, T., and
Sekiguchi, M., 1986, Structure and function of _dnaQ_ and
mutD mutators of _Escherichia coli_, _Mol. Gen. Genet._,
205:9.

Villani, G., Boiteux, S., and Radman, M., 1978, Mechanism of
ultraviolet light-induced mutagenesis; extent and
fidelity of _in vitro_ DNA synthesis on irradiated
templates, _Proc. Natl. Acad. Sci. USA_, 75:3037.

Woodgate, R., Bridges, B.A., Herrera, G., and Blanco, M.,
1987, Mutagenic DNA repair in _Escherichia coli XIII_.
Proofreading exonuclease of DNA polymerase III holoenzyme is not operational during UV mutagenesis, _Mutation
Research_, 183:31.

CENTROMERE SEPARATION: EMERGING RELATIONSHIP WITH ANEUPLOIDY

Baldev K. Vig[1] and Neidhard Paweletz[2]

[1]Department of Biology, University of Nevada
Reno, NV 89557, USA

[2]Institute of Cell and Tumor Biology
German Cancer Research Center
D-6900 Heidelberg 1, FRG

INTRODUCTION

It is well established that the centromeres of various chromosomes in a given genome at metaphase/anaphase junction separate in a species specific sequence (Vig and Rattner, 1989). In humans, for example, the centromeres of chromosomes 18, 17 and 2 separate before do others and all acrocentrics carrying large quantity of r-DNA separate the last. Similarly, amongst the three pairs of chromosomes of the plant <u>Crepis capillaris</u> the largest chromosome is the first to split at its centromere while the second largest separates the last. These examples indicate that centromere separation is not dependant upon the relative length of the chromosome in the genome nor it correlates with the position of the centromere in the chromosome.

This sequential separation (or splitting) of the centromeres is independent of the application of spindle arrestants or hypotonic treatment and is not tissue specific (see Vig et al., 1989). For instance, Chinese hamster (Figueroa and Vig, 1983) and frog (Belcheva et al, 1980) cells show similar sequence of separation whether treated or not with colcemid or hypotonic solutions. The bone marrow and spermatogonial cells of the Chinese hamster (Singh et al., 1977) show identical sequence of separation as do the cells from human lymphocytes, bone marrow, amniotic fluid (Mehes,1975; Bajnockzky and Buhler, 1984) and chorionic villae (Mehes, personal communication). The long-term cultures of cells grown in artificial media also express strict sequence of separation (see, e.g., Vig, 1983; Vig and Zinkowski, 1986; Zinkowski et al., 1986).

CONTROL OF CENTROMERE SEPARATION

This strict maintenance of sequential separation may point to some sort of strict genetic control and might have some evolutionary significance. At first sight, the sequence of separation in some organisms appears to correlate well with the quantity of pericentric heterochromatin (Fig. 1). In various species and sub-species of mouse the chromosomes showing the least amount of pericentric heterochromatin seperate the earliest; those with increasing quantities separate later (Vig, 1982). Similar situation is found in the wood lemming (<u>Myopus schisticolor</u>). In cattle, the X chromosome is almost entirely devoid of any heterochromatin and it separates the earliest. A correlation between the timing of centromere separation and the quantity of pericentric heterochromatin is also apparent for all but one chromosome in the Indian muntjac (Gerlach et al., 1984) and potorus (Vig, 1981). At the surface, thus, it appears that the timing of centromere separation is controlled by the quantity of pericentric heterochromatin.

However, the separation sequence in some organisms defies this correlation. In the Chinese hamster, for example, the heaviest C-bands are found in the pericentric region of the X and Y chromosomes. Yet these chromosomes separate in the middle of the overall sequence defined for the genome. The chromosomes separating last in the genome are numbers 5, 6 and 7, these carry only moderate quantity of pericentric heterochromatin. Chromosomes 3 and 4 carry about similar quantities of pericentric heterochromatin as do, say, nos. 5 or 8 and yet the former separate earlier than the latter. Similar 'irregularities' are encountered for the human genome in which, for example, chromosomes number 1, 9 and 16 separate earlier than the Y, 21, 22, 13, 14 and 15 inspite of much larger quantities of pericentric heterochromatin in the former. These studies show that whereas pericentric heterochromatin may be a factor in determining the timing of centromere separation, its quantity is not the sole determinant.

A rather close correlation emerges when one considers the DNA composition of pericentric heterochromatin and the timing of centromere separation. Pericentric heterochromatin, in most instances, is made up of repetitive DNA, its composition being characteristic of the species. As discussed elsewhere (Vig, 1987) there emerges a rather tight link between the quantity of repetitive DNA and the timing of centromere separation.

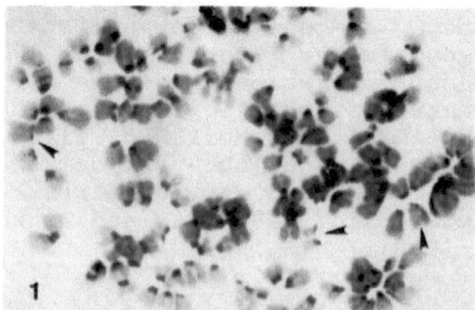

Fig. 1. A C-banded mouse cell showing clear separation of centromeres with small C-bands (arrow heads). Note lack of separation of centromeres associated with large C-bands.

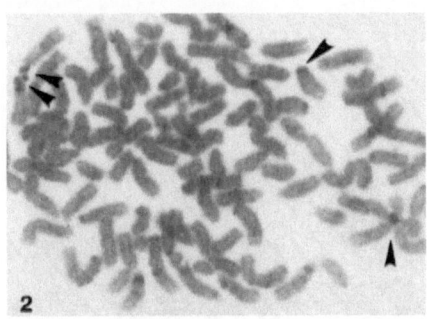

Fig. 2. A mouse cell showing BrdU incorporation during
the last few minutes of S-phase. The thymidine
analog is located in the centromeric regions of
only a few chromosomes. (From Broccoli et al.,
1989; with permission)

These correlates can be demonstrated for Chinese hamster,
mouse, human, wood lemming and domestic pig, among others
(Vig, 1987).

REPLICATION OF PERICENTRIC/CENTRIC REGION

Since the separation of a centromere is a post-DNA
replication phenomenon, the timing of separation might depend
upon the timing of replication of pericentric repetitive DNA.
It stands to logic that larger blocks of repetitive DNA would
take longer to complete replication than would smaller blocks
and, hence, one would observe the hierarchy of centromere
separation dependant upon the timing of completion of replication of pericentric/centric DNA. We, therefore, labeled the
mouse chromosomes with 5-bromodeoxyuridine (BrdU) or 5-iodo-
deoxyuridine (IudR) during last part of S phase. The results
were informative in that the centromeres which completed
replication earlier also separated earlier at metaphase. It
appears that the pericentric region completes its replication
during late S and that the replication fork moves right into
the centromere proper. This creates a differential in the
timing of completion of replication and separation of the
centromeres (Vig and Broccoli, 1988). Since DNA replication
and centromere separation are distanced by several hours,
apparently some sort of post-replication 'maturation' is
involved before the centromere separates. Our studies also
show that the mammalian centromere replicates during the late
S phase (Fig. 2). This is in contrast to the data on yeast
which indicates that the yeast centromeres complete replication during early S (McCarrol and Fangman, 1988).

MULTICENTRIC CHROMOSOMES AND PREMATURE CENTROMERE SEPARATION

A correlation between the quantity of repetitive DNA in
the pericentric region and the timing of centromere separation
is, however, not without some rather interesting exceptions.
One such situation is encountered in dicentric and multicentric chromosomes found in long term cultures of brain tumor

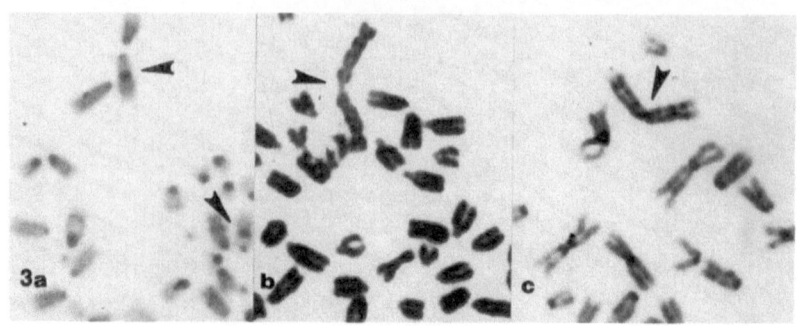

Fig. 3. Parts of mouse cells showing (a) two dicentrics in brain tumor cells, (b) an octacentric in L-cell and (c) location of pericentric heterochromatin associated with the various centromeres in the octacentric. The active centromeres in the dicentrics are those located terminally; those in the octacentric are in the middle. Arrowheads point to inactive centromeres in the dicentric but to active centromere(s) in the octacentric.

cells and L-929 cells of mouse origin. The former carry between one and three dicentric chromosomes which have one centromere placed acrocentrally and the other somewhere in the middle of the arm (Fig. 3a). At metaphase these appear as normal acrocentrics; however, at prophase the two primary constrictions are quite evident. Both centromeres have associated pericentric heterochromatin (Vig and Zinkowski, 1986). Quantitatively the two blocks appear equal; rarely one may be larger than the other. The L-cells carry one rather large chromosome which appears to carry as many as eight centromeres (Vig, 1984b). At metaphase this chromosome looks like any ordinary biarmed chromosome as if it originated by Robertsonian translocations. It is at prophase that one observes three primary constrictions in each arm and two located in the middle of the chromosome (Fig. 3b). It was proposed that all centromeres except one in these dicentric as well as multicentric chromosomes undergo premature separation at prophase (Fig. 3c). These prematurely separating centromeres are inactive or accessory centromeres. The prophase separation is not in conformity with the usual timing of centromere separation for normal centromeres which separate at the onset of anaphase/late metaphase. These prematurely separating centromeres may be associated with larger blocks of pericentric heterochromatin, and, hence, of repetitive DNA, than are the active centromeres of some monocentrics and yet the former separate before do the latter.

But what about the replication of pericentric region surrounding the inactive centromeres which separate prematurely. BrdU incorporated cells show that the DNA in the vicinity of inactive centromeres replicates before does the DNA associated with the active centromeres (Fig. 4). This is rather surprising because the repetitive DNA in the pericentric regions of active as well as accessory centromeres in

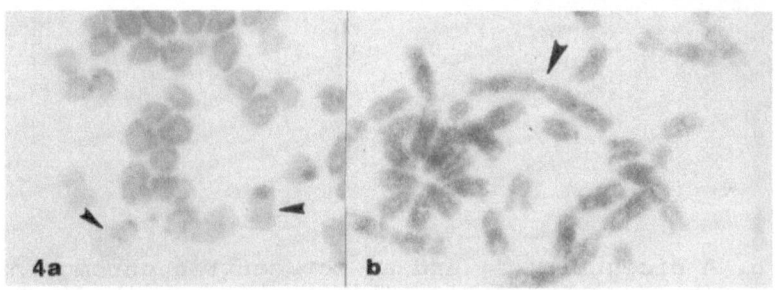

Fig. 4. BrdU incorporated chromosomes in mouse cells.
(a) Brain tumor cells showing replication of
the DNA in the active centromere in the
dicentric. The inactive centromeres (arrowheads) had completed replication before the
addition of BrdU. (b) Localization of BrdU in
the middle, active centromere (arrow head) in
the octacentric in L-cells. (From Vig and
Broccoli, 1988, with permission)

mouse is similar in its base composition. The data, thus,
suggest that it may not be the nucleotide sequence of the DNA
but some other property which determines its timing of
replication. The exact answer to the question is not at hand
but factors like those controlling initiation of replication,
extent of methylation etc may be the elements dictating the
timing of initiation and completion of DNA synthesis. Our
recent studies indicate that the pericentric repetitive DNA
associated with prematurely separating centromeres also
initiates replication earlier - sometime during early S.
Hence, it may not be the differential rate of DNA synthesis in
the two blocks of heterochromatin which separates them in
their timing of replication but differential onset of initiation of synthesis.

KINETOCHORE FORMATION IN MULTICENTRICS

For sake of clarity of the discussion which follows, a
distinction between the terms centromere and kinetochore is
warranted. Centromere is that part of the chromosome which is
called primary constriction. It is an integral, constitutive
component of the chromosome and is made up of nucleo-protein
complex. The kinetochore, however, may be defined as a
facultative organelle which is apparently composed of several
proteins (CENP-A, B, C, D, E etc; see Vig and Rattner, 1989)
deposited at the site of the centromere. The kinetochore binds
to spindle microtubules and presents a trilamellar, globular
or some similar structure under the electron microscope. Up
until recently the only means of visualizing the kinetochore
was electron microscopy. However, with the discovery of
antikinetochore, antibodies present in the serum of sclerod-
erma (var CREST) patients, it has been possible to view the
kinetochore proteins at the light microscope level(Moroi et
al., 1980).
Monocentric chromosomes, as a rule, express only one site
of kinetochore proteins juxtaposed at the centromere.

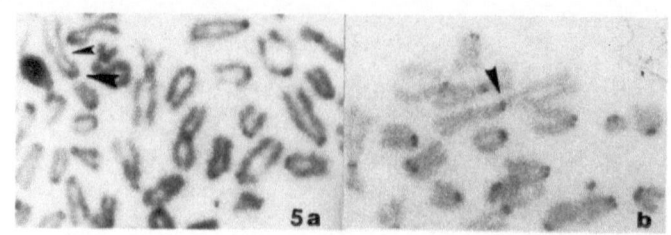

Fig. 5. A dicentric (a) and an octacentric chromosome
(b) treated with antikinetochore antibody. Each
shows only one kinetochore at the site of the
active centromere. Small arrowhead in (a)
points to inactive centromere.

Multicentric chromosomes which have more than one active
centromere, however, exhibit kinetochore proteins at every
centromeric site (see later). Multicentric chromosomes which
undergo premature centromere separation generally exhibit only
one site of deposition of kinetochore proteins (Fig. 5). This
is the site of the active centromere. This situation has been
encountered for the dicentric chromosomes found in the brain
tumor cells of mouse (Vig and Zinkowski, 1986) as well as for
the octacentric found in the mouse L-929 cells (Zinkowski et
al., 1986). In essence, therefore, such multicentric chromosomes are functionally monocentric. This observation can
explain a lack of binding of spindle microtubules at sites
other than the one of active centromere, since it is presumed
that spindle microtubules bind to the kinetochore and not to
the centromere. There are, however, reports in which accessory centromeres occasionally respond to the antikinetochore
antibody. These sites exhibit reduced binding of the antibody
indicating reduced quantity of one or more protein components,
e.g., in human dic(X;X) (Merry et al., 1985) and dic(13;13)
(Earnshaw et al., 1989).

MULTICENTRIC CHROMOSOMES LACKING PREMATURE CENTROMERE SEPARATION

Whereas stable multicentric chromosomes exhibit premature
centromere separation, a lack of kinetochore formation and
early replication of the pericentric/centric DNA, there exist
unstable multicentric chromosomes which fail to display any of
these properties (Fig. 6a). Instability of these chromosomes,
however, may result in the genesis of new multicentric
chromosomes in every cell generation. One example is a SV40
transformed cell line of rat endothelial origin (Diglio et
al., 1983). These chromosomes create bridges and centric
fragments. The latter, upon rejoining and DNA replication,
result in the genesis of new multicentric chromosomes. These
cells show three types of anaphase aberrations; (1)typical
chromatid bridges resulting from breakage-fusion-bridge cycle
(McClintock, 1941), (2) separation and proper mitotic migra-
tion of only one centromere during anaphase leaving the other
centromeres unseparated and (3) total lack of separation of
the centromeres into two units so that the two centromeres in
a dicentric migrate to the opposing poles of the spindle
creating a chromosome-type bridge. Breakage of the chromatids
near the site of the first unsplit centromere in (2), rejoin-

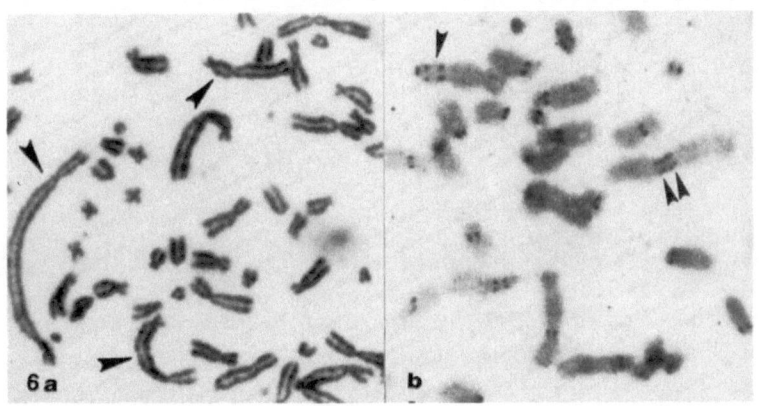

Fig. 6. Parts of two SV40 transformed rat cerebral
endothelial cells showing (a) multiple dicentrics and (b) the positions of kinetochores in
di- and multi-centric chromosomes. Note in (b)
kinetochore dots separated by interchromatin
segments (arrowhead) and four adjacently
located, almost coalescing kinetochores (double
arrowheads). ((b) is reproduced from Broccoli
et al., 1989, with permission)

ing of the broken ends, delayed splitting of the unsplit
centromere and its replication in the ensuing S phase would
create new multicentrics with increased number of centromeres
(Vig et al., 1989b). A break in the sister chromatids of the
chromosome bridge described in (3), consequent rejoining,
centromere splitting and subsequent replication of the
chromosome would result in the production of a new dicentric.
Should breakage occur at the same position in various cell
cycles the dicentric would apparently look like a stable
component of the genome. The genesis of these new multicentrics and details of the cell line have been previously
published (Vig and Paweletz, 1988). The intriguing fact about
these cells is the delayed, post-anaphase separation of some
centromeres in contrast to the premature separation of
accessory centromeres in stable multicentrics. The centromeres eres separating at post-anaphase stage exhibit the
binding of spindle microtubules.

If the absence of kinetochore proteins indicates inacti-
vity of the inactive centromeres, then multicentric chromosomes in rat should exhibit kinetochore proteins at every
centromeric site. This is borne out by the application of
antikinetochore antibody to these chromosomes (Broccoli et
al., 1989). Not only certain chromosomes exhibit two or more
kinetochores placed distally from each other but also there is
evidence of close localization of the centromeres resulting in
coalescence of closely placed kinetochores (Fig. 6). In some
chromosomes several centromeres are placed close to each other
resulting in the formation of a compound centromere. This
results in the kinetochore proteins being deposited in a long
stretch along the length of the chromosome. Electron microscopic analysis of these regions (Fig. 7) has shown the

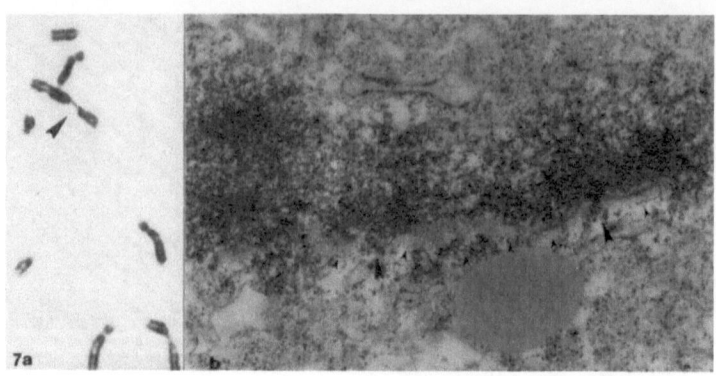

Fig. 7. (a) An elongated, compound kinetochore in a rat
        cerebral endothelial cell as seen under the
        light microscope (arrowhead). (b) The compound
        centromere under the electron microscope
        displays as many as six kinetochores (small
        arrowheads), some of which are separated by
        inter centromeric chromatin (large arrowheads).
        (Vig et al., 1989, with permission)

presence of up to six kinetochores at these sites (Vig et al.,
1989). Even though these are the first observations for the
cells grown in tissue culture, this phenomenon finds equivalence in nature in the long neck like structure of the
centromere of the X chromosome of Indian muntjac (Brinkley et
al., 1984).

In contrast to the prematurely separating inactive
centromeres, the functional centromeres separating at post-
anaphase stage and their pericentric regions also show
characteristic property of DNA replication in the late S
phase. When rat Bl cells are fed with BrdU and analysed for
DNA synthesis, all centric/pericentric regions show incorporation and DNA synthesis having occurred simultaneously. Though
these observations on kinetochore protein deposition, centromere activity (as studied by spindle microtubule binding) and
late S phase DNA replication indicate the expected properties
of the active centromeres, these do not provide any clue to
the question why certain centromeres separate at post-anaphase
stage and not at their proper time at meta-anaphase junction.

CENTROMERE SEPARATION, DNA REPLICATION AND KINETOCHORE FORMATION
IN SOMATIC CELL HYBRIDS

Of late we have studied the relative timing of centromere
separation in the two genomes in a quasi-stable mouse-human
somatic cell hybrid. This cell line, HR61, carries about a
dozen human chromosomes. In this hybrid all human chromosomes
separate at their centromeres before such separation initiates
in the mouse genome. In contrast to the observations of
Graves and Zelesco (1987), we found no evidence of overlap in
the timing of separation of the two genomes (Vig and Athwal,
1989). Also, the DNA replication studies show that all human
chromosomes complete their replication, including the pericen-

tric and centromeric regions, ahead of the replication of the mouse component. This is in contrast to the earlier suggestions (e.g., Kao and Puck, 1970) that the retained genome (mouse in HR61) replicates earlier in the cell cycle than the segregant genome (human).

Generally, in the quasi stable hybrid HR61 all chromosomes show kinetochore proteins. There are, however, isolated instances in which one or more chromosomes do not respond to the antikinetochore antibody indicating a lack of these proteins at the site of the centromere (Fig. 8a). The frequency of kinetochore lacking chromosomes is dramatically increased in some cells (Fig. 8b) and is reminiscent of lack of such activity of the accessory centromeres and double minutes (Haff and Schmid, 1988). It is likely that this lack of kinetochore proteins leads to lack of spindle fiber attachment to the chromosome and, hence, elimination of such chromosomes in the somatic cell hybrids. This suggestion is in conformity with the frequency of kinetochore-less micronuclei in the quasi-stable population. It is worth mentioning that the kinetochore lacking micronuclei do not appear to have originated from acentric fragments since upon Hoechst 33258 treatment of some kinetochore-less micronuclei we find evidence for the presence of blocks of pericentric heterochromatin present in mouse chromosomes.

CENTROMERE SEPARATION, ANEUPLOIDY AND HUMAN DISEASE

Errors of centromere separation are emerging as a correlate between nondisjunction and human diseases. More than a decade ago Fitzgerald et al., (1975) reported premature separation of the X chromosome in elderly women. Certain cells contained as many as three copies of this chromosome which appeared as pairs of 'rail road tracks' - the two chromatids lying parallel to each other without any connection. It was concluded that premature centromere separation,

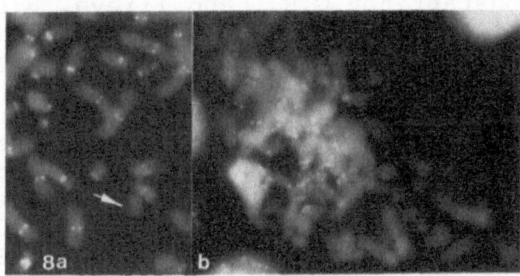

Fig. 8. (a) Part of a mouse-human hybrid cell treated with antikinetochore antibody. Note all chromosomes except one exhibiting kinetochores. Apparently, the chromosomes lacking kinetochore (arrowhead) would fail to attach to spindle microtubule. (b) A mouse cell in culture in which several chromosomes appear to have no kinetochores.

somehow, resulted in unequal segregation of this chromosome. It had been previously shown that the X chromosome in man is not the first one to split at its centromere (Vig, 1983). Thus this out-of-phase separation may be the cause of nondisjunction. Similarly, a G-group chromosome in a woman suffering from chronic myelogenous leukemia has been reported to show out-of-phase separation and aneuploidy (Vig, 1984a). More recently, correlates between premature centromere separation and aneuploidy have been recorded for chromosome 21 and a C-group chromosome (Fitzgerald, 1987). Bajnoczky and Mehes (1988) have also reported a similar correlation for human chromosomes 18 and 21 in a family exhibiting trisomy for these chromosomes.

In other human diseases, errors of centromere separation are beginning to emerge. These include centromere puffing in Robert's syndrome (Tompkin et al., 1979; German, 1979) Alzheimer disease (Moorehead and Heyman, 1983), acute myelogenous leukemia (Sandberg, 1980) and non-lymphocytic leukemias (Gallo et al., 1984; Shairaishi et al., 1982). Out-of-phase centromere separation has also been reported in Kleinfelter syndrome, ataxia talengiectasia and pregnancy complications (Buhler et al., 1987). Suggestions have been made that nondisjunction based on errors of centromere separation are genetically controlled (Madan et al., 1987) and that this dominantly inherited trait may be responsible for miscarriages in some cases (Gabbaron, 1986). In summary, there appears to be emerging a relationship between errors of centromere separation, human diseases and various trisomies.

REFERENCES

Bajnoczky, K. and Buhler, E.M., 1984, Sequence of centromere separation in cultured human amniotic cells, Acta Biol., Hung., 34:107.
Bajnoczky, K., and Mehes, K., 1988, Parental centromere separation sequence and aneuploidy in the offspring. Hum. Genet., 78:286.
Belcheva,R.G., Konstantinov,G.H., and Ilieva, H.L., 1980, Sequence of centromere separation in the mitotic chromosomes of Rana ridibunda, CR Bulg. Acad. Sci., 33:1689.
Buhler, E.M., Fessler, R., Beutler,C., and Gargano, G., 1987,Incidental findings of double minutes (DM), single minutes (SM), homogeneously staining regions (HSR), premature chromosome condensation, and premature centromere division (PCD), Ann. Genet., 30:75.
Brinkley, B.R., Valdivia, M.M., Tousson, A, and Brenner, S.L., 1984, Compound kinetochores of Indian muntjac: evolution by linear fusion of unit kinetochores. Chromosoma 91:1.
Broccoli, D., Paweletz, N., and Vig, B.K., 1989, Sequence of centromere separation: characterization of multicentric chromosomes in a rat cell line, Chromosoma 98, 13.
Diglio, C.A., Wolff, D.E., and Meyers, P., 1983, Transformation of rat cerebral endothelial cells by Rouse sarcoma virus, J. Cell. Biol., 97:15.

Earnshaw, W.C., Ratrie III, H., and Stetten, G., 1989,
        Visualization of centromere proteins CENP-B and CENP-C
        on a stable dicentric chromosome in cytological
        spreads, <u>Chromosoma</u>, 98:1.
Figueroa, M.L., and Vig, B.K., 1983, sequence of centromere
        separation: a lack of colcemid effect on the Chinese
        hamster genome, <u>Cytogenet. Cell Genet.</u>, 36:627.
Fitzgerald, P., 1987, Premature centromere division and
        aneuploidy, <u>in</u>: "Aneuploidy Part A", B.K. Vig and
        A.A.Sandberg, ed., Alan R. Liss, New York.
Fitzgerald, P.H., Pickering, A.F., Mercer, J.M., and Miethke,
        D.M., 1975, Premature centromere division: a mechanism
        of nondisjunction causing X chromosome aneuploidy in
        somatic cells of man, <u>Ann. Hum. Genet.</u>, 38:417.
Gabbaron, J., Jimenez, A., and Glover, G., 1986, Premature
        centromere division dominantly inherited in a subfer-
        tile family, <u>Cytogenet. Cell Genet.</u>, 43:69.
Gallo, J.H., Misawa, S., and Testa, J.R., 1984,Centromere
        spreading in acute non-lymphocytic leukemia, <u>Cancer
        Genet. Cytogenet.</u>, 12:105.
Gerlach, B., Sulleder, E., Hauke, M., Harms, H., Schmid, M.,
        and Aus, H.M., 1984, Application of a high resolution
        TV-microscope system to estimate the sequence of
        centromere separation in muntjac chromosomes, <u>Cyto-
        metry</u>, 5:562.
German, J., 1979, Roberts syndrome. I. Cytological evidence
        for a disturbance in chromatid pairing, <u>Clin. Genet.</u>,
        16:441.
Graves, J.A.M. and Zelesco, P.A., 1987, Chromosome segregation
        from cell hybrids. V. Does segregation result from
        asynchronous centromere separation? <u>Genome</u>, 30:124.
Haff, T., Schmid, M., 1988, Analysis of double minute and
        double minute-like chromatin in human and murine tumor
        cells using antikinetochore antibodies, <u>Cancer Genet.
        Cytogenet.</u>, 30:73.
Kao, F.T., and Puck, T., 1970, Genetics of somatic mammalian
        cells: linkage studies with human-Chinese hamster cell
        hybrids, <u>Nature,</u> 228:329.
Madan, K., Lindout, D., Palan, A., 1987, Premature centromere
        division (PCD): a dominantly inherited cytogenetic
        anomaly, <u>Hum. Genet.</u>, 71:193.
McCarrol, R.M., and Fangman, W.L., 1988, Time of replication
        of yeast centromeres and telomeres, <u>Cell</u>, 54:505.
McClintock, B., 1941, Spontaneous alterations in chromosome
        size and form in <u>Zea</u> <u>mays</u>, <u>Cold Spring Harb. Symp.
        Quant. Biol.</u>, 9:72.
Mehes, K., 1975, Non-random anaphase segregation of mitotic
        chromosomes, <u>Acta Genet. Med. Gemellol.</u>, 24:175.
Merry, D.E., Pathak, S., Hsu, T.C., and Brinkley, B.R., 1985,
        Antikinetochore antibodies: use as probes for inactive
        centromeres, <u>Am. J. Hum. Genet.</u>, 37:425.
Moorhead, P., and Heyman, A., 1983, Chromosome studies of
        patients with Alzheimer disease, <u>Am. J. Med. Genet.</u>,
        14:545.
Moroi, Y., Peebles, C., Fritzler, M.J., Steigerwald,J. and
        Tan, E.M., 1980, Autoantibody to centromere (kineto-
        chore) in scleroderma sera, <u>Proc. Nat. Acad. Sci., USA</u>,
        77:1627.
Sandberg, A.A., 1980, "Chromosomes in Human Cancer and
        Leukemia", Elsevier, New York.

Shiraishi, Y., Taguchi, H., Niiya, K., Shiomi, F., Kikukawa, K., Kubonishi, S., Ohmura, T., Hamawaki, M., and Ueda, N., 1982, Diagnostic and prognostic significance of chromosome abnormalities in marrow and mitogen response of lymphocytes of acute lymphocytic leukemia, Cancer Genet. Cytogenet., 5:1.

Singh, J.R., and Miltenberger, H.G., 1977, The effect of cyclophosphamide on the centromere separation sequence in the Chinese hamster spermatogonia, Hum. Genet., 39:359.

Tompkins, D., Hunter, A., and Roberts, M., 1979, Cytogenetic findings in Roberts-SC phocomelia syndrome, Clin. Genet., 5:1.

Vig, B.K., 1981, Sequence of centromere separation: an analysis of mitotic chromosomes from long-term cultures of Potorus, Cytogenet. Cell Genet., 31:129.

Vig, B.K., 1982, Sequence of centromere separation: role of centromeric heterochromatin, Genetics, 102:795.

Vig, B.K., 1983, Sequence of centromere separation: occurrence, possible significance and control, Cancer Genet. Cytogenet., 8:249.

Vig, B.K., 1984a, Out-of-phase separation of a G-group chromosome in a woman with chronic myelogenous leukemia, Cancer Genet. Cytogenet., 12:167.

Vig, B.K., 1984b, Sequence of centromere separation: orderly segregation of multicentric chromosomes in mouse L-cells, Chromosoma, 90:39.

Vig, B.K., 1987, Sequence of centromere separation: a possible role for repetitive DNA, Mutagenesis, 2:155.

Vig, B.K., and Athwal, R.S., 1989, Centromere separation: separation in a quasi-stable mouse-human hybrid, Chromosoma, (in press).

Vig, B.K., and Broccoli, D., 1988, Sequence of centromere separation: differential replication of pericentric heterochromatin in multicentric chromosomes, Chromosoma, 96:311.

Vig, B.K., and Paweletz, N., 1988, Sequence of centromere separation: generation of multicentric chromosomes in a rat cell line, Chromosoma, 96:275.

Vig, B.K., and Rattner, J.B., 1989, Centromere, kinetochore and cancer, CRC Critical Rev. Oncogen., (in press).

Vig, B.K., Sternes, K.L., and Paweletz, N., 1989, Centromere structure and function in neoplasia, Cancer Genet. Cytogenet., 43:151.

Vig, B.K., and Zinkowski, R.P., 1986, Sequence of centromere separation: a mechanism for orderly separation of dicentrics, Cancer Genet. Cytogenet., 23:347.

Zinkowski, R.P., Vig, B.K., and Broccoli, D., 1986, Characterization of kinetochores in multicentric chromosomes, Chromosoma, 49:243.

GENETIC ANALYSIS OF GENOTOXIC EFFECTS ON CHROMOSOMES AND CELL

DIVISION IN LOWER EUKARYOTES

Etta Kafer[1] and Andreas Kappas[2]

[1]McGill University
 Biology Department
 Montreal, Canada H3A 1B1

[2]National Research Center "Democritus"
 Institute of Biology
 153 10 Athens, Greece

INTRODUCTION

When test systems use lower eukaryotes for the identification of aneugenic agents, three major questions arise: (i)
How relevant are such tests? i.e., in molecular terms, are the
vulnerable targets sufficiently conserved that similar effects
can be expected in higher organisms? (ii) How informative and
conclusive are the tests developed in lower eukaryotes? i.e.,
can genetic tests identify induced primary aneuploidy and
distiguish such events from other effects that lead to
selection of secondary malsegregants? (iii) How repeatable
are the results obtained? i.e., can such tests easily be used
for routine analysis, and are results easy to interpret and
able to give useful information when multiple effects are
induced?

GENETIC ANALYSIS OF TARGETS FOR THE INDUCTION OF ANEUPLOIDY

A variety of approaches has produced information about
targets of agents that induce aneuploidy and about their
evolutionary conservation. The genetic methods use two
different types of mutations which can reveal cellular
components important for normal segregation: (a) resistance
mutations, and (b) mutations in cell cycle and DNA repair
genes. Their analysis can complement investigations that use
chemicals as probes (Liang and Brinkley, 1985).

Resistance Mutations

Specific targets of chemicals which interfere with
chromosome segregation have in many cases been identified or
confirmed by genetic and molecular analysis of mutations that
confer resistance. Resistant strains are often obtained in
response to treatment with inhibitors of mitosis and chromo-

Mechanisms of Environmental Mutagenesis-Carcinogenesis, Edited by
A. Kappas, Plenum Press, New York, 1990

some segregation, and similar resistance in different species indicates conservation of target proteins, the best-known one being tubulin. Such mutants can also confirm or identify common targets, if they show cross-resistance to several agents; e.g., β-tubulin mutants in CHO cells were found to be resistant to colcemid as well as griseofulvin (Cabral et al., 1980), or in Aspergillus not only to benomyl, but also to pFPA (p-fluorophenylalanine; Morris and Oakley, 1979). In some cases, such resistant strains show reduced growth in the absence of the inhibitor, so that suppressor mutations can be selected. Some of these have been found to occur in genes coding for interacting proteins; e.g., suppressors of benomyl-resistant β-tubulin mutants in Aspergillus were identified as mutations in the tubC gene which apparently codes for an alternate form of β-tubulin (May et al., 1985).

Mutations in Cell Cycle and DNA Repair Genes

Mutations of both these types, which enabled cloning of the corresponding genes, is providing evidence for conservation of proteins that are essential for normal chromosome segregation (as recently discussed by Parry and Parry, 1988, in a sequel to an interesting comparison of mitosis in fungi and higher eukaryotes).

Cell cycle and nondisjunction mutants. Mutations in many genes essential for cell cycle or mitosis, or chromosome segregation in meiosis, have been isolated as temperature sensitive lethals in several fungal species. Some of these mutant strains show large increases in aneuploidy or malsegregation under certain conditions (e.g., Morris et al., 1982; Hartwell and Smith, 1985). A block in mitosis, and release by shifting to permissive temperature, may well lead to nonsynchronous recovery and rather nonspecifically cause malsegregation. However, only some of the cell cycle arrest mutants show this effect. In addition, mutants which do not block mitosis, but more specifically cause aneuploidy or malsegregation, have also been identified and isolated (e.g., Upshall and Mortimer, 1984; Mecks-Wagner et al., 1986). Among these, cold-sensitive lethals which destabilize protein complexes at permissive temperature have been especially valuable (Thomas and Botstein, 1986). While, so far, the function of only a few of the corresponding gene products has been identified, conservation of proteins has been demonstrated in several cases by cross-species complementation or postulated on the basis of recognizable homologies at the DNA level. The latter was found, e.g., for yeast topoisomerases and especially for their active sites in a wide ragne of organisms, from bacteria to yeast and mammals (Lynn et al., 1989). Their relatedness to mammalian enzymes was further evident from tests with antitopoisomerase drugs which were effective also in yeast, once permeable strains became available (Nitiss and Wang, as reported by Resnick, 1989). Similarly, primary cell cycle control genes were found to be homologous not only between different yeasts, but also between fungal and vertebrate genes and proteins (Featherstone, 1989).

DNA repair defective mutants. Mutations in certain DNA repair genes which reduce recombination also increase nondisjunction. These are of special interest when effects are not restricted to meiosis but also found in mitotic cells (e.g., as in the case of <u>rad52</u> in budding yeast; Schild and Mortimer, 1985; or the <u>rec</u>⁻mutations, <u>uvsC</u> and <u>E</u>, in Aspergillus). While homology to genes in higher organisms has not yet been found in most of these cases, wide-ranging relatedness has been demonstrated for several genes or enzymes; e.g., the recombination endo-exonuclease which is reduced to low levels in <u>rad52</u> strains (Chow and Resnick, 1988) shows immunological crossreactivity not only to the <u>recC</u> protein of <u>E. coli</u>, but also endonucleases in higher eukaryotes (Fraser et al., 1986).

Considering the obvious differences in various structural components for mitosis, and especially the finding of unusual intranuclear mitoses in fungi, the molecular conservation of the cellular components responsible for chromosome segregation is indeed remarkable (as concluded by Parry and Parry, 1988).

GENETIC EVIDENCE FOR INDUCED ANEUPLOIDY: YEAST vs. ASPERGILLUS

Genetic tests for induction of mitotic aneuploidy in fungi use mainly the yeast <u>Saccharomyces cerevisiae</u> or the filamentous fungus <u>Aspergillus nidulans</u>. Several methods and different types of approaches have been developed in each species to demonstrate aneuploidy. It seems, therefore, worthwhile to clarify some of the differences between the two systems, and also to identify the similarities which may be hard to see for outsiders, especially because of the differencies in terminology. Since available meiotic data from fungi have recently been reviewed by Bond (1987) and discussed in the light of results obtained in Sordaria, meiotic tests will not be included in the following discussion which will focus on the following three aspects: a) Tolerance and evidence for aneuploidy; b) genetic tests for induced mitotic malsegregation; and c) tests which actually identify aneuploids after treatment of diploid, or haploid, strains.

<u>Tolerance and Evidence for Aneuploidy</u>

At first sight it seems obvious that yeast has a much greater tolerance for aneuploidy than Aspergillus. However, the difference is in fact not very large, once it is taken into account that yeast has at least twice as many chromosomes as Aspergillus and that, on average, these carry at most half as many dosage-sensitive genes.

<u>Yeast aneuploids</u>. The most straight-forward information about aneuploids in yeast comes from the analysis of triploids which, after meiotic segregation, produce 15-20% viable ascospores. As expected, many of these are aneuploids, disomic for several chromosomes, as first shown by Parry and Cox (1970). Disomy for up to 5 chromosomes was identified in test crosses to strains with markers on many chromosomes, but certain combinations of extra chromosomes could be recovered much more frequently than others, either because distribution of homologues was not random or, more likely, because certain types were more viable. In a similar approach, Campbell and Doolittle (1987) aimed at identifying the original genotype of ascospores from triploids using crosses to tester strains with markers on 16 chromosomes. They demonstrated that the original ascospore colonies resulted from independent and random distribution of each set of 3 homologues, but that most

aneuploids suffered extensive loss of extra homologues during analysis. They concluded, that "repeated and non-identical chromosome losses" resulted in more rapidly growing cells, because "multiply-disomic configurations retard efficient cell proliferation".

Similar conclusions were reached by investigators working with systems that produced aneuploid ascospores from diploids. For example, Klapholz and Esposito (1982) analysed meiotic products of diploids homozygous for a meiotic rec⁻ mutant, spoll, which causes aneuploidy. Their results clearly showed that for aneuploid types, disomic for specific chromosomes, the frequency of recovery and the stability was correlated. Similarly, Louis and Haber (1989) found that the most viable aneuploids are usually nearly haploid (or nearly diploid), and that certain but not other doubly disomic types are inviable.

Aspergillus aneuploids, similarities to yeast. Comparing these features of aneuploidy in yeast with those of Aspergillus, the similarity which is rarely mentioned is rather striking; the general instability and varying growth rate of disomics, and the random loss of extra chromosomes which produces more competitive types, is a typical feature of Aspergillus aneuploids. Furthermore, as in yeast, certain specific aneuploids are recovered more frequently than others (Fig. 1). Among 2n+1 trisomics, which often grow almost as well as diploids, types with distinctly reduced conidiation, are most easily identified (e.g., 2n+1 trisomic for chromosome VI). On the other hand among disomics, which all grow less well than haploids, recoveries are high for types that either grow relatively well (e.g., n+1 for IV or VI), or conidiate relatively well (especially n+1 for III), and as in yeast only n+2 with certain pairs of extra chromosomes are viable. Since in yeast, 5 or 6 extra chromosomes appear to be the upper limit for reasonable growth of disomics, n+2 or 3 would be the equivalent in Aspergillus, and is indeed found (e.g., Kafer, 1984).

Aspergillus aneuploids, differences. While yeast aneuploids show various features which are also found in Aspergillus, there is no doubt that qualitative differences exist. A major one is, that triploids of yeast, are not relatively stable but also fertile, while Aspergillus triploids are sterile. Furthermore, the latter are unstable enough to constitute an excellent source of mitotic aneuploids (up to 30% among conidia) and even haploids are easily obtained from triploids. In contrast, diploids of Aspergillus are quite stable and haploids are very rare segregants (10^{-4} to 10^{-5} among conidia). The crucial difference to yeast however is, that Aspergillus haploids have a selective advantage over diploids and therefore, in spite of low frequencies, can produce prominant sectors in various situations of stress, or after chromosome loss which reduces imbalance.

Evidence for aneuploidy or chromosome malsegregation. A considerable difference in procedure arises from the fact that certain disomics of yeast are stable enough to be used for various tests (enumerated by Whittaker et al., 1988) which includes tests for chromosome loss. Even the corresponding 2n-1 types persist long enough in yeast, that at least a

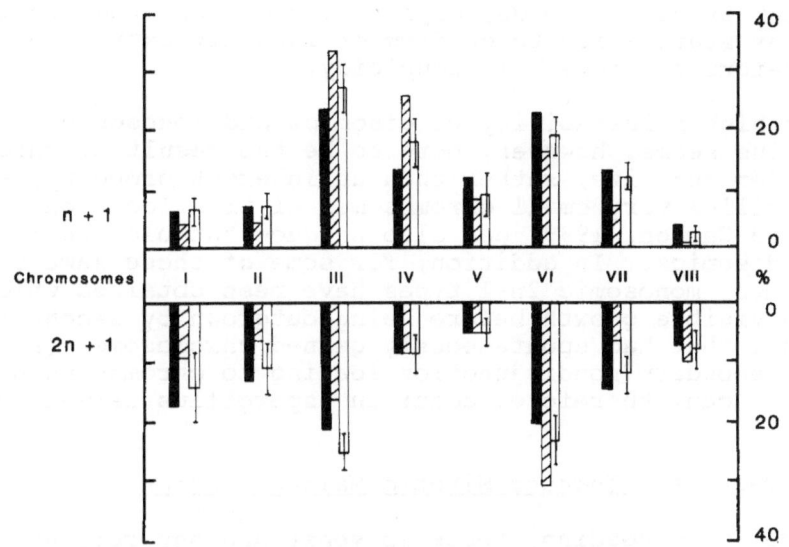

Fig. 1. Typical frequencies of recovery for di- and
trisomies obtained for each of the 8 chromosomes in hyperhaploid and hyperdiploid
segregants of Aspergillus. Relative percent
for each of the 8 types are shown for aneuploids induced by two types of different
compounds:
Benomyl, chloral hydrate, pFPA (total
, n+1: 228, 2n+1: 262);
Bleomycin, MMS and Botran (total
, n+1: 119, 2n+1: 70); and
average (±SE) for all recently
, identified aneuploids (total >600).

fraction can be identified as monosomics when they are
selected in tests for induced aneuploidy (see below). In
contrast, 2n-1 types are not normally found in Aspergillus and
haploid products are frequently obtained. The latter appears
not to be the case in yeast, since even after treatment with
effective aneugens like MBC (methyl benzimidazole-2-yl-
carbamate, the active ingredient of benomyl) and multiple
chromosome loss from diploids, haploid products were not even
mentioned (Wood, 1982). However, expression of recessive
markers simultaneously on many chromosomes was frequent enough
to permit mitotic mapping of genes.

A further obvious difference derives from the mycelial
growth of Aspergillus which, combined with the low competi-
tiveness of heterokaryons (and mononucleate conidia), distinguish Aspergillus not only from yeast but also from Neurospora. These features, added to the reduced growth and
conidiation of aneuploids, result in the recognizable phenotypes of n+1 and 2n+1 aneuploids of Aspergillus (Kafer and
Upshall, 1973). Simple disomics, and some of the trisomics
with one specific extra chromosome, can be recognized

visually, especially if the aneuploid center is replated. Such phenotypic classification, therefore, can replace cytological analysis which is impossible in these fungi. For aneuploids from heterozygous diploids, euploid sectors can then be tested for marker segregation to confirm or identify extra homologues and therefore the level of aneuploidy.

The higher instability of disomics and monosomics in Aspergillus seems, however, more to be the result of chromosome number and size, rather than an inherent property, since in Aspergillus very small chromosomes of translocation strains (Brody and Carbon, 1989) can also produce "stable" well-growing disomics. In addition, for some of these same small chromosomes, monosomic 2n-1 types have been obtained which produced visible growth before being outgrown by secondary segregants that had spontaneously gained chromosomes (Kafer, 1976). Secondary nondisjunction leading to chromosome gain in monosomics can, therefore, occur in Aspergillus as well as in yeast.

Genetic Tests for Induced Mitotic Malsegregation

Under this heading, tests in yeast and Aspergillus are compared which are used to measure increases in the frequency of chromosomal malsegregation, but which do not attempt to demonstrate aneuploidy. A similar comparison has been published in a recent review (Zimmermann, 1987, section on "Mitotic Aneuploidy in Fungi") and aspects clarified there will not be discussed further.

Tests using selection by recessive resistance and colour. Genetic tests for mitotic malsegregation in yeast generally select rare segregants that express certain recessive markers after treatment of heterozygous diploid cells in liquid medium (Parry et al., 1979; Resnick et al., 1986). Since spontaneous frequencies are very low (10^{-5} or less), selection of segregants is essential. The corresponding test in Aspergillus involves treatment of germinating conidia and plating onto selective media, or growth on selective medium containing the chemical to be tested (Gualandi and Morpurgo, 1984; Kafer et al., 1986). In both species, malsegregants are selected in the first instance by resistance, and potential malsegregants identified by colour. Tests for additional markers, present in coupling on the selected homologue, make it possible to estimate the frequency of types which are homozygous for all linked markers, and to judge the likelihood of their origin by chromosome malsegregation rather than by coincidence of more than one event of recombination. For compounds that only induce malsegregation, such tests are economical and reliable.

For testing with yeast, various chromosomes with satisfactory markers for selection on opposite arms are available. This is not the case in Aspergillus, where the only selective markers on chromosomes with colour mutations on the opposite arm require selection with compounds that on their own are able to induce aneuploidy (pFPA and pimaricin, Morpurgo et al., 1979; de Bertoldi et al., 1980). Such tests are therefore mainly used to check for induced recombination (and non-selective tests are used for measurements of chromosomal segregation instead, or in addition).

In both species, when the selective methods are used, mainly euploid segregants expressing all recessive markers of one chromosome are identified, and these are used as a measure of induced malsegregation. In yeast, the induced event is chromosome loss and homozygous diploid segregants apparently arise from 2n-1 "precursor" aneuploids by "upregulation", i.e., spontaneous nondisjunction. The original monosomic aneuploids can be, but only occasionally are identified. In selective tests of Aspergillus, precursor aneuploids have never been recovered and it is assumed (but has never been demonstrated) that 2n malsegregants arise from secondary chromosome loss in 2n+1 types which contain two copies of the homologue carrying the recessive resistance mutation. On the other hand, 2n-1 types are assumed to lead only to the production of haploid segregants which usually predominate (e.g., de Bertoldi et al., 1980).

It has to be kept in mind that in both species such systems, which basically select for chromosome loss and identify euploid segregants homozygous for whole chromosomes, do not in fact give evidence for induced primary aneuploidy. Various other effects that damage chromosomes can also lead to a selective advantage for products of secondary spontaneous nondisjunction. If it is important to know the origin of induced malsegregants, follow-up tests have to be used to support the conclusion of increased aneuploidy.

Nonselective tests for euploid malsegregants in Aspergillus. A number of tests have been developed in this species, which avoid selective media and use recessive colour mutations and tests for linked nutritional markers instead. Such visual identification of induced effects is very satisfactory when frequencies are high, but may require large samples when levels of induction are low (e.g., when brief exposure of conidia in liquid is used: Crebelli et al., 1986; Kafer, 1986). More frequently such tests therefore use longer exposure during growth on media which contain the test chemicals (called "Plate Tests" by Morpurgo et al., 1979). In the latter case, colour sectors from small patches in growing colonies have to be purified, tested and classified. If significant increases are obtained exclusively for chromosomal-type segregants, results are taken to indicate induced aneuploidy (e.g., Kappas et al., 1974).

Tests which Identify Aneuploids Induced in Well-marked Diploid and in Haploid Strains

Selection for extra copies of dosage sensitive markers in yeast. A genetic test which selects for induced aneuploids that are identified on the basis of increased gene dosage has been developed in yeast during the last few years (Whittaker et al., 1988; summarized by Resnick et al., 1989). Two suitable mutants located on the same chromosome in coupling are used, namely a temperature-sensitive nutritional mutation (arg4-8) and a low-level copper resistance mutation (cup1). Such strains show a growth advantage on certain selective media when copy number of the corresponding homologue increases. Hyperploidy can therefore be selected using one marker and confirmed using the other (by objective tests at specific temperatures and concentrations). Meiotic (disomic) as well as mitotic (tri- and tetrasomic) products can give clear evidence of aneuploidy. Unfortunately other genetic types (especially revertants of arg4-8) are also selected and have to be distinguished and eliminated from counts. These and other minor problems (like reduced growth due to coincidental aneuploidy of other chromosomes or "petite" mutations) complicate such tests. So far, not much validation of this system has been published (apart from a small sample using a well-marked diploid and treatment with MBC; Whittaker et al., 1988). The usefulness and performance of the test with problem compounds can therefore not yet be judged.

Nonselective tests for reciprocal products in yeast. Another theoretically attractive method has been proposed and used for tests of EMS by Parry et al., (1979). It involves plating of treated cells from one of the commonly used diploids (e.g., D6) onto non-selective medium on which complementary products of nondisjunction (or reciprocal crossing over) can produce half-sectored red-while colonies which can be analysed further. Tests of the white sectors for linked recessive markers are able to identify potential 2n-1 types and companion red sectors can then be analysed to get evidence of the complementary 2n+1 products. However, except in cases where frequencies are high, this nonselective test is unlikely to be practical; it has apparently not been used for any of the recently reported tests (Parry and Parry, 1988).

Nonselective tests for aneuploids from well-marked Aspergillus diploids. Similar conclusive evidence for induced aneuploidy can be obtained in Aspergillus when germinating conidia are treated and plated onto complete medium. In this case, coloured half-colonies are only found as reciprocal products of mitotic crossing over after treatments which induce recombination. In contrast, after induced nondisjuction, trisomics (2n+1 for a single or several chromosomes) are the main products recovered. These can be recognized by slightly abnormal growth patterns and can be identified by testing of euploid sectors that segregate for colour mutations and linked markers (as described above, section on "Aspergillus aneuploids, differences"). In samples with relatively low frequencies of aneuploids, certain trisomics with very characteristic phenotypes can even be classified without further testing. Complex types which segregate for markers on many chromosomes, can be replated if analysis is desirable, and usually will produce recognizable tri- and disomics in abundance. In combination, these visual and genetic tests conclusively identify primary aneuploids resulting from chromosome gain during the first mitotic division of germinating conidia.

In contrast, if the test compounds (also) have clastogenic or recombinogenic effects, other types of imbalanced segregants are produced and more detailed analysis is required for identification of whole-chromosome aneuploidy. For reliable genetic identification of induced types, detailed analysis of small but random samples of informative abnormal colonies, is the best approach. This involves replating of aneuploid centers and testing of their euploid sectors homozygous for well-marked chromosomes. Only if chromosomal segregation predominates in all cases, is aneuploidy indicated. On the other hand, if crossover sectors are (also) frequent in some of the analysed abnormal types, clastogenic effects and partial aneuploidy is much more likely. This is confirmed by the observation, that each replated colony shows a different appearance (being the result of a unique chromosomal mutation) and typical trisomic phenotypes are rare.

Results from such tests are conclusive in most cases but, like in the two preceding methods for yeast, the extra steps for identification of aneuploids sound complicated and make this method not very attractive to those interested in the development of easy and simple routine tests. One additional diagnostic finding is, that clastogens equally induce unbalanced sectoring types in quiescent as well as dividing nuclei, while aneugens (and recombinogens) are effective only if germinating conidia with dividing nuclei are treated (as shown in Table 1 for an experiment in which conidia of a well-marked diploid were treated with ethyl acetate; methods as in Kafer, 1984). Similar differences can be observed also in yeast when stationary vs. long-phase cells are treated.

Treatment of haploid strains of Aspergillus. When haploid strains are treated, one very obvious difference is found: only malsegregation can produce viable imbalanced products, while clastogenic effects produce mainly lethal events (some imbalanced, relatively stable types are also found, but these do not show frequent euploid sectors). Therefore, induction of increased levels of abnormal, frequently sectoring colonies in haploid germinating conidia is conclusive evidence for induced primary aneuploidy. Furthermore, the majority of such aneuploids are expected to be relatively simple disomic types which can often be identified visually. Genetic analysis of such aneuploids is unnecessary (and is, of course, not possible). The only drawback of this test is, that such types are usually not very frequent, so that fairly large samples are required. Also, relatively low plating densities have to be used which permit growth and sector formation of aneuploid colonies and lead to good recovery or even visual identification of aneuploids in the original platings. Tests using haploids are presumably also feasible with the system of Whittaker et al., (1988) in yeast, but no results with haploid strains have been published so far.

STANDARDIZATION OF TESTS, UNEXPECTED RESULTS AND OPEN QUESTIONS

In this section certain aspects of mitotic tests and of recent results in yeast and Aspergillus wil be compared: a) standardization of protocols, modified procedures with different effects on results in the two species;: b) the question of metabolic activation for aneugens; c) testing of chemicals with multiple effects and identification of primary vs. secondary induced aneuploidy.

Standardization of Protocols

Tester strains. Considering the protocols for yeast which have actually been used to test large numbers of chemicals, it is obvious that methods of exposure are relatively uniform and that efforts are being made to use basically similar tester strains (Resnick et al., 1986). However, as found for Aspergillus (Kafer et al., 1986) the choice between simple vs.

Table 1. Ethyl Acetate Effects on Survival and Aneuploid Frequencies of Diploid, Quiescent or Germinating, Conidia of Aspergillus

Dose of ethyl acetate Concentration %	Treatment h	Total No. of colonies	Percent survival of Q	Percent survival of P	Abnormal colonies %	Abnormal colonies [No]
Q: Quiescent conidia (in buffer at 30°)						
0	–	937	100		1.2	[11]
1.25	2	146	100		2.0	[3]
2.5	2	384	100		1.6	[6]
5	2	163	29		1.8	[3]
6.5	2	114	2		0.9	[1]
P: Germinating conidia[b] (in liquid complete medium at 37°)						
Germinating control						
0	0[b]	93	50[c]		1	[1]
	1.5[d]	148	51[c]—>100		3.4	[5]
1.25	2[b]	399		100	3	[12]
1.5	2	510		67	3.7	[19]
2.0	2	310		6	4.5	[14]
2.5	2	70		1	7.1	[5]

[a] All replated and shown to be typical tri- or disomic aneuploids.
[b] Hours of treatment after 3 hours of preincubation.
[c] Reduced "survival", i.e. less colony forming units, as a result of clumping preceding germination.
[d] Untreated conidia germinate after 1.5 h further incubation (total about 4.5 h).

58

informative test diploids requires a compromise and agreement on the best strains or procedures is difficult, especially if large data bases have already been collected with slightly varying strains and protocols. The following observations illustrate how certain variables can unexpectedly influence results, how minor differences can be important, and how unforeseen results can influence methods of testing.

Media. An interesting and unexpected finding has recently been reported by Taylor-Mayer et al., (1988) who demonstrated profound influence on frequencies of induced segregants when yeast cells were exposed in supplemented minimal vs. peptone-containing complete medium (the latter is normally used). Increasing standardization of media is especially needed for Aspergillus tests, since different minimal and complete media are used in each laboratory, not only during exposure, but also for growth (and the ingredients for the "standard" media of Pontecorvo et al., 1953, are no longer available).

Stage and age of starting culture. It is probably general knowledge that freshly grown cells (of yeast, or conidia of Aspergillus) show more uniform vigorous growth, but also that mature but not old cells show better survival. In yeast, this higher resistance to the lethal effects of some but not other mutagens was found to be correlated with levels of induced recombination (Davies and Parry, 1976). Standardization of starting inoculum for the cultures which will be treated is therefore also required (but such details are rarely even mentioned).

Cold storage and cold treatment. In yeast, starting cultures are stored while samples are being assayed for spontaneous frequencies to avoid "jackpots" and obtain reliable control data (e.g., Whittaker et al., 1988). This useful procedure turns out to be difficult in Aspergillus when germinating conidia need to be treated. Cold storage (4-6°C) apparently affects the stability of control cells, and increases levels of "spontaneous" aneuploids if storage lasts >36 hrs (Table 2). In contrast, treated cells show little additional increase of aneuploidy even after storage for several days (see Table 2, bottom half). On the other hand, shorter (e.g., overnight periods) of cold treatment which enhance induction of aneuploids by aceton or ethyl acetate in yeast (Zimmermann and Mayer, 1984; Zimmermann et al., 1985), have not produced any singificant effects in controls or acetone-treated cells of Aspergillus (Table 3).

Insolubility and solvent effects. Since various solvents cause aneuploidy (e.g., not only acetone, but also ethanol), treatment with water-insoluble compounds at effective concentrations can become very difficult. In such cases, DMSO is a frequently used solvent. However, DMSO has been found to increase aneuploids in some systems (Fulton and Bond, 1984; Resnick et al., 1986), while being negative in others (e.g., in Drosophila, Traut, 1983). DMSO also has created problems in some tests by enabling induction of aneuploidy by certain other agents which on their own were inactive (e.g., pFPA in Neurospora; Griffiths, 1982; or botran in Aspergillus; Kafer, 1989), or by suppressing the effects of certain active

Table 2. Effects of Cold Storage on Survival and Aneuploid Frequencies in Platings of Germinating Diploid Conidia of Aspergillus, Controls and Treatments with Aneugens

Averages of 5 platings (± SE) or of 2-3 platings (samples of >1500 conidia plated)

Preincubation h (37°)	Survival (percent)[a] Days of storage at 4°					Frequency of aneuploids (% of survivors) Days of storage at 4°				
	0	1.5	2	3-5	7-8	0	1.5	2	3-6	7-8
Untreated										
0	100[a]	101	101 ±7	92	94 ±6	0.8 ±1	1.0	1.0 ±0.3	1.3	1.3 ±0.1
2	98 ±8	80	80 ±10	74	53 ±2	1.5 ±0.2	0.8	1.1 ±0.4	1.6	1.6 ±0.6
3	68[b] ±8	50	35 ±12	25	28 ±10	1.5 ±0.6	2.0	4.3 ±1.8	6.9	9.3 ±1.8
4 – 4.5	42[b] ±6	19	13 ±4	16	8.3 ±3	1.7 ±0.4	1.8	6.1 ±2.2	9.3	13.0 ±3
Treated with pFPA (5 mM) for 5-6 hours (37°) after preincubation[a]										
2	42[a]					12			15	
3	30					19			22	
Treated with chloral hydrate (10 mM) for 3 vs 2 hours (after 2 vs 3 h preincubation)										
2	16[a]	13		14		5		22	21	
3	12	8		8		16		21	37	
Treated with benomyl (6 µg/ml) for 6.5 hours during germination										
	9[a]			8.5		78		75		

[a] Survival of untreated conidia is calculated relative to "day 0, quiescent" control (i.e., 0 h preincubation), while survival of treated, germinating, conidia is based on "day 0, germinating" control (i.e., 4-4.5 h preincubation).

[b] while reduced survival of "day 0, control" (after 3-4.5 h preincubation) is due to clumping of conidia (see Table 1), further reduction after storage does not result from increased clumping.

Table 3. Acetone Induced Aneuploidy in Diploid and Haploid Germinating Conidia of Aspergillus and Lack of Increase by Cold Treatment

Strain Ploidy	Acetone concentration %	Pregermination in CM at 37° h	Treatment at 37° in CM + acetone h	After cold storage[b] h	Sample size No	Survival %	Frequencies of abnormal colonies %	Factor of change[a] Acetone effect	Cold effect
2n	0	0			559	100	1.9±0.5	1[a]	
		3			448	86±13	1.3±1.1		
		4.5			874	46±8.7	2.3±1.2		
		4.5		(0.4)	669	21±7.9	2.4±1.0		1.0
	3	3	1-2		1593	24±7.4	7.3±2.6	3	
		3	1.5	0.8	1540	23±6.9	4.0±1.1		0.55
	6	3	1-2		541	10±8	18±6.9	8	
		3	1	1.0	759	4±3	12		0.66
	9	3	1-2		228	0.6±0.3	28±3	12	
		3	1	1.0	197	0.2	27		0.96
n	0	0			1318	100	0.6±0.4	1[a]	
		3			1255	91±8	0.5±.03		
		4.5			1975	59±7	0.5± 0.2		
	1.5	3	1 or 2		1652	45±6.5	0.5±.04	1	
	3-4	3	2		2299	56±8	1.5±0.3	3	
	6	3	2		1616	20±1.3	4.0±1.4	8	
	8.9	3	2		510	1.0±0.8	8.5±6	17	

[a]Increase of aneuploids (=tested abnormals) compared to control, and relative frequencies after cold treatment.
[b]Some samples were stored overnight in ice (="cold treatment") and reincubated ≈1 h next day.

61

chemicals (e.g., nocodazole and ethyl acetate in yeast; Mayer and Goin, 1987).

Chronic exposure in plate tests of Aspergillus. A further aspect that will require standardization in Aspergillus is the method of inoculation for exposure in "plate tests" (mentioned above). Such exposure can be very suitable for stable, poorly soluble chemicals and several variants of inoculation are used, each with certain advantages. Plating conidia onto or in overlays of media which contain chemicals, has the advantage that aneuploid colonies may be visually recognizable, but it results in variable plating densities and variable colony size. An alternative method is to use a standardized number of transfers which controls available space for growth and determines maximal colony size (Kappas, 1978; Demopoulos et al., 1982, as used for tests summarized in Table 4).

Metabolic Activation for Aneugens

The question whether any chemicals might be modified by the human metabolism to produce aneugens from inactive compounds has recently arisen (discussed e.g., by Albertini et al., 1988). Certainly there seems to be little evidence at this time that exogenous sources of metabolic activity would be appropriate in meiotic tests (Parry and Parry, 1988). However, when the expected effects are unknown, addition of S9 mix to plates may detect altered effects of chemicals and add information about possible modifications by the human metabolism. Such unexpected modifications by S9 mix of the effects on mitotic segregation in Aspergillus diploids during treatment with econazole and diethylstilbestrol (DES) are shown in Table 4 (methods as described by Kappas, 1978, and Demopoulos et al., 1982). Different results for DES have also been obtained for different procedures with yeast by Albertini et al., (1988) and so far no explanation for the various contradictory data has emerged.

Tests for Primary vs. Secondary Induced Aneuploidy

A variety of observations (including the ones described above) demonstrate that chromosomal nondisjunction, and even more so chromosome loss, are not independent of other genotoxic effects. The following aspects will be discussed briefly: (i) effects of inhibitors of recombination on nondisjunction, and crossing over in aneuploids; (ii) effects of clastogens on nondisjunction; (iii) chemicals with multiple effects and the question of separate vs. interrelated mechanisms.

Recombination and aneuploidy. Lack of recombination is often correlated with increased nondisjunction (most prominently in rec⁻ mutants, like rad52 mentioned above). Correlated effects on these two processes also have been observed in other organisms, especially for meiosis. In the case of disomics in yeast, mitotic recombination in the centromere region, but not in regions distant from centromeres, were found to be associated with "restoration" of haploidy, i.e., with chromosome loss (Esposito et al., 1982). This is different from mitotic crossing over in diploids which is not concentrated around the centromere in yeast (as found

Table 4. Effects of Metabolic Activation (S9 mix) on Frequency of Segregants (Crossovers or Malsegregants) in Diploids of Aspergillus Exposed to Diethylstilbestrol (DES) and Econazole during Colony Growth

Compound	Concentration (µg/ml)	Toxicity[a]		Number of mitotic segregants per 100 colonies[b]							
				Total (all types)		-S9			+S9		
		-S9	+S9	-S9	+S9	Crossovers	Malsegregants Total (2n+n)		Crossovers	Malsegregants Total (2n+n)	
Control	0	-	0	28	21	18	10	(2+8)	14	7	(3+4)
DES	0.4	2	3	28[c]	43	-[c]	-[c]		35	8	(1+7)
	0.5	5	5	16	49	-	-		37	12	(0+12)
	0.75	9	11	17	41	-	-		30	11	(1+10)
	1.0	13	15	20	45	-	-		33	12	(2+10)
	1.5	16	20	20[c]	44	-[c]	-[c]		36	8	(0+8)
Econazole	0.01	26	2	59	35	24	35	(5+30)	30	5	(1+4)
	0.02	40	4	51	47	13	38	(12+26)	40	7	(0+7)
	0.03	48	6	38	44	3	35	(4+31)	36	8	(1+7)
Benomyl	0.15	18	-	180	-	30	150	(70+80)	-	-	
	0.3	32	-	282	-	27	235	(87+148)	-	-	

a Reduction of colony diameter (% of control), 3 days after inoculation.
b Average of 3-6 experiments, >100 colonies analysed for each chemical concentration.
c Totals unchanged, types not identified (presumably similar to control).

in Drosophila and Aspergillus, possibly related to differences in centromere structure?). For Aspergillus, evidence for high levels of mitotic crossing over in the centromere region of aneuploids, especially disomics, has been observed for ethanol-induced aneuploids (Kafer, 1984) and was demonstrated for disomics by Assinder et al., (1986; if one takes into account the recent correction of the genetic map by Arst, 1988). Therefore, in some cases, recombination may appear to be increased when euploid sectors from aneuploid segregants are tested. However, careful analysis of individual aneuploid types which produce crossover sectors can demonstrate, that in such cases restoration of euploidy and better growth is not a consequence of recombination. On the other hand, the latter is typically found when colonies heterozygous for chromosomal aberrations produce crossover sectors that eliminate distal deletions (or nondisjunctional types from the same colony). The euploid, homozygous normal, products recovered in such cases are basically complementary to the homozygous mutant types observed in certain retinoblastoma tumor cells (Hansen and Cavenee, 1988).

Clastogens and nondisjunction. Among the indirect genotoxic effects that can be important for meiotic aneuploidy are translocations. These aberrations, even when balanced but heterozygous, cause large increases of meiotic nondisjunction of the heterostructural homologues. Similar increases of secondary nondisjunction occur also in vegetative cells of Aspergillus, and presumably yeast, but frequencies are modified by selection and therefore are hard to judge. In addition, unbalanced aberrations induced by clastogens can indirectly increase spontaneous nondisjunction, especially if this leads to restitution of euploidy. Tests which can identify any of these primary effects, are therefore, useful for distinctions between induced aneuploidy and secondary spontaneous events of nondisjunction plus selection.

Multi-effect chemicals and multi-purpose tests. Many compounds tested in Aspergillus have shown increases of all types of genetic segregants, crossovers as well as chromosomal types. In such cases, testing of euploid segregants is inadequate for identification of induced aneuploidy. Detailed analysis of unbalanced primary products, on the other hand, clearly demonstrates that such results can be caused by more than one type of primary effect. In Aspergillus, malsegregants induced by γ-rays could be identified as secondary spontaneous products from nuclei with chromosomal aberrations, as expected for this clastogenic agent (Kafer, 1986; however, the recent intriguing results in yeast by Resnick et al., 1989, reopen the question of radiation-induced aneuploidy). On the other hand, unexpectedly, clastogenic as well as aneugenic effects were found after treatments with MMS (methylmethane sulfonate; Kafer, 1988). Apparently genotoxic targets other than DNA are likely for this and other alkylating agents, so that chromosomal aberrations as well as aneuploids may be induced (Bonatti et al. 1986; Abbondandolo, 1989). Similarly, in yeast, hydroxyurea appears to induce aneuploidy in addition to other effects (Mayer et al., 1986). This could be related to the observed inhibitory effect of hydroxyurea on DNA replication, in addition to its mutagenic and recombinogenic activity. The results of this detailed and careful study are very suggestive of dual effects, but in the sample shown "monosomics" were increased only at the highest dose (from 0.2% to 1% of selected segregants). At that dose reciprocal crossing over in nonselected samples was also highly increased (>15-fold; and the more relevant measurements of recombination close to the centromere were carried out only at a lower dose). Actual identification of aneuploids would therefore have been of special interest.

REFERENCES

Abbondandolo, A., Bonatti, S., Cercignani, G., De Ferrari, M., Ipata, P.L., Rocco, M., Tozzi, M.G., and Viaggi, S., 1989, The role of nucleotide pool alkylation in the induction of numerical chromosome aberrations, in: "Aneuploidy, Part C: Mechanisms of Origin", B.K. Vig and M.A. Resnick, eds., Alan Liss, New York.

Albertini, S., Friederich, U., and Wurgler, F.E., 1988, Induction of mitotic chromosome loss in the diploid yeast Saccharomyces cerevisiae D61.M by genotoxic carcinogens and tumor promotors, Environ. Mol. Mutagen., 11:497.

Arst, Jr., H.N., 1988, Localisation of several chromosome I genes of Aspergillus nidulans: implications for mitotic recombination, Mol. Gen. Genet., 213:545.

Assinder,S.J., Giddings, B., and Upshall, A., 1986, Mitotic crossing over in chromosome I disomics of Aspergillus nidulans, Mol. Gen. Genet., 202:382.

Bonatti, S., De Ferrari, M., Pisano,V., and Abbondandolo, A., 1986, Cytogenetic effect by alkylated guanine in mammalian cells, Mutagenesis, 1:99.

Bond, D.J., 1987, Mechanisms of aneuploid induction, Mutation Res., 181:257.

Brody, H., and Carbon, J., 1989, Electrophoretic karyotype of Aspergillus nidulans, Proc. Natl. Acad. Sci., USA, 86:6260.

Cabral, F., Sobel, M.E., and Gottesman, M.M., 1980, CHO mutants resistant to colchicine, colcemid or griseoful-vin have an altered β-tubulin, Cell, 20:29.

Campbell, D.A., and Doolittle, M.M., 1987, Coincident chromo-somal disomy in meiotic dyads from triploid yeast, Curr. Genet., 12:569.

Chow, T.Y.K., and Resnick, M.A., 1988, An endo-exonuclease activity of yeast that requires a functional RAD52 gene, Mol. Gen. Genet., 211:41.

Crebelli, R., Bellincampi, D., Conti, G., Conti, L., Morpurgo, G., and Carere, A., 1986, A comparative study on selected chemical carcinogens for chromosome malsegre-gation, mitotic crossing-over and forward mutation induction in Aspergillus nidulans, Mutation Res., 172:139.

Davies, P.J., and Parry, J.M., 1976, The induction of mitotic gene conversion by chemical and physical mutagens as a function of culture age in the yeast, Saccharomyces cerevisiae, Mol. Gen. Genet., 148:165.

De Bertoldi,M., Griselli, M., and Barale, R.,1980, Different test systems in Aspergillus nidulans for the evaluation of mitotic gene conversion, crossing-over and non-disjunction, Mutation Res., 74:303.

Demopoulos, N.A., Kappas, A., and Pelecanos, M., 1982, Recombinogenic and mutagenic effects of the antitumour antibiotic bleomycin in Aspergillus nidulans, Mutation Res., 102:51.

Esposito, M.S., Maleas, D.T., Bjornstad, K.A., and Bruschi, C.V., 1982, Simultaneous detection of changes in chromosome number, gene conversion and intergenic recombination during mitosis of Saccharomyces cerevisiae: spontaneous and ultraviolet light induced events, Curr. Genet., 6:5.

Featherstone, C., 1989, The complexities of the cell cycle, Trends Biochem., 14:85.

Fraser, M.J., Chow, T.Y.K., Cohen, H., and Koa, H., 1986, An immunochemical study of Neurospora nucleases, Biochem. Cell Biol., 64:106.

Fulton, A.M., and Bond, D.J., 1984, Dimethylsulfoxide induces aneuploidy in a fungal test system, Mol. Gen. Genet., 197:347.

Griffiths, A.J.F., 1982, Short-term tests for chemicals that promote aneuploidy, in: "Chemical Mutagens, Principles and Methods for their Detection", Vol. 7, F.J. de Serres and A. Hollaender, eds., Plenum Press, New York.

Gualandi, G., and Morpurgo, G., 1984, Methods for detecting the induction of mitotic chromosomal misdistribution in Aspergillus nidulans, in: "Handbook of Mutagenicity Test Procedures", B.J. Kilbey, M. Legator, W. Nichols and C. Ramel, eds., Elsevier, Amsterdam.

Hansen, M.F., and Cavenee, W.K., 1988, Retinoblastoma and the progression of tumor genetics, Trends Genet, 4:125.

Hartwell, L.H., and Smith, D., 1985, Altered fidelity of mitotic chromosome transmission in cell cycle mutants of S. cerevisiae, Genetics, 110:381.

Kafer, E., 1976, Mitotic crossing over and nondisjunction in translocation heterozygotes of Aspergillus, Genetics, 82:605.

Kafer, E., 1984, Disruptive effects of ethyl alcohol on mitotic chromosome segregation in diploid and haploid strains of Aspergillus nidulans, Mutation Res., 135:53.

Kafer, E., 1986, Tests which distinguish induced crossing-over and aneuploidy from secondary segregation in Aspergillus treated with chloral hydrate or γ-rays, Mutation Res., 164:145.

Kafer, E., 1988, MMS-induced primary aneuploidy and other genotoxic effects in miotic cells of Aspergillus, Mutation Res., 201:385.

Kafer, E., 1989, Botran and bleomycin induce crossing over, and bleomycin also increases aneuploidy in diploid strains of Aspergillus, Mutation Res., (submitted).

Kafer, E., Scott, B.R., and Kappas, A., 1986, Systems and results of tests for chemical induction of mitotic malsegregation and aneuploidy in Aspergillus nidulans, Mutation Res., 167:1.

Kafer, E., and Upshall, A., 1973, The phenotypes of the eight disomics and trisomics of Aspergillus nidulans, J. Heredity, 64:35.

Kappas, A., 1978, On the mechanisms of induced somatic recombination by certain fungicides in Aspergillus nidulans, Mutation Res., 51:189.

Kappas, A., Georgopoulos, S.G., and Hastie, A.C., 1974, On the genetic activity of benzimidazole and thiophanate fungicides on diploid _Aspergillus nidulans_, _Mutation Res._, 26:17.

Klapholz, S., and Esposito, R.E., 1982, A new mapping method employing a meiotic rec⁻ mutant of yeast, _Genetics_, 100:387.

Liang, J.C., and Brinkley, B.R., 1985, Chemical probes and possible targets for the induction of aneupoidy, _in_: "Aneuploidy. Etiology and Mechanisms", V.L. Dellarco, P.E., Voytek and A. Hollaender, eds., Plenum Press, New York.

Louis, E.J., and Haber, J.E., 1989, Non recombinant meiosis I nondisjunction in _Saccharomyces cerevisiae_ induced by tRNA ochre suppressors, _Genetics_, 123:81.

Lynn, R.M., Bjornsti, M.A., Caron, P.R., and Wang, J.C., 1989, Peptide sequencing and site-directed mutagenesis identify tyrosine-727 as the active tyrosine of _Saccharomyces cerevisiae_ DNA topoisomerase I, _Proc. Natl. Acad. Sci. USA_, 86:3559.

May, G.S., Weatherbee, J.A., Gambino, J., Tsang, M.L.S., and Morris, N.R., 1985, Identification and function of beta tubulin genes in _Aspergillus nidulans_, _in_: "Molecular Genetics of Filamentous Fungi", W.E. Timberlake, ed., Alan Liss, New York.

Mayer, V.W., and Goin, C.J., 1987, Aneuploidy induced by nocodazole or ethyl acetate is suppressed by dimethyl sulfoxide, _Mutation Res._, 187:31.

Mayer, V.W., Goin, C.J., and Zimmermann, F.K., 1986, Aneuploidy and other genetic effects induced by hydroxyurea in _Saccharomyces cerevisiae_, _Mutation Res._, 160:19.

Meeks-Wagner, D., Wood, J.S., Garvik, B., and Hartwell, L.H., 1986, Isolation of two genes that affect mitotic chromosome transmission in _S. cerevisiae_, _Cell_, 44:53.

Morpurgo, G., Bellincampi, D., Gualandi, G., Baldinelli, L., and Crescenzi, O.S., 1979, Analysis of mitotic nondisjunction with _Aspergillus nidulans_, _Envir. Health Perspect._, 31:81.

Morris, N.R., Kirsch, D.R., and Oakley, B.R., 1982, Molecular and genetic methods for studying mitosis and spindle proteins in _Aspergillus nidulans_, _Methods Cell Biol._, 25:107.

Morris, N.R., and Oakley, C.E., 1979, Evidence that p-fluorophenylalanine has a direct effect on tubulin in _Aspergillus nidulans_, _J. Gen. Microbiol._, 114:449.

Parry, E.M., and Cox, B.S., 1970, The tolerance of aneuploidy in yeast, _Genet.Res._, 16:333.

Parry, E.M., and Parry, J.M., 1988, Induced aneuploidy in fungi and higher eukaryotes, _in_: "Aneuploidy, Part B: Induction and Test Systems", B.K. Vig and A.A. Sandberg, eds., Alan Liss, New York.

Parry, J.M., Sharp, D., and Parry, E.M., 1979, Detection of mitotic and meiotic aneuploidy in the yeast _Saccharomyces cerevisiae_, _Environ. Health Perspect._, 31:97.

Pontecorvo, G., Roper, J.A., Hemmons, L.M., Macdonald, K.D., and Bufton, A.W.J., 1953, The genetics of _Aspergillus nidulans_, _Adv. Genet._, 5:141.

Resnick, M.A., 1989, Aneuploidy, _Genome_, 31:469.

Resnick, M.A., Mayer, V.W., and Zimmermann, F.K., 1986, The detection of chemically induced aneuploidy in <u>Saccharomyces cerevisiae</u>: an assessment of mitotic and meiotic systems, <u>Mutation Res.</u>, 167:47.

Resnick, M.A., Skaanild, M., and Nilsson-Tillgren, T., 1989, Lack of DNA homology in a pair of divergent chromosomes greatly sensitizes them to loss by DNA damage, <u>Proc. Natl. Acad. Sci. USA</u>, 86:2276.

Schild, D., and Mortimer, R.K., 1985, A mapping method for <u>Saccharomyces cerevisiae</u> using <u>rad52</u>-induced chromosome loss, <u>Genetics</u>, 110:569.

Taylor-Mayer, R.E., Mayer, V.W., and Goin, C.J., 1988, Effect of treatment medium on induction of aneuploidy by nocodazole in <u>Saccharomyces cerevisiae</u>, <u>Environ. Mol. Mutagen.</u>, 11:323.

Thomas, J.H., and Bolstein, D., 1986, A gene required for the separation of chromosomes on the spindle apparatus in yeast, <u>Cell</u>, 44:65.

Traut, H., 1983, The solvent dimethyl sulfoxide (DMSO) does not induce aneuploidy in oocytes of <u>Drosophila melanogaster</u>, <u>Environ. Mutagen</u>, 5:273.

Upshall, A., and Mortimore, I.D., 1984, Isolation of aneuploid-generating mutants of <u>Aspergillus nidulans</u>, one of which is defective in interphase of the cell cycle <u>Genetics</u>, 108:107.

Whittaker, S.G., Rockmill, B.M., Blechl, A.E., Maloney, D.H., Resnick, M.A., and Fogel, S., 1988, The detection of mitotic and meiotic aneuploidy in yeast using a gene dosage selection system, <u>Mol. Gen. Genet.</u>, 215:10.

Wood, J.S., 1982, Genetic effects of methyl benzimidazole-2-yl-carbamate on <u>Saccharomyces cerevisiae</u>, <u>Mol. Cell. Biol.</u>, 2:1064.

Zimmermann, F.K., 1987, Genetic evidence for aneuploidy in fungi, <u>in</u>: "Aneuploidy, Part A: Incidence and Etiology", B.K. Vig and A.A. Sandberg, eds., Alan Liss, New York.

Zimmermann, F.K., and Mayer, V.W., 1984, Induction of aneuploidy by oncodazome (nocodazole), an antitubul agent, and acetone, <u>Mutation Res</u>., 141:15.

Zimmermann, F.K., Mayer, V.W., Scheel, I., and Resnick, M.A., 1985, Acetone, methyl ethyl ketone, ethyl acetate, acetonitrile and other polar aprotic solvents are strong inducers of aneuploidy in <u>Saccharomyces cerevisiae</u>, <u>Mutation Res</u>., 149:339.

DNA REPAIR AND THE RECOMBINATION BARRIER BETWEEN DIVERGENT (HOMOLOGOUS) CHROMOSOMES

Michael A. Resnick[1] and Torsten Nilsson-Tillgren[2]

[1]Yeast Genetics/Molecular Biology Group
National Institute, Environmental Health
Sciences
Research Triangle Park, NC, 27709, U.S.A.

[2]Institute of Genetics
University of Copenhagen, Denmark

INTRODUCTION

Recombinational interactions between homologous chromosomes are important in meiosis and are generally considered relevant to chromosomal segregation. In mitotic cells sister chromatid recombination is a common response to DNA damage and it appears to have repair function. Using the model eukaryote Saccharomyces cerevisiae, we have pursued the role and mechanisms of recombination in repair, meiotic development and chromosome segregation or stability (Resnick, 1987; Nilsson-Tillgren et al., 1986; Kielland-Brandt et al., 1983). This yeast provides an opportunity to examine these processes in synchronized mitotic and meiotic cells, identify mutants, obtain relevant proteins as well as understand the relationship between mitotic and meiotic functions.

The role for recombination in the repair of DNA double-strand damage has been well-established in yeast (Resnick, 1976, and summarized in Resnick, 1987). Many genes have been identified that are required for the repair of double-strand breaks (DSBs) induced by ionizing radiation (Game, 1983). Mutants defective in DSB repair have been shown to be defective in repair of double-strand damage induced by proralens and bleomycin (Magana-Schwencke et al., 1982; Chanet et al., 1985; Moore, 1978). Several of these recombination genes are dispensable in mitotic cells but essential during meiotic development due to defects in recombination (Game et al., 1981; summarized in Resnick, 1987).

Double-strand breaks can be repaired via recombination in a variety of eukaryotic systems and may account for genomic rearrangements and even antibody diversity (Kucherlapati and Smith, 1988). We are addressing the importance of DSBs and consequences to genomic stability when they remain unrepaired.

Mechanisms of Environmental Mutagenesis-Carcinogenesis, Edited by
A. Kappas, Plenum Press, New York, 1990

Results with yeast have demonstrated that DSBs in rad52 mutants, which lack repair, can act as dominant lethal lesions (Ho and Mortimer, 1973; J. Nitiss and M. Resnick, unpublished results). However, the dominant lethal consequence is likely to be due to a defect in an intermediate step in recombination (Resnick et al., 1986).

DSB DAMAGE IN HOMOLOGOUS CHROMOSOMES

We have taken a unique approach to investigating the genetic and biochemical consequences of DSBs (and double-strand damage) in the genome of wild type (for repair) yeast. The system that we have developed involves using diploid strains containing 15 (of 16) homologous chromosome pairs and one pair of DNA divergent (homologous) chromosomes. Because double-strand damage induced in homologous chromosomes of G-1 cells can be repaired via recombination, it is possible to focus specifically on damage and repair between in the homologous chromosomes. Since the homologous chromosomes examined are those for which monosomy can be tolerated, we can detect both breakage and loss.

The homologous chromosome pairs that have been investigated are chromosomes III and V (procedures for developing such strains and the genetics are described in Nilsson-Tillgren et al., 1981; and Holmberg, 1982). The homologous chromosomes were obtained from the related yeast Saccharomyces carlsbergensis using individual chromosomes transfer techniques. For chromosome III (370-kb), the "left-half" of the chromosome is divergent and the right-half is homologous; chromosome V (~570-kb) is entirely divergent. Based on DNA hybridization studies, the level of homology in the divergent regions is 80-90% within genes and decreases in the noncoding DNA. Since meiotic recombination as measured by tetrad analysis is absent in the divergent regions and normal in the homologous region, the DNA divergence is clearly a block to recombination (see Nilsson-Tillgren, et al., 1986; Holmberg, 1982; and Resnick et al., 1989 for a further discussion of these strains).

Genetic consequences of radiation-induced DNA damage could be monitored by using chromosomes that were well-marked genetically on both sides of the centromere. The following is an example of the chromosomes III (Resnick et al., 1989); the chromosomes are divergent from the left end up to the mating type locus and homologous beyond MAT:

his4-x	+		MAT@	+	S. carlsbergensis
his4-y	leu2	MATa	thr4		S. cerevisiae

To determine the effects of DSBs, G-1 cells (>95%) were exposed to ionizing radiation in a Shepherd Cesium source and plated to fully nutrient plates. Changes in phenotype were determined by replica-plating to medium diagnostic for the genetic markers on chromosomes III or V. All experiments were performed at survival levels greater than 50%. For chromosome III, the his4-x/his4-y alleles are complementing. A genetic change is indicated by colonies that are histidine requiring. The chromosome V genetic signal is the recessive canavanine resistance marker.

In strains containing a homologous chromosome III or V pair, low nonlethal doses (<10krad) of ionizing radiation were highly effective in causing chromosome loss (Resnick et al., 1989). As many as 10 to 15% of colonies arising from the irradiated cells were aneuploid for one or the other homoeologous chromosome. The genetic observation was confirmed by whole chromosome gel (OFAGE) electrophoresis. Induction of aneuploidy for chromosomes III and V was proportional to the size of the divergent region. The level of aneuploidy among colonies from irradiated cells containing homologous chromosomes was 20-50 fold less. Based on the reported results for chromosomes III (Resnick et al., 1989) and unpublished results with V (M. Resnick and T. Nilsson-Tillgren), we have concluded that in the absence of precise homology, the recombinational repair of DSBs is ineffective. The absence of repair leads to profound chromosomal consequences, i.e., aneuploidy. In addition we have found that there is a relatively high frequency of chromosome III translocations. Nearly 1% of the genetic "signalling"events are associated with translocation. We also conclude that unrepaired DSBs are not lethal lesions although they can lead to lethality either through some incompleted chromosome metabolic processing in mutants such as rad52 and possibly through the production of translocations.

Strains containing homologous chromosomes, therefore, provide a highly sensitive system for genetically detecting DSBs due to ionizing radiation and characterizing the consequences of such damage. It is likely that the resolution of the system can be extended down to a few rad and that it can be used for identifying and characterizing other types of double-strand damage.

DSB INDUCED RECOMBINATION BETWEEN HOMOLOGOUS CHROMOSOMES

Although there is a good correlation between the induction of DSBs and aneuploidy for chromosomes III (cf. Figure 2 in Resnick et al., 1989) and V (unpublished), less than half the DSBs in the divergent DNA appear to cause aneuploidy. The efficiency of DSB induction was estimated from observations with total chromosomal DNA ($4 \times 10-4$ DSBs/kb/krad; Resnick and Martin, 1976). Since translocations can be induced and since the apparent efficiency is much less than 100%, we considered that there may be a limited amount of recombinational repair between the divergent DNAs even though there was less than 80-90% homology. To investigate this idea, two approaches were taken. The first involved the use of completely divergent DNA and the second examined intragenic recombination between heteroalles of a gene in the homologous chromosomes.

Completely divergent human HeLa DNA (370 kb) was contained in a novel yeast vector provided by Dr. Philip Hieter. The stability and transmission of this artificial human chromosome fragment in yeast could be followed genetically using a marker system that involved the accumulation of a precursor in adenine synthesis (Hieter et al., 1985). Less than 5% of the vector contained yeast sequences. It was

expected that there would be no opportunity for recombinational DSB repair in the single copy artificial chromosome in
G-1 cells. As anticipated, chromosome loss was induced by
ionizing radiation; however, the efficiency was nearly a
factor of two over that for a pair of homologous yeast
chromosomes. The presence in cells of a pair of human
artificial chromosomes with the same HeLa fragment suppressed
the induced loss. We conclude that for the yeast homologous
chromosomes there is a limited capability for DSB recombinational repair whereas there is no opportunity for repair of
breaks in a single HeLa containing yeast artificial chromosomes.

We directly confirmed the recombinational repair between
divergent chromosomes by detecting the induction of prototrophs between his4 noncomplementing heteroalleles in the
divergent portion of chromosome III. In terms of meiotic
recombination, such alleles are relatively inactive as
compared to intragenic recombination between homologous
chromosomes. Low radiation doses (<10 krad) increased
recombination 20-30 fold over the low spontaneous level. A
mutation in the gene PMS1 whose product is involved with
mismatch correction (Williamson et al., 1985; Kramer et al,
1989) resulted in a further 3-4 fold enhancement (M. Resnick,
Z. Zgaga,S. Fogel and T. Nilsson-Tillgren, submitted).

CONCLUSIONS

Our results have demonstrated that DNA divergence
normally prevents recombination in mitosis and meiosis. As a
general consideration such divergence may be a contributing
factor in speciation. However, DNA damage and DSBs in
particular can act as initiating lesions to overcome the
recombinational barrier. These observations are important for
understanding mechanisms by which translocating might occur.
Possibly they form between related DNA sequences during
recombinational repair.

The recombinational repair between divergent sequences
might also lead to novel forms of genes. We are currently
investigating this idea as well as the genetic control of
systems that normally prevent recombination between homoeologous chromosomes. The importance of recombination between
divergent DNAs has recently been demonstrated for crosses
between two bacterial species Salmonella typhimurium and
Escherichia coli (Rayssiguier et al., 1989). Exchange of
genetic material only occurs in mutants defective for particular mismatch ocrrection genes. For the case of yeast we have
shown that DNA lesions can also overcome recombinational
barriers presented by divergent DNAs.

REFERENCES

Chanet, R., Cassier, C., and Moustacchi, E., 1985, Genetic
        control of the bypass of mono-adducts and of the repair
        of cross-links photoinduced by 8-methoxypsoralen in
        yeast, Mutat. Res., 145:145.

Game, J.C., 1983, Radiation sensitive mutants and repair in yeast, in: "Yeast Genetics", J.F.T. Spencer, D.M. Spencer, and A.R.W. Smith, eds., Springer Verlag, New York.

Game, J.C., Zamb, T.J., Braun, R., J., Resnick, M.A., and Roth, R., 1981, The role of radiation (rad) genes in meiotic recombination in yeast, Genetics, 94:51.

Hieter, P., Mann, C., Snyder, M., and Davis, R.W., 1985, Mitotic stability of yeast chromosomes: a colony color assay that measures nondisjuction and chromosome loss, Cell, 40:381.

Ho, K., and Mortimer, R.K., 1973, Induction of dominant lethality by X-rays in a radiosensitive strain of yeast, Mutat. Res., 20:45.

Holmberg, S., 1982, Genetic differences between Saccharomyces carlsbergensis and S. cerevisiae II. Restriction endonuclease analysis of genes in chromosome III, Carls. Res. Comm., 47:233.

Kucherlapati, R., and Smith, G.R., eds., 1988, "Genetic Recombination", American Society for Microbiology, Washington D.C.

Kielland-Brandt, M.C., Nilsson-Tillgren, T., Petersen, J.G.L., Homberg, S., and Gjermansen, C., 1983, Approaches to the genetic analysis and breeding of brewer's yeast, in: "Yeast Genetics", J.F.T. Spencer, D.M. Spencer, and A.R.W. Smith, ed.s, Springer Verlag, New York.

Kramer, B., Kramer, W., Williamson, M.S., and Fogel, S., 1989, Heteroduplex DNA correction in Saccharomyces cerevisiae is mismatch specific and requires functional PMS genes, Molec. Cell. Biol., 9: 4432.

Magana-Schwencke, N., Chanet, J.A.P., Chanet, R., and Moustacchi, E., 1982, The fate of 8-methoxypsoralen photoinduced cross-links in nuclear and mitochondrial yeast DNA: comparison of wild-type and repair-deficient strains, Proc. Natl. Acad. Sci. U.S.A., 70:1722.

Moore, C.W., 1978, Response of radiation-sensitive mutants of Saccharomyces cerevisiae to lethal effects of bleomycin, Mutat. Res., 51:165.

Nilsson-Tillgren, T., Gjermansen, C., Kielland-Brandt, M.C., Petersen, J.G.L., and Holmberg, S.,1981, Genetic differences between Saccharomyces carlsbergensis and S. cerevisiae. Analysis of chromosome III by single chromosome transfer, Carls.Res. Comm., 46:65.

Nilsson-Tillgren, T., Gjermansen, C., Holmberg, S., and Petersen, J.G.L., Kielland-Brandt, M.C., 1986, Analysis of chromosome V and the ILV1 gene from Saccharomyces carlsbergenesis, Carlsberg Res. Comm., 51.

Raysiggeuer, C., Thaler, D.S., and Radman, M., 1989, The barrier to recombination between Escherichia coli and Salmonella typhimurium is disrupted in mismatch-repair mutants, Nature, 342:396.

Resnick, M.A., 1976, The repair of double-strand breaks in DNA: a model involving recombination, J. Theoret. Biol., 59:97.

Resnick, M.A., 1987, Investigating the genetic control of biochemical events in meiotic recombination, in: "Meiosis", P. Moens, ed., Academic Press.

Resnick, M.A. and Martin, P., 1976, The repair of double-strand breaks in the nuclear DNA of Saccharomyces cerevisiae and its genetic control, Mol. Gen. Genet., 143:119.

Resnick, M.A., Nitiss, J., Edwards, C., and Malone, R.E.,
        1986, Meiosis can induce recombination in RAD52 mutants
        of Saccharomyces cerevisiae, Genetics, 113:531.
Resnick, M.A., Skaanild, M., and Nilsson-Tillgren, T., 1989,
        Lack of DNA homology in a pair of divergent chromosomes
        greatly sensitizes them to loss by DNA damage, Proc.
        Nat. Acad. Sci. USA, 86:2276.
Williamson, M.S., Game, J.C., and Fogel, S., 1985, Meiotic
        gene conversion mutants in Saccharomyces cerevisiae. I.
        Isolation and characterization of pms1-1 and pms1-2,
        Genetics, 110:609.

CONSEQUENCES OF ALTERING TUBULIN LEVELS IN YEAST

Brant Weinstein and Frank Solomon

Department of Biology and Center for Cancer
Research
Massachusetts Institute of Technology
Cambridge, Massachusetts 02139, U.S.A.

INTRODUCTION

Microtubules are the central component of the mitotic and
meiotic spindles, and their proper assembly and function is
vital to the fidelity of chromosome segregation. Twenty-five
years of microtubule research has provided the fundamentals
for understanding microtubule function. All microtubules are
composed primarily of two different kinds of proteins, alpha
tubulin and beta tubulin. The genes for these proteins have
been isolated and sequenced from a wide variety of organisms.
Alpha and beta tubulin form a heterodimer, which serves as the
subunit of the microtubule polymer. The properties of the
heterodimer are fairly well understood, and _in vitro_ assembly
experiments have defined parameters and factors that control
the intrinsic ability of tubulin to self-assemble. Nontubulin
proteins that bind to microtubules have been identified, and
are candidates for elements that may control microtubule
assembly, organization, and function in the cell. Biochemical
approaches have been complemented by genetic analyses of
microtubule function, demonstrating the essentiality of the
tubulin proteins and dissecting their functional properties.
Nontubulin proteins required for microtubule function have
also been identified by mutational analysis, and some of them
may turn out to be the same as the proteins identified
biochemically.

The yeast _Saccharomyces cerevisiae_ has proven a
particularly useful model system for genetic and molecular
genetic analyses of microtubules. It has simple and well
defined microtubule structures that change in a reproducible
way throughout the cell cycle (which is itself well defined by
cellular morphology). There are only two genes for alpha
tubulin and a single gene for beta tubulin, in contrast to
higher eukaryotes which generally contain many of both. The
tubulin genes have all been cloned and sequenced, and as
mentioned above mutational analysis has yielded information on
the functional properties of the tubulins as well as potential
microtubule associated proteins. We have chosen to study
microtubules in yeast for many of the reasons noted above, but
we have undertaken to probe quantitative rather than qualita-

tive aspects of microtubule assembly and function. In previous
studies gene disruption was used to entirely eliminate
tubulins or other proteins implicated in microtubule function
from the cell, and mutational analysis was used to introduce
novel altered forms of the tubulins or the other proteins into
the cell. We have asked instead how changes in the levels of
the tubulins affect microtubule assembly and function, posing
a number of specific questions:

- Living cells contain heterodimeric tubulin in two
  biochemically separable pools; assembled into microtubule structures and in unassembled soluble form. The
  relative amount of tubulin in each varies in a reproducible way throughout the cell cycle. What are the
  consequences of changes in the relative or absolute
  amounts of heterodimer in these pools ? What are the
  mechanisms the cell uses to control its levels of
  heterodimer ?

- The levels of alpha and beta tubulin are always
  equivalent and tubulins have never been observed in
  monomeric form in vivo. Why is this apparent homeostasis maintained in cells ? What would the consequences
  be of imbalances in the levels of alpha and beta
  tubulin relative to one another ? Do the individual
  tubulin proteins have any unique properties ? What are
  the mechanisms that the cell uses to maintain the
  equivalence of tubulin monomers ?

We have sought to address these questions through two
general sorts of approaches in yeast. First, we have constructed strains with moderate increases or decreases in the
dosages of tubulin genes by eliminating gene copies or
introducing extra gene copies into yeast cells. The levels of
tubulin mRNA and tubulin protein are analyzed, as well as the
phenotypic consequences to the cell. These experiments were
the contribution of Wendy Katz, Vida Praitis, Lorraine Pillus,
and Peter Schatz, who was a joint student with David Botstein.
Secondly, we have transiently overexpressed both of the
tubulin gene products using a strong inducible promoter,
GAL10, and again studying the consequences to the cell of
exposure to excess tubulins. These results lead to a number of
possible answers to the questions posed above.

CONSEQUENCES OF ALTERATIONS IN TUBULIN LEVELS IN YEAST

Reducing Tubulin Gene Dosages

In the process of cloning and characterizing the two
yeast alpha tubulin genes, TUB1 and TUB3, we observed that
elimination of the two copies of TUB3 from a diploid yeast
cell made the cells sick, while elimination of one of the two
copies of TUB1 was lethal. We also found that the product of
the TUB1 gene accounted for the majority of the alpha tubulin
in wild type yeast cells and that the two genes were functionally interchangeable at sufficient levels of expression
(Schatz et al., 1988). These results were assumed to indicate
that substantial reduction in heterodimeric tubulin levels

could not be tolerated by yeast cells. Subsequent experiments demonstrated that this original interpretation was incorrect.

A diploid "hemizygote" yeast strain was constructed with one of the two copies of the yeast beta tubulin gene, TUB2, eliminated (manuscript in press). Unlike the hemizygotes for TUB1 or TUB3, these grew normally, sporulated well, and had no detectable increase in the frequency of chromosome transmission errors. Their microtubule structures also appeared normal by immunofluorescence staining. Examination of the levels of the tubulin polypeptides in these cells by western blotting showed that the levels of beta tubulin were reduced to 43% of the wild type level. These cells would therefore be capable of forming only 43% as much heterodimer as a wild type cell, and yet they were phenotypically normal (with what appeared by inspection to be normal amounts of assembled microtubules). This level of heterodimer is even less than a TUB1 hemizygote would be expected to produce. Therefore, reduction in heterodimer levels was apparently not the problem with the alpha tubulin hemizygotes, leading to the conclusion that the deleterious defect in alpha tubulin hemizygotes was due rather to the presence of excess, unheterodimerized beta tubulin. And since beta tubulin hemizygotes do contain an excess of alpha tubulin (the TUB1 gene product is present at 81% of the level in wild type cells), a small excess of this protein apparently does not cause the same deleterious effect as excess beta tubulin.

Increasing Tubulin Gene Dosages

The conclusion that excess, unheterodimerized beta tubulin (but not alpha tubulin) was harmful was given further support by experiments in which extra copies of tubulin genes were introduced into yeast cells. Haploid yeast cells were constructed with duplications of the TUB2 gene or extra TUB1 genes on a low copy number plasmid. The strain into which extra copies of the TUB1 gene had been introduced showed an approximately 30% increase in alpha tubulin levels, and no increase in beta tubulin. As in the case of the beta tubulin hemizygotes, the accumulation of excess alpha tubulin in these strains has no detectable phenotype. Surprisingly, the cells into which an extra copy of the TUB2 gene had been introduced also appeared to grow fairly normally. However, examination of tubulin protein and RNA levels in these cells revealed that although these cells did in fact contain extra beta tubulin, this increase in beta tubulin had been accompanied by a corresponding increase in alpha tubulin as well, in spite of the fact that the construction had been designed only to increase the gene dosage of the beta tubulin gene. We suggest that cells compensated for increases in the TUB2 dosage by the generation of aneuploid cells containing an extra copy of chromosome 13, which bears the two alpha tubulin genes. A similar phenomenon was observed while attempting to obtain diploids hemizygous for the TUB1 gene. All of the strains isolated that bore the marker for the TUB1 deletion turned out to be triploid for chromosome 13, resulting in the restoration of TUB1 to its full diploid gene dosage (Schatz et al., 1988). We are currently performing matings to haploids with marked chromosome 13 to test this hypothesis.

Inducible Overexpression of Tubulin Proteins

The experiments we have performed using yeast strains with stable alterations in tubulin gene dosage are consistent with the conclusion that the presence of beta tubulin in excess of alpha tubulin is harmful to cells, but that the converse is not true. Since all of the strains that we were able to study in this way were either unaffected by the alterations in their tubulin gene set (i.e., constructions resulting in excess alpha tubulin) or in some way compensated for their alterations (probably by further alterations through the generation of aneuploids), we have not been able to directly observe from these experiments how excess beta tubulin causes its toxic effects. For this we turned to a transient tubulin overexpression system.

We constructed high copy number plasmids with the TUB1 or TUB2 genes transcriptionally fused to the galactose-inducible promoter GAL10, and introduced these plasmids into diploid yeast cells. When grown on glucose- or raffinose-containing media, the plasmid-encoded tubulin genes were not expressed and the cells were entirely normal. When galactose was added to cultures of these cells growing in raffinose media, however, the plasmid-encoded tubulin gene products were strongly induced (approximately 25-fold within 8 hours after galactose addition) and a variety of phenotypes were elicited, depending on the plasmid(s) the strain carried. Cells overexpressing beta tubulin in this manner underwent division arrest, dramatic loss of viability, rapid depolymerization of all normal microtubule structures, and at relatively high levels of overproduced tubulin formed novel beta tubulin-enriched structures. Cells overexpressing alpha tubulin also underwent cell division arrest, but unlike the beta tubulin overproducing cells they did not undergo dramatic loss of viability, did not depolymerize their microtubules, and did not form any structure from the overproduced protein.

Models for Microtubule Depolymerization by Excess Beta Tubulin

In trying to understand what the phenotypes of cells transiently overexpressing tubulins could tell us about the nature of the defect in the alpha hemizygote and extra-beta haploid, we focused on the microtubule depolymerization phenotype for a number of reasons. Microtubule loss occurs at a fairly low level of beta tubulin overproduction; 50% of the cells in an induced culture have lost all of their microtubules by the time beta tubulin has accumulated to only about a two-fold excess. Microtubule loss is also the only one of the three beta-tubulin specific phenotypes that is clearly related to normal microtubules. The dramatic loss of viability may occur for secondary reasons and the ability of beta tubulin to form structures may be unrelated to its capacity to interact with normal microtubule structures. Two general sorts of models suggest themselves for how overproduction of beta tubulin could effect depolymerization of microtubules. Excess beta could interact with components necessary for microtubule assembly, either the tubulin heterodimers or proteins whose function is required for microtubule stability, to reduce their effective concentration in the cell. Alternatively, excess beta could interact directly with assembled microtubules or microtubule organizing centers to inhibit microtubule assembly. Either of these interactions could result in depolymerization of microtubules, which are inherently dynamic structures.

One model previously suggested (Burke et al., 1989) postulates that loss of microtubules in cells overproducing beta tubulin is a direct result of the formation of the novel beta tubulin-enriched structures. According to this hypothesis these structures, although composed predominantly of beta tubulin, act as a low energy "sink" for normal alpha/beta tubulin heterodimers and thus result in the depolymerization of normal microtubules. A prediction of this hypothesis would be that the alpha tubulin in the cell should be relocalized into the novel beta tubulin-enriched structures coincidentally with the disappearance of normal microtubule structures. This prediction is not borne out by our data. Kinetic experiments show that 50% of the cells in a beta tubulin-overproducing culture have lost their microtubules by 30 minutes after galactose addition. The novel beta tubulin-enriched structures are not observed at all prior to approximately 100 minutes after galactose addition, and there is clearly no quantitative reassortment of the alpha tubulin into any structure detectable by immunofluorescence until more than an hour after loss of microtubules. Furthermore, in cells which have been transferred to glucose media after a brief induction with galactose, reappearance of normal microtubules occurs in many cells that still retain beta structures. We cannot eliminate the possibility, however, that excess beta tubulin interacts with alpha-beta heterodimers to form small, nonproductive complexes (such as beta-beta-alpha trimers). These complexes would have to be incapable of adding onto the beta-enriched structures, and analogous complexes would not be formed by either alpha tubulin (to make, for example, alpha-alpha-beta trimers) or derivatives of beta tubulin.

Another model would propose that excess beta leads to the formation of beta-beta dimers, which sequester beta tubulin to produce a deficiency of alpha-beta heterodimers. In order for quantitative removal of beta tubulin from the heterodimer pool to take place, the dissociation constant for formation of beta-beta homodimers must necessarily be not substantially larger than that for alpha-beta heterodimers, since loss of microtubules occurs when beta has accumulated to three times its normal level or less. However, making the dissociation constants for beta-beta dimers and heterodimers roughly equivalent requires that there be a substantial amount of beta-beta dimer present even when the levels of beta and alpha tubulin are equivalent. There is no evidence for beta homodimers in normal cells.

Beta tubulin could also interact with non-tubulin microtubule associated factors necessary for microtubule stabilization. In vitro studies have shown that microtubule-associated proteins from higher eukaryotes are capable of binding to peptides derived from the tubulin proteins (Littauer et al., 1986). To test whether we could reproduce some of the phenotypes associated with overproduction of beta tubulin by overproduction of a sub-portion of the beta molecule, we engineered a series of derivatives of the TUB2 gene for overproduction in yeast. Derivatives were constructed encompassing the 5' or 3' halves of beta tubulin, the 5' or 3' three-quarters of beta tubulin, lacking only the last 12 amino acids ("Tailless" beta tubulin), or containing a stop codon a few amino acids downstream of the initiator methionine (so that abundant message but not protein would be produced). These derivatives were put onto high-copy plasmids under the control of the galactose promoter and subjected to the same sorts of analyses described for the full-length tubulin gene products. Only the largest of these derivatives, the "Tailless" TUB2 lacking the divergent last 12 amino acids, had any phenotypic effect on the cells overproducing it. This derivative, it should be noted, is capable of supporting normal microtubule function. Cells in which the wild type chromosomal copies of TUB2 have been replaced with the tailless gene are virtually wild type in all respects, including vegetative growth, sporulation, and mating ability. The only detectable phenotype of this replacement is a hypersensitivity to the antimicrotubule drug benomyl (Katz and Solomon, 1988). The peptides that strongly bound MAP2 and tau in the experiments of Littauer et al. were all derived from the C-terminal end of beta tubulin. Since both 3'1/2 and 3'3/4 derivatives of beta tubulin were overproduced to high levels in our experiments, loss of microtubules and other phenotypes associated with beta tubulin overproduction may not be explainable by this model.

Excess beta tubulin could also be physically interacting with microtubules or microtubule organizing centers (in yeast, the spindle pole bodies). In the former case, addition of unheterodimerized beta tubulin would occur at microtubule ends, preventing further elongation by heterodimeric tubulin and resulting in depolymerization of the microtubule. In the latter case, free beta tubulin would bind to the spindle pole bodies during normal depolymerization/repolymerization of microtubules, preventing addition of heterodimeric tubulin and repolymerization of the microtubules. Rigorous tests of these models would require in vitro experiments assessing the ability of purified excess yeast beta tubulin to interfere with assembly of purified yeast tubulin heterodimers, perhaps with purified yeast spindle pole bodies as well. The components for such experiments are unavailable at present.

Although it is attractive to postulate that the microtubule loss phenotype is causally related to the dramatic loss of viability and to the toxicity of beta tubulin in the experiments on modest changes in gene dosage, we have at present no evidence for such a link.

REGULATION OF TUBULIN LEVELS IN YEAST

Autoregulation in Higher Eukaryotes

A large body of work has been performed on the regulation of tubulin levels in higher eukaryotic cells. These experiments have elucidated an autoregulatory mechanism whereby cells adjust the rate of degradation of their tubulin messenger RNA according to the level of unassembled tubulin in the cytoplasm. This effect was first observed by Ben-Ze'ev and Penman (1979), who performed two dimensional gels on extracts from ^{35}S methionine pulse-labelled cells after treatment with the antimicrotubule drugs colchicine or vinblastine. They found that colchicine, which depolymerizes microtubules and increases the size of the unassembled tubulin pool, resulted in a specific decrease in the synthesis of the tubulins. Vinblastine, which sequesters tubulin and thus decreases the size of the unassembled pool, had the opposite effect. They further found that the ability of isolated total RNA from these same cells to yield tubulin proteins in _in vitro_ translation reactions also correlated with the expected changes in the unassembled heterodimer pool. Cleveland, et al., (1981) demonstrated the same phenomena using specific antitubulin antisera and DNA probes. Caron et al., (1985) and Pittinger and Cleveland (1985) both showed that a specific reduction in tubulin protein synthesis and tubulin mRNA levels in response to microtubule depolymerizing drugs could occur in enucleated "cytoplasts" as well as in intact cells. Their results indicated that the primary level of control for the autoregulatory effect in most cells was degradation of messenger RNA, not synthesis. Caron et al. (1985) showed using lower concentrations of drug to only partially depoly- merize cellular microtubules that a direct correlation could be shown between the amount of tubulin in polymer and the level of tubulin synthesis. The results of Caron et al. also provided support for the notion that the autoregulatory control mechanism senses the level of tubulin in the unas- sembled heterodimer pool rather than in microtubules. They found that tubulin mRNA degradation was more sensitive to depolymerization of microtubules in a cell type which con- tained 85% of its tubulin as assembled tubulin (and thus 15% in the unassembled pool) than in another cell type which contained 47% of its tubulin in the assembled pool. This is the result one might expect if the quantity of tubulin in the unassembled pool is being sensed, since depolymerization would cause a much greater increase in tubulin synthesis in the former cell type than in the latter. Other papers from Don Cleveland's lab (Gay et al., 1987; and Yen et al., 1988) showed that the cis-acting site for autoregulatory control resided within the first 13 coding nucleotides of the tubulin mRNA's. Their results further indicated that the control mechanism in fact recognized the translation product of the first 12 nucleotides of tubulin mRNA's, the peptide sequnce MREI. This tetrapeptide motif is highly conserved amongst tubulins (it is present at the amino terminus of the yeast tubulins) and is not found at the amino termini of any other protein yet sequenced. Cleveland's group was also able to demonstrate using protein synthesis inhibitors, polysome gradient analysis, and 3' truncated tubulin translation products that the autoregulatory mechanism requires that the mRNA be bound by ribosomes and capable of synthesizing a translation product of greater than 41 amino acids (Pachter et al., 1987; and Yen et al., 1988).

Tubulin Heterodimer Regulation in Yeast

We have used yeast cells containing altered dosages of tubulin to study tubulin regulation. As we mentioned, yeast tubulins do have the conserved MREI sequence at their amino termini, and the sequence from yeast beta tubulin is capable of substituting for higher eukaryotic autoregulatory sequences in transient expression assays, so one might expect the autoregulatory mechanism to be conserved as well. Our results so far cast some doubt upon this, however. Diploid hemizygotes with one of their two copies of beta tubulin deleted contain only 43% as much total tubulin heterodimer as wild type diploid cells. Although direct measurement of the proportions of this tubulin in the assembled and unassembled pools by biochemical fractionation has been difficult in the hemizygotes, the appearance of the microtubule structures in these cells by immunofluorescence is indistinguishable from wild type cells, suggesting that there has been little or no deletion of assembled tubulin. If this is so, then there has of necessity been a reduction in the unassembled tubulin pool in these cells of between 50% and 100% without any increase in tubulin synthesis (since total tubulin level decreased in direct correlation to the gene dosage). This result suggests that there is no "up-regulation" of tubulin synthesis in yeast cells in response to decreased levels of unassembled tubulin. However, the evidence for up-regulation is not compelling in higher eukaryotes either. Treatment of higher eukaryotic cells with vinblastine or other antimicrotubule drugs which dramatically decrease the levels of unassembled tubulin give at most a two-fold increase in the levels of tubulin mRNAs, and in most experiments much less or even no increase at all. Most of the experiments in higher eukaryotes involved increases in the unassembled tubulin pool rather than decreases. We have performed genetic analogs of these experiments as well, by making modest increases rather than decreases in tubulin gene dosage. In haploid yeast cells containing extra copies of the genes for both beta tubulin (integrated into the genome) and alpha tubulin (on a low copy number plasmid), messenger RNA levels for both genes are increased to levels that mirror their gene dosages. However, there is little or no increase in either tubulin polypeptide. Thus it appears that although there has been a down-regulation of tubulin levels in these cells, in striking contrast to higher eukaryotes it has occurred at the level of the proteins rather than their messenger RNA's. We are currently investigating further the nature of tubulin heterodimer regulation in yeast.

Tubulin Monomer Regulation in Yeast

The regulation experiments in higher eukaryotes concern control of tubulin levels in response to levels of tubulin heterodimer. They do not address the problem of how it is that cells maintain equivalence of the two components of these heterodimers, alpha and beta tubulin. We have begun to study this problem, again using yeast with small stable changes in their tubulin gene dosages. These experiments have been difficult to perform for cells with extra gene dosages of beta tubulin or reduced gene dosages of alpha tubulin (in either case resulting in an excess dose of beta over alpha) because of the toxicity of beta tubulin described above. However, the results are reasonably clear for cells manipulated to contain excess gene dosage of alpha tubulin. In the diploid beta tubulin hemizygotes the levels of tubulin messenger RNA in the cell accurately reflect the gene dosages. That is, there is a normal complement of alpha tubulin message and one-half the normal level of beta tubulin message. Beta tubulin protein is also reduced to approximately half the normal level. The levels of the TUB1 alpha tubulin gene product are at approximately 81% of normal. These results suggest first of all that there is no regulation at all at the level of messenger RNA. They also indicate that there is no up-regulation of beta tubulin protein synthesis (or reduction in beta tubulin degradation) to increase the levels of beta tubulin to match those of alpha tubulin. There does appear to be some degree of down-regulation of alpha tubulin protein, although this regulation appears to be incomplete (since the levels of alpha tubulin were only reduced to 81% of wild type, not 50%). Similar sorts of results have been obtained in the haploid cells containing extra copies of the TUB1 alpha tubulin gene. These cells also contain tubulin messenger RNA levels that accurately reflect the dosages of the genes encoding them, again indicating a lack of regulation at the RNA level. They have normal levels of beta tubulin protein but levels of alpha tubulin 30% in excess of normal. This 30% increase in TUB1 protein is substantially less than the approximately three-fold increase in the TUB1 message, and again indicates that there has been down-regulation of the levels of the alpha tubulin protein. We have not yet determined for either the beta tubulin hemizygotes or the cells bearing excess alpha tubulin genes whether the down regulation of alpha tubulin protein that we observe occurs through decreased translation or increased degradation.

Our findings with respect to maintenance of equivalence in alpha tubulin and beta tubulin levels are quite analogous to the results we have described for regulation at the level of tubulin heterodimer. That is, there appears to be no regulation at all at the level of tubulin mRNA and that regulation of tubulin protein levels occurs solely by down-regulation of protein levels in some manner and not by their up-regulation. A caveat has to be attached to these results, as mentioned above, since we cannot test message and protein levels in the symmetrical case in which beta tubulin gene dosages are engineered to be in excess of alpha tubulin gene dosages (i.e., the diploid alpha tubulin hemizygote and haploid beta tubulin duplication).

Regulation of Tubulin Assembly in Yeast

In addition to regulating the total levels of their tubulin polypeptides, eukaryotic cells must also regulate the relative proportion of this tubulin present in assembled form, as well as the structural and functional properties of the assembled tubulin. All of these properties are constantly in flux in growing cells, and one would expect that tight control would need to be exerted to prevent errors in chromosome transmission. Although it has been postulated for some cells that the expression of differential tubulin isotypes can help to determine these properties (Cleveland, 1987), this clearly cannot be the case in yeast which has only a single beta tubulin gene and two functionally interchangeable alpha tubulin genes (Schatz et al., 1986). In yeast cells the properties described above must be determined by nontubulin microtubule-associated factors, and the search for these factors is already being pursued genetically and biochemically by many labs. One consequence of the reduction in tubulin heterodimer levels in the diploid beta tubulin hemizygotes that we have constructed is that there has necessarily been a dramatic shift in the distribution of heterodimer from the unassembled to the assembled pool (since, as described above, the amount of assembled tubulin was probably unchanged). We would therefore postulate that there has been an induction or increase in the levels of factor(s) stabilizing assembled microtubules in these cells. Since they grow well and are amenable to further analysis it might be possible to genetically or biochemically identify such factors, for example by simply identifying proteins whose levels are increased in the hemizygotes relative to wild-type cells. Two dimensional gel analysis of proteins that associate with microtubules in yeast has already led to the identification of a number of potential microtubule-stabilizing factors (ref Pillus), and these would be attractive candidates for factors whose levels might be elevated in the hemizygotes.

SUMMARY AND CONCLUSIONS

We have used two different approaches to study the consequences of alterations in the levels of the tubulin proteins, and to assess how normal levels of these proteins are maintained. We have exploited the powerful molecular genetic techniques available in yeast to construct cells containing either moderate, stable increases or decreases in tubulin gene dosage or inducibly overexpressed high levels of the tubulin gene products. The results of both approaches suggest that unheterodimerized beta tubulin has an intrinsic capacity to interfere with the function of normal microtubules that is not possessed by alpha tubulin. Furthermore, the inducible overexpression experiments reveal that only a two-fold excess of beta tubulin in vivo causes depolymerization of normal microtubules, an effect not duplicated by alpha tubulin. Based on a number of considerations which we discuss, we propose that this destabilization may be the result of physical interaction of unheterodimerized beta tubulin with the ends of microtubules or with microtubule organizing centers.

We find no evidence for regulation of tubulin messenger RNA by changes in the unassembled tubulin heterodimer pool, in contradiction to the results from higher eukaryotes, although we do find evidence for down-regulation at the level of the tubulin proteins. We do not yet understand the reason for the discrepancy between our results and previous reports of autoregulatory control of tubulin mRNA levels. Our experiments do have the advantage of having been performed in phenotypically normal cells (the TUB2 hemizygote) without the use of antimicrotubule drugs. We also find that yeast cells are capable of dramatically adjusting the relative proportions of tubulin present in the assembled and unassembled pools to cope with alterations in total tubulin levels. We hope to extend these results to determine the mechanism of this regulation and identify the nontubulin factors involved.

Finally, our results make the important point that the assembly and function of microtubules are quite sensitive not only to qualitative but also to quantitative defects in the tubulins (and perhaps other minor microtubule associated proteins as well). Defects in the regulation of synthesis and stability of microtubule constituents should be considered as possible sources of aneuploidy in eukaryotic cells.

REFERENCES

Ben-Ze'ev, A., Farmer, S.R., and Penman, S., 1979, Mechanisms of regulating tubulin synthesis in cultured mammalian cells, Cell, 17:319.

Burke, D., Gasdaska, P., and Hartwell, L, 1989, Dominant effects of tubulin overexpression in Saccharomyces cerevisiae, Molecular and Cellular Biology, 9:1049.

Caron, J.M., Jones, A.L. and Kirschner, M.W., 1985, Autoregulation of tubulin synthesis in hepatocytes and fibroblasts, The Journal of Cell Biology, 101:1763.

Caron, J.M., Jones, A.L., Rall, L.B. and Kirschner, M.W., 1985, Autoregulation of tubulin synthesis in enucleated cells, Nature, 317:648.

Cleveland, D.W., 1987, The multitubulin hypothesis revisited: what have we learned?, The Journal of Cell Biology, 104:381.

Cleveland, D.W., Lopata, M.A., Sherline, P., and Kirschner, M.W., 1981, Unpolymerized tubulin modulates the level of tubulin mRNAs, Cell, 25:537.

Gay, D.A., Yen, T.J., Lau, J.T.Y., and Cleveland, D.W., 1987, Sequences that confer β-tubulin autoregulation through modulated mRNA stability reside within exon 1 of a β-tubulin mRNA, Cell, 50:671.

Katz, W., and Solomon, F., 1988, Diversity among β-tubulins: a carboxy-terminal domain of yeast β-tubulin is not essential in vivo, Molecular and Cellular Biology, 8:2730.

Littauer, U.Z., Giveo, D., Thierauf, M., Ginzburg, I., and Postingl, H., 1986, Common and distinct tubulin binding sites for microtubule-associated proteins, Proc. Natl. Acad. Sci. U.S.A., 83:7162.

Pachter, J.S., Yen, T.J., and Cleveland, D.W., 1987, Autoregulation of tubulin expression is achieved through specific degradation of polysomal tubulin mRNAs, Cell, 51:283.

Pittinger, M.F., and Cleveland, D.W., 1985, Retention of autoregulatory control of tubulin synthesis in cytoplasts: demonstration of a cytoplasmic mechanism that regulates the level of tubulin expression, The Journal of Cell Biology, 101:1941.

Schatz, P.J., Solomon, F., and Botstein, D., 1986, Genetically essential and nonessential a-tubulin genes specify functionally interchangeable proteins, Molecular and Cellular Biology, 6:3722.

Schatz, P.J., Solomon, F., and Botstein, D., 1988, Isolation and characterization of conditional-lethal mutations in the TUB1 a-tubulin gene of the yeast Saccharomyces cerevisiae, Genetics, 120:681.

Yen, T.J., Gay, D.A., Pachter, J.S., and Cleveland, D.W., 1988, Autoregulated changes in stability of polyribosome-bound β-tubulin mRNAs are specified by the first 13 translated nucleotides, Molecular and Cellular Biology, 8:1224.

Yen, T.J., Machlin, P.S., and Cleveland, D.W., 1988, Autoregulated instability of β-tubulin mRNAs recognition of the nascent amino terminus of β-tubulin, Nature, 334:580.

A GENETIC ASSAY USING RODENT/HUMAN HYBRID CELLS TO EVALUATE THE GENOTOXIC EFFECTS OF CHEMICALS FOR MULTIPLE ENDPOINTS

Shahbeg S. Sandhu[1], Ramadevi Gudi[2] and
Raghbir S. Athwal[2]

[1]Genetic Toxicology Division, U.S. Environmental
 Protection Agency, Research Triangle Park
 NC 27711, USA

[2]Department of Microbiology and Molecular
 Genetics, New Jersey Medical School UMDNJ
 185 S. Orange Avenue, Newark, NJ 07103, USA

INTRODUCTION

Since the mid-seventies, a large number of chemicals have been tested in a variety of short-term _in vitro_ and _in vivo_ test systems (STTS). It is apparent that all chemicals do not induce a similar spectrum of genotoxic effects in STTSs. Some chemicals induce predominantly gene mutations and/or chromosome aberrations while certain chemicals or classes of chemicals, in addition to interacting with cellular DNA, also induce genomic alterations by interfering with the normal functioning of cellular organelles associated with cell division.

Recently there has been some controversy on the utility of STTS for evaluating carcinogenic hazards. Two shortcomings for most of the short-term bioassays have contributed to this controversy: (1) most of the STTSs have been developed to detect one endpoint, i.e., gene mutation or chromosome aberration, and (2) genetic measurements particularly for cytogenetic effects have been based on the use of relatively high doses of test chemicals, making it difficult to extrapolate from the observed effects to low doses to which humans are normally exposed.

Since it is not practical to eliminate from our environment all the chemicals that are capable of inducing genotoxic effects in STTS, it is prudent to develop test systems that are capable of yielding information for ranking chemicals for their relative hazards to human health and environmental integrity. The STTS that can be used for evaluating the chemicals for multiple genetic endpoints at relatively low doses may be useful for assessing the toxic potential of environmental chemicals.

Mechanisms of Environmental Mutagenesis-Carcinogenesis, Edited by
A. Kappas, Plenum Press, New York, 1990

In this article, we report on the development of a test system using rodent-human monochromosomal hybrid cells for simultaneously evaluating the potential of a chemical to induce mutations, chromosome aberrations, and aneuploidy at relatively low dose levels.

The utility of hybrid cells for evaluating the mutagenic effects (Waldren and Puck, 1987), aneuploidy (Athwal and Sandhu, 1985; Sandhu et al., 1988), and clastogenic effects (Gudi et al., 1989) has been demonstrated. These assays are relatively more sensitive than the classical cytogenetic analysis for evaluating induced genetic effects.

CELL LINES

The monochromosomal hybrid cell lines used in our assay system contained a single marker human chromosome incorporated into mouse cells (Fig. 1). The development of these cell lines has been described by Athwal et al., (1985b) and is briefly discussed in this volume by Athwal and Kaur. One of these cell lines, R3-5, was used by Athwal and Sandhu (1985) for developing both a simplified cytogenetic assay and selection assay for aneuploidy (Sandhu et al., 1988). Another cell line, R2-12, containing human chromosome 5, was employed for developing a genetic assay for clastogenicity (Gudi et al., 1989). The important feature of these cell lines is the presence of a dominant selectable marker, Ecogpt, integrated into human chromosomes 2 and 5. Ecogpt is a bacterial gene that codes for the enzyme xanthine-guanine phosphoribosyl-transferase (XGPRT) which catalyzes the formation of xanthine monophosphate (XMP) in the salvage pathway of purine biosynthesis. The XGPRT is analogous to mammalian enzyme hypoxanthine guanine phosporibosyl transferase (HGPRT) and is normally not present in the mammalian cells. A single human chromosome with Ecogpt was transferred into mouse cells deficient for hgprt. Thus, the genotype of the hybrid cells is hgprt$^-$/Ecogpt$^-$.

The presence of a human chromosome in the hybrid cells does not confer selective advantage/disadvantage and is not required for cell survival. Thus, the genotoxic effects are measured for a chromosome that is not associated with the genes required for cell survival.

The cultivation of MHC in the medium containing mycophenolic acid (25 µg/ml) and xanthine (70 µg/ml) [MX medium] assures the presence of the human chromosome in every cell. The cells are routinely cultivated at 37°C in 10% CO^2 in Dulbecco's modified Eagle's Minimum Essential Medium (DMEM) supplemented with 10% fetal calf serum.

GENETIC ASSAY FOR ANEUPLOIDY

The classical microscopic methods for analyzing the ability of chemicals to induce aneuploidy are labor intensive, relatively insensitive, and require high treatment doses to yield statistically meaningful results. However, if large cell populations can be analyzed either by automation or by genetic selection techniques, the utility of test systems for

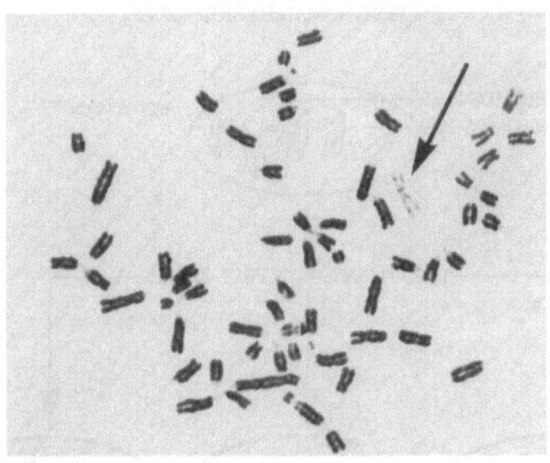

Fig. 1. Metaphase spread of a monochromosome hybrid
cell (R3-5), stained by G-11 method, showing a
differentially stained human chromosome.

evaluating genetic effects of chemicals at relatively low dose
levels can be significantly enhanced. Genetic selection
assays have been successfully exploited in the development of
test systems for mutagenicity using microbial cells and
mammalian cells in culture. Our efforts have been to develop
assay systems for assessing the ability of chemicals to induce
aneuploidy based on the principle of differential growth of
cells in selective medium.

The rationale of the method is: when R3-5 cells with a
genotype hgprt$^-$/Ecogpt$^+$ are cultured in MX medium, every
growing cell contains a transferred human chromosome. XGPRT
is analogous to the mammalian HGPRT enzyme and can also
utilize the purine analogue 6-TG. In the purine biosynthetic
pathway, the monophosphate of 6-TG inhibits the enzyme
phoshoribosyl pyrophosphate amido transferase and therefore is
toxic to the cells. Thus, cells with the hgprt$^-$/Ecogpt$^+$
genotype (R3-5) cannot grow in the medium containing 6-TG
(TG-medium). The loss of the human chromosome from R3-5 cells
results in the hgprt$^-$/Ecogpt$^-$ genotype, and these cells are
capable of growth in TG medium. Treatment of R3-5 cells with
an aneuploidy inducer would generate a population of cells
containing 0, 1, or 2 human chromosomes (Fig. 2). The cloning
efficiency of this cell population in TG medium would provide
the frequency of cells which have lost the human chromosome.
The increase in the frequency of chromosome loss in cultures
treated with a test chemical is used as an index for the
potential of a chemical to induce aneuploidy.

The feasibility of the assay was illustrated by using two
model compounds, i.e., Colcemid and nocodazole, and the data
on the frequency of chromosome loss as determined by the
cloning efficiency of cells in 6-TG medium is shown in Fig. 3.

Both chemicals yielded a dose-related response for

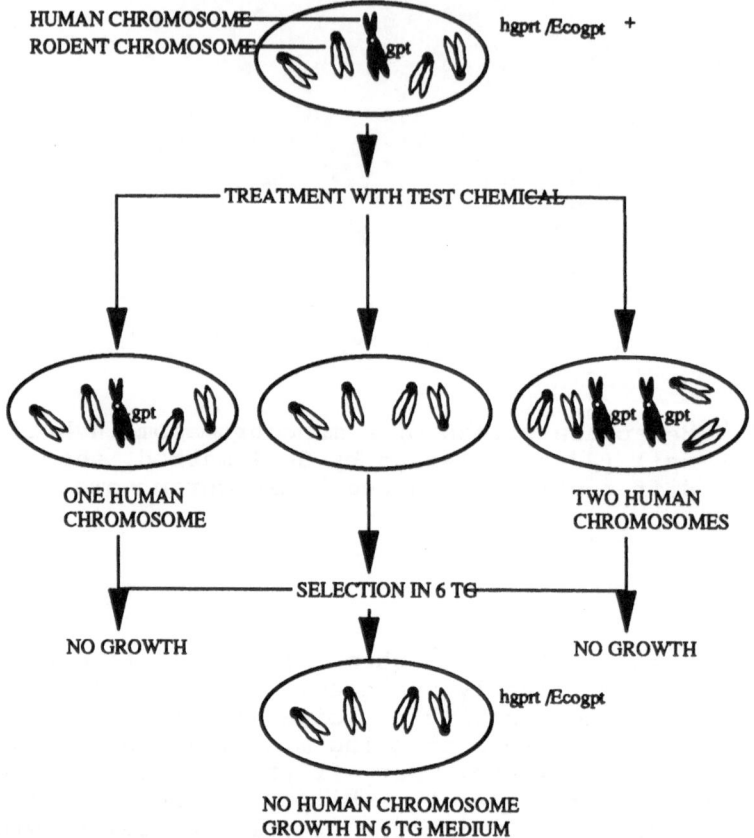

Fig. 2. Experimental approach for genetic assay to
measure chromosome loss in rodent/human cell
line R3-5.

chromosome loss. The chromosome loss in the control cultures
was rather high (10.3%) and in the treated cultures reached as
high as 61.9% at 0.016 μg/ml for nocodazole and 62.9% at 0.032
μg/ml for Colcemid. High frequency of loss in the untreated
cultures reflects the instability of chromosome 2 in the
hybrid cells. This instability may in fact be utilized as an
asset to enhance the power of the test to detect the potential
genotoxic effects of test chemicals at low concentrations. The
absence of human chromosome 2 in the cells isolated from the
colonies grown in 6-TG medium after treatment with 0.02 μg/ml
of Colcemid was confirmed by microscopic analyses of 200
metaphases and by slot-blot analyses of DNA from the same
population of cells for the presence of human DNA (data not
shown).

GENETIC ASSAY FOR CLASTOGENICITY

This assay is based on the cloning efficiency of cells in

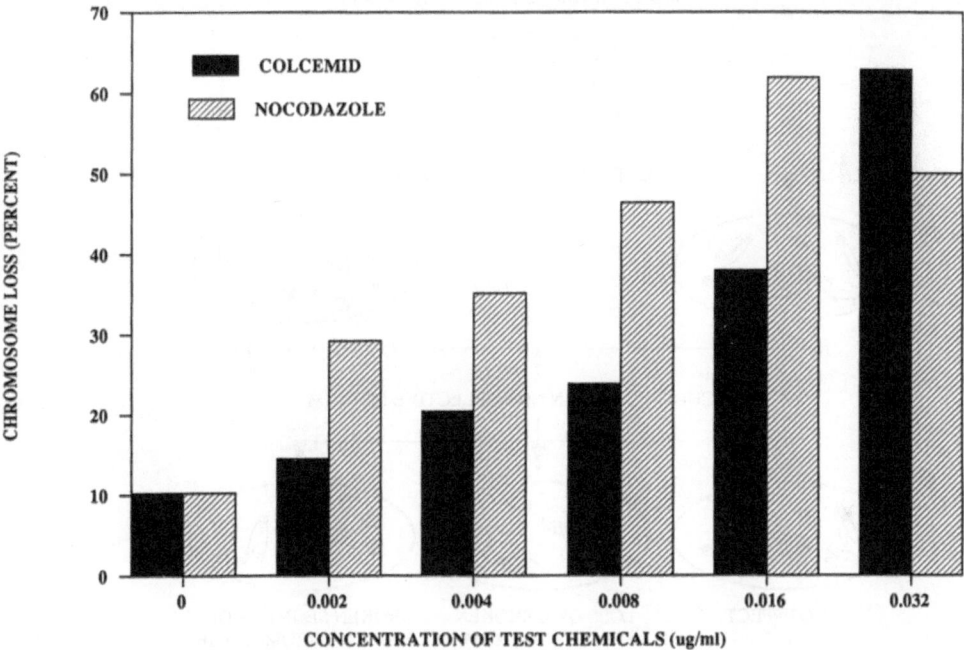

Fig. 3. Chromosome loss from the monochromosomal hybrid cell line R3-5, measured by cloning efficiency after treatment with Colcemid and nocodazole.

a medium that simultaneously selects for two genetic markers present on the same human chromosome, i.e., for the retention of one and loss of another when separated by a break in the chromosome. For this assay, mouse-human hybrid cell line R12-2 containing a single human chromosome 5 is utilized. In addition to the integrated marker, Ecogpt, this chromosome carries a gene for sensitivity to diptheria toxin [DT^S]. Human cells are sensitive to a concentration of 10^{-13} M DT, whereas mouse cells are resistant to 10^{-7} M DT. The sensitivity to DT is expressed as a dominant phenotype so that the hybrid cells are sensitive to 10^{-13} M DT (Athwal et al., 1985a). The growth of cells in M X medium containing DT (10^{-13} M) permits the quantitation of cells that have lost a specific segment of human chromosome 5 containing DT^S gene, separated by a break between DT^S and Ecogpt. The principle of the assay is shown in Fig. 4.

The feasibility of the assay system was illustrated using x-irradiation as a model clastogen. The details of the experimental protocol were published earlier (Gudi et al., 1989); only a summary of the data is presented in Fig. 5.

Following radiation, the frequency of chromosome breaks approximately doubled (7.25%) at 25 rads as compared to the

MONOCHROMOSOMAL HYBRID CELL

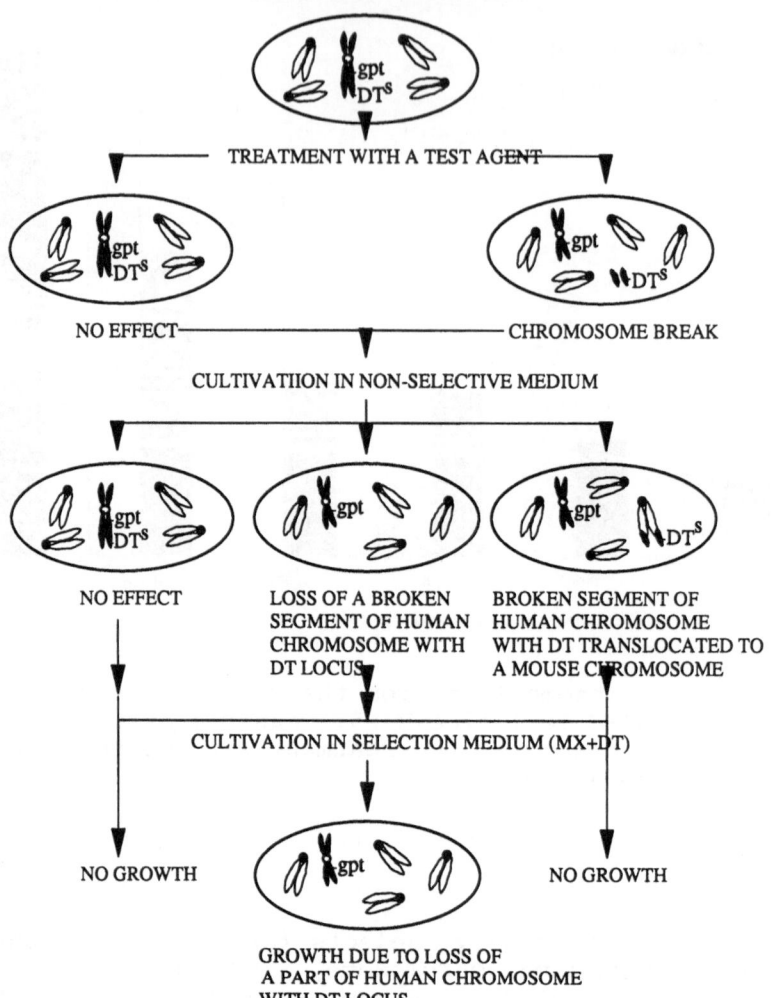

TREATMENT WITH A TEST AGENT

NO EFFECT — CHROMOSOME BREAK

CULTIVATIION IN NON-SELECTIVE MEDIUM

NO EFFECT

LOSS OF A BROKEN
SEGMENT OF HUMAN
CHROMOSOME WITH
DT LOCUS

BROKEN SEGMENT OF
HUMAN CHROMOSOME
WITH DT TRANSLOCATED TO
A MOUSE CHROMOSOME

CULTIVATION IN SELECTION MEDIUM (MX+DT)

NO GROWTH

NO GROWTH

GROWTH DUE TO LOSS OF
A PART OF HUMAN CHROMOSOME
WITH DT LOCUS

Fig. 4.   Experimental approach and rationale of the
genetic assay for detecting clastogenicity
using monochromosomal hybrid cells (R12-2).

control (3.50%) and increased to 34.4% at 200 rads. The cells
exposed to X-rays were also analyzed cytogenetically to
determine breaks in the human chromosome. Cytogenetic
analysis of 100 metaphases was performed 24 h after irradiation. Different categories of breaks and configurations in
the human chromosome were pooled to calculate the frequency of
chromosome aberrations. A comparison of the frequencies of
chromosome aberrations detected by cytogenetic and by selection assays is presented in Fig. 6. The dose-related increase
in the frequency of breaks observed in the selection assay was
comparable to the dose response obtained from cytogenetic
analyses. At each dose of the radiation, higher levels of
chromosome damage, observed in the cytogenetic data in
comparison with the selection assay, may be attributed to the
target size. In the cytogenetic analysis, modifications along

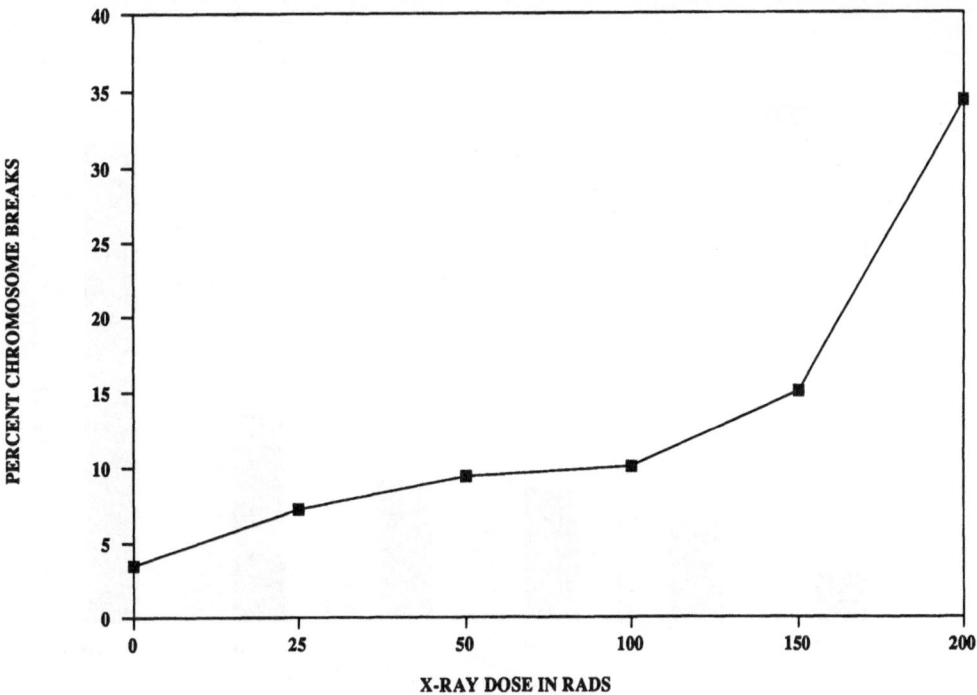

Fig. 5. Frequency of chromosome breaks in R12-2 cells induced by X-irradiation determined by cloning efficiency of cells cultivated in a M X medium containing DT.

the entire length of chromosome 5 were scored, whereas the selection assay measures the breaks that occur only in the region between Ecogpt and DT genes. Moreover, this assay measures only those breaks in which the broken segment of the chromosome has actually been lost from the cell. A similar assay system using human-hamster hybrid cells in which human chromosome 11 has been transferred into hamster cells has been shown by Waldren and Puck (1987) to be ten times more sensitive for detecting chromosome aberrations as compared to the classical cytological observations using CHO cells.

UNIFIED GENETIC ASSAY

The feasibility of using hybrid cell lines for detecting aneuploidy and clastogenicity has been demonstrated. A unified assay in which three genetic endpoints, i.e., gene mutation, chromosome breaks, and aneuploidy, can be detected in a single experiment using three different media is currently being evaluated. For this assay, cell line R12-2 containing human chromosome 5 with two genetic markers, i.e., Ecogpt and DT[S], will be used. We are presently evaluating this assay with a variety of carefully selected model compounds, each eliciting a genotoxic effect by a different mechanism.

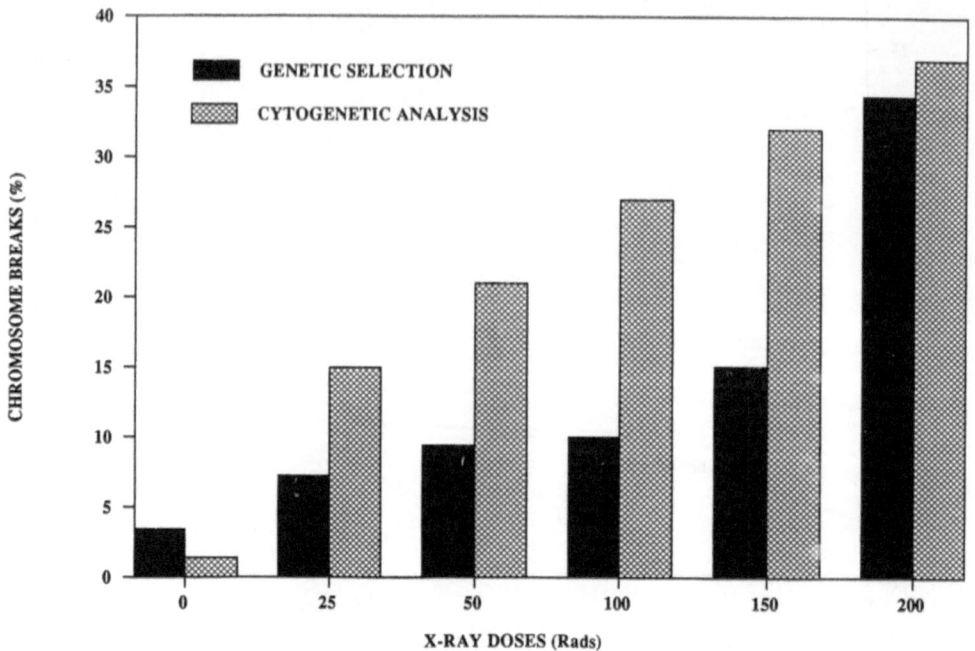

Fig. 6. Comparison of X-irradiation induced chromosome breaks hybrid cells (R12-2) measured by genetic selection and cytogenetic analysis.

ACKNOWLEDGMENTS

Although the research described in this article has been supported by the United States Environmental Protection Agency through Cooperative Agreement number CR 815926 to New Jersey Medical School, UMDMJ, Newark, NJ, it has not been subjected to Agency review and therefore does not necessarily reflect the views of the Agency and no official endorsement should be inferred. Mention of trade names or commercial products does not constitute endorsement or recommendation for use.

This project was in part funded by the New Jersey Hazardous Substances Management Research Center, N.J. Institute of Technology, Newark, NJ 07102.

We are grateful to Dr. B.S. Gill and Ms. Velva Milholland for their assistance in the preparation of this manuscript.

REFERENCES

Athwal, R.S., and Sandhu, S.S., 1985, Use of a human x mouse hybrid cell line to detect aneuploidy induced by environenmental chemicals, Mutation Res., 149:73.

Athwal, R.S., Searle, B.M., and Jansons, V., 1985a, Diptheria toxin sensitivity in a monochromosomal hybrid containing human chromosome 5, <u>J. Heredity</u>, 76:329.

Athwal, R.S., Smarsh, M., Searle, B.M., and Deo, S., 1985b, Integration of a dominant selectable marker into human chromosomes and transfer of the marked chromosome to mouse cells by microcell fusion, <u>Somatic Cell Mol., Genet.</u>, 11:177.

Gudi, R., Sandhu, S.S., and Athwal, R.S., 1989, A genetic method to quantitate induced chromosome breaks using a mouse/human monochromosomal hybrid cell line: Identification of potential clastogenic agents, <u>Mutation Res.</u>, 225:149.

Sandhu, S.S., Gudi, R., and Athwal, R.S., 1988, A genetic assay for aneuploidy: quantitation of chromosome loss using a mouse/human monochromosomal hybrid cell line, <u>Mutation Res.</u>, 201:423.

Waldren, C.A., and Puck, T.T., 1987, Steps toward experimental measurement of total mutations relevant to human disease, <u>Somatic Cell Mol. Genetics</u>, 13:411.

ANEUPLOIDY IN HUMANS

Judith H. Ford

Genetics Department
The Queen Elizabeth Hospital
Woodville, South Australia, 5011
Australia

ANEUPLOIDY: THE MOST DRAMATIC MUTATION IN HUMANS

Aneuploidy, the change in chromosome number by the addition (trisomy) or loss (monosomy) of a single chromosome, has a most dramatic effect on the cells and organisms in which it occurs. It is extremely harmful because it changes the gene dosages and creates genetic imbalances of a large number of genes simultaneously.

In human pregnancy, aneuploidy is responsible for about 35% of miscarriages and for about half the chromosome abnormalities (about 0.3%) at birth. It probably also accounts for some early embryonic loss prior to implantation as monosomy is rarely observed except for monosomy X and some trisomies have never been observed in miscarriages. A conservative estimate would be that aneuploidy occurs in about 10% of conceptions.

When the whole individual or aborted embryo is aneuploid, the divisional error leading to the chromosomal imbalance has usually occurred during meiosis. Data gained from a study of heritable chromosomal heteromorphisms has indicated that the majority of errors are of maternal origin and are due to errors in meiosis I. The estimated contribution of paternal meiotic error for chromosome 21 is 25% but the rate of paternal contribution does not exceed 12% for any other chromosome (Hassold, 1985).

All individuals have some aneuploid cells which have been generated by mitotic error. If the error occurs in one of the first cleavage divisions then a high proportion of cells will be aneuploid and the individual will be described as a mosaic. In the majority of people, however, the errors will be restricted to less than 5% of cells.

Mosaicism usually confers some or all of the characteristics of a full trisomy. For example, a mosaic trisomy 21 will be similar to but less affected than a full trisomy 21 and a mosaic Klinefelter syndrome may have reduced but not complete infertility. The effect of lower levels of aneuploidy is largely unpredictable and will vary according to the frequencies of aneuploid cells found in the various tissues.

Aneuploidy and Tumours

That aneuploidy leads to the progression of tumours has been demonstrated by numerous investigations on leukaemias and cancers. Indeed, the extent and types of aneuploidy can often be used to stage the disease. It also seems likely that aneuploidy may contribute to the initiation of some tumours. Evidence for its role in initiation is diverse. For example, trisomy 21 individuals have increased rates of reticuloen-dothelial neoplasms (Harnden and O'Riordan, 1973) and aneuploid inducing chemicals also induce cell transformation in vitro (Barrett et al., 1985). The evidence from retinoblas-toma and Wilm's tumour families show that these tumours are induced by the induction of hemizygosity or homozygosity at specific loci. Although these tumours usually involve aneuploidy for only a part of a chromosome, which is probably generated by a different mechanism to full aneuploidy, the genetic effects of full aneuploidy are equivalent at any gene locus and are thus also likely to be tumorogenic.

Ageing and Aneuploidy

Aneuploidy cells appear more frequently in older people, particularly in older females. Several studies in the 1960's attempted to perform population studies of this phenomenon (Jacobs and Court-Brown, 1966, Hamerton et al., 1965, Mattevi and Salzano, 1975). The studies were limited by the technology available at this time but concluded that chromosome loss was the most significant feature of age-related changes. The increased frequency of loss which occurred with increasing age in females fitted a cubic curve, with the most rapid change occurring after age 45. The frequency of error then pla-teaued. By comparison, the change in males fitted a linear regression with increasing age. In both sexes, increases in hyperploidy were either subtle or absent.

Some studies attempted to correlate the incidence of aneuploid cells with various physiological disorders. Nielsen et al. (1968) studied cells from normal controls and patients with senile dementia and found the highest level of hypo-ploidy in the senile dementia group. Other studies supported an association between a deterioration in mental functioning and increased hypodiploidy (e.g. Bettner and Jarvik, 1971).

Few population studies of aneuploidy have been performed since the advent of chromosome banding techniques. One study performed by Ford and Russell (1985) on 16 healthy women within the reproductive age range gave some interesting results. Ford and Russell examined the ratio of hypoploidy to hyperploidy, for X chromosomes and autosomes, in a large number of cells per subject. Their results showed that both X chromosomes and autosomes exhibited age-related changes but suggested that the changes were of different types. X chromo-somes were found to show a ratio of loss to gain of about 3:1 at all ages, but the frequency of total aneuploidy, loss plus gain, showed a significant increase with age. By contrast, there was no clear association between the frequency of autosomal aneuploidy and age. The ratio of loss to gain, however, was significantly greater in the younger women aged

21-35 years than in the older women aged 36 to 50 years. This suggested that the mechanism of error for autosomes altered with age.

Mechanisms of Aneuploidy Induction

It is usually stated that all trisomies are generated by nondisjunction and that most monosomies are also generated as a consequence of this process. Some authors recognise that monosomies may also be generated by chromosome loss following lagging. Several other mechanisms have been proposed to explain misdivision but many authors have found it convenient to use the term nondisjunction in a general way to indicate any defect of the division process which gives rise to aneuploidy (e.g. Bond and Chandley, 1983). Although perhaps convenient, this approach is certain to mask the real mechanisms of aneuploidy and will not be used here.

Meiosis and Mitosis

The elements of meiosis and mitosis which seem to be critical to the proper segregation of chromosomes and chromatids are illustrated in Figures 1A and 1B. In meiosis I, chiasmata are usually formed between the chromosomes in a bivalent and even the smallest chromosomes usually have at least one chiasma. At metaphase the bivalents are orientated on the spindle such that the centromeres of the two chromosomes are orientated to the opposite poles of the spindle. The point of equilibrium between the two chromosomes seems to be established by the position of the proximal chiasma. At anaphase, the chromosomes move to the opposite poles of the spindle.

In meiosis, the vulnerable stages would thus appear to be:

- chromosome pairing or synapsis,
- formation of chiasmata,
- spindle formation,
- attachment and orientation of centromere-kinetochore complexes to the appropriate spindle poles,
- resolution of chiasmata to allow segregation.

Four different elements could thus potentially be involved in error,(i) the chromosome pairing mechanism, (ii) the mechanism of the formation and/or resolution of the chiasmata, (iii) the structure and hence function of the kinetochore-centromere complex, (iv) the structure and function of the spindle.

The second meiotic division can be regarded as meiotic mitosis. Thus mitosis and meiosis II can be assessed together. In mitosis the vulnerable stages would appear to be:

- chromosome replication,
- spindle formation,
- attachment and orientation of the centromere-kinetochore complexes to the appropriate spindle poles,
- separation of the sister-chromatids to allow segregation.

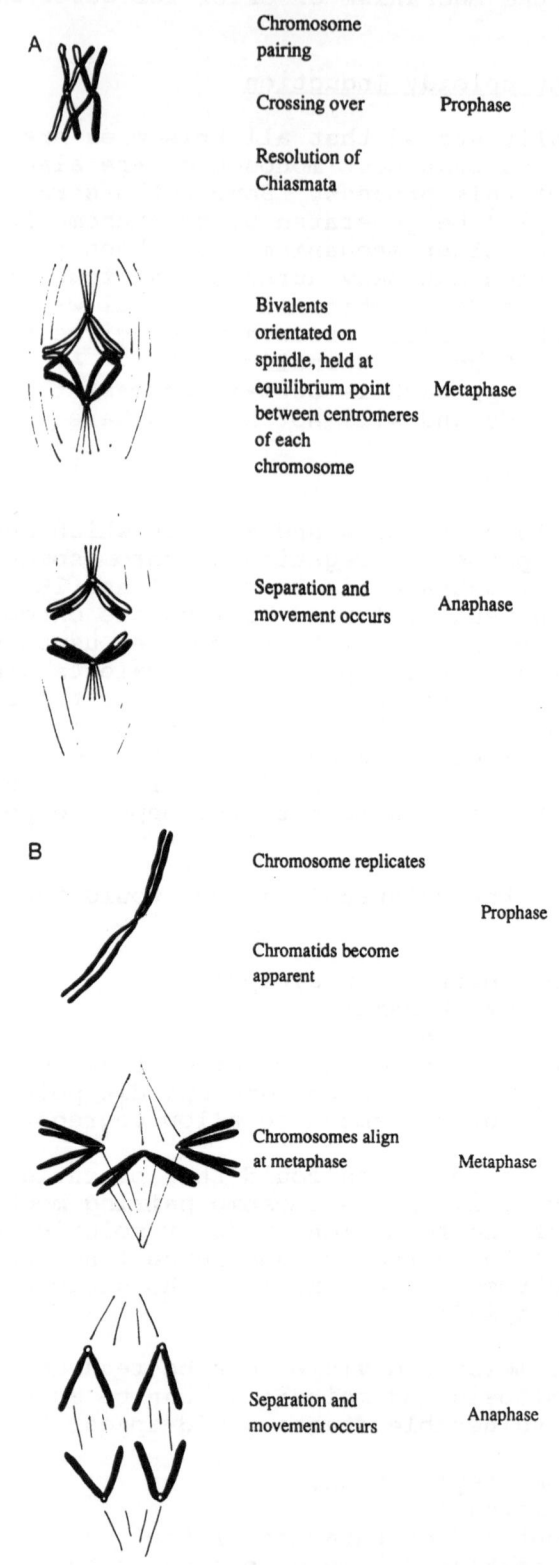

A

Chromosome
pairing

Crossing over   Prophase

Resolution of
Chiasmata

Bivalents
orientated on
spindle, held at
equilibrium point   Metaphase
between centromeres
of each
chromosome

Separation and
movement occurs   Anaphase

B

Chromosome replicates

                       Prophase

Chromatids become
apparent

Chromosomes align
at metaphase   Metaphase

Separation and
movement occurs   Anaphase

Fig. 1.   A, Meiosis; B, Mitosis.

Four different elements could thus potentially be involved in error, (i) the rate of synthesis of the various elements including the chromosome replication, (ii) the mechanism of the separation of the sister chromatids, (iii) the structure and hence function of the centromere-kinetochore complex, (iv) the structure and function of the spindle.

There are elements which are common to both meiosis and mitosis and some that are not. Meiotic chromosome replication has not been mentioned as a potentially error-prone stage because in female mammals this occurs so many years before reproduction. In males, however, this could perhaps cause some discordance in timing of the meiotic events. Even so, it is likely that it would be observed as pairing failure rather than segregation failure.

The next section of this paper presents the results obtained recently in our laboratory in studies of mitosis. This will be followed by a discussion of the recent data available from studies of meiotic segregations. The aims of this comparison are to identify the mechanisms which are common to both processes and to recognize the sources of the errors which are not.

Aneuploidy in Human Mitosis

Actively growing lymphocyte cultures can be briefly treated with warm hypotonic solution in order to induce a wave of anaphases (Ford and Congedi, 1987) which are rare in untreated preparations. Rapid fixation results in a preparation with up to 30% of the cells at anaphase or early telophase. The first experiments in this study recorded the progress of cells through division and the differences between groups of subjects of different ages. In the later experiments, the behaviour of individual chromosomes was studied by hybridizing tritium-labelled, chromosome-specific alpha satellite sequences to the preparations.

MATERIALS AND METHODS

Subjects

The subjects were all females who were either volunteers from the hospital personnel or older subjects who had agreed to participate in a South Australian longitudinal study on aging. The studies on individual chromosomes were performed on a sub-group of the subjects. Two of these subjects were in their early twenties, 23 and 25; two were in their forties, 42 and 47; three were in their sixties, 62, 64 and 64 and one was 82. Studies of chromosomes X, 17 and 18 were completed on the four younger subjects but because of unsatisfactory cell growth, the studies on the older subjects were incomplete. Investigations on the X chromosome were completed on two subjects who were both aged 64. Chromosomes 17 and 18 were investigated on the 62 and 82 year olds.

To test for differences between the subjects in the first experiments, the subjects were divided into different age-groups. For the latter experiments, where there were too few

subjects for this treatment, the subjects were divided into
two groups, namely pre-menopausal and post-menopausal and the
results within the groups pooled. The first group included
subjects SP, RS and HF. All other subjects were in the second
group.

Cell Cultures

All the studies were performed on phytohaemagglutinin-
treated whole blood cultures in which the T-lymphocytes were
stimulated to divide. These were cultured in R.P.M.I. medium
with 20% fetal calf serum at 37°C water bath. No microtubule
inhibitor such as colchicine was added. After 72 hours of
culture, the medium was rapidly decanted from the tubes and
replaced with 2.5 mls. of prewarmed 0.075 M KCl. The cells
were exposed to the hypotonic solution for 5 minutes to
induce anaphase. After this time, 7.0 mls. of 3:1 methanol:
acetic acid fix solution was added to the tubes which were
immediately plunged into an ice bucket. They were allowed to
remain in the ice for about 10 minutes before they were
centrifuged and the fluid replaced with 100% fixative. Three
further changes of fixative were made over a 30 minute period.
The cells were spread onto dry, clean slides and allowed to
air dry.

For _in situ_ hybridization studies, the slides were
hybridized with tritium labelled DNA probes specific to the
pericentromeric alpha satellite region of chromosome X, 17 or
18. Probes TRX and TRI7 (Choo et al., 1987) and L1.84
(Devilee et al., 1986) which were kindly donated by Drs. Choo
and Devilee were used for chromosomes X, 17 and 18 respectively. After hybridization, the slides were coated with
photographic emulsion and exposed for one or two weeks. They
were then developed and stained with Giemsa stain.

Method for Scoring Slides Hybridized with Chromosome Specific Probes

In these preparations, metaphases often retain their ring
formation and anaphases are often seen as two segregating
rings (Fig. 2). For each cell scored, the position of the
labelled chromosomes or chromatids were recorded on the sheet.
Prior to scoring, the slide labels were covered and the slides
were given an arbitrary code. All scoring was done blind.

RESULTS

Changes in Cell Division with Age

Tables 1-3 show the changes that occur in cell division
with age. These are changes that affect cells in general and
might be expected to affect all chromosomes.

It is easily seen that progression from metaphase to
anaphase is inhibited with increasing age. Moreover anaphase
appears to be less efficient. This suggests that there is
loss of efficiency of spindle function with age.

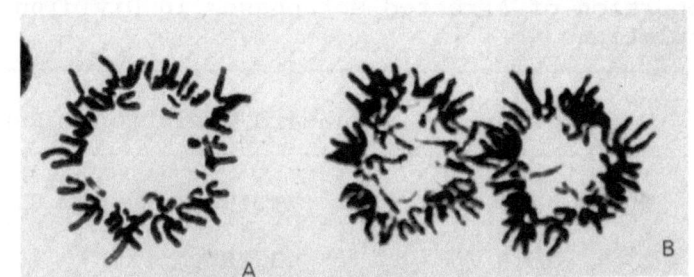

Fig. 2.   A, Typical ring metaphase seen in non-
colchicine treated preparations.
B, Typical early anaphase. Segregating
chromatids remain in ring orientations.

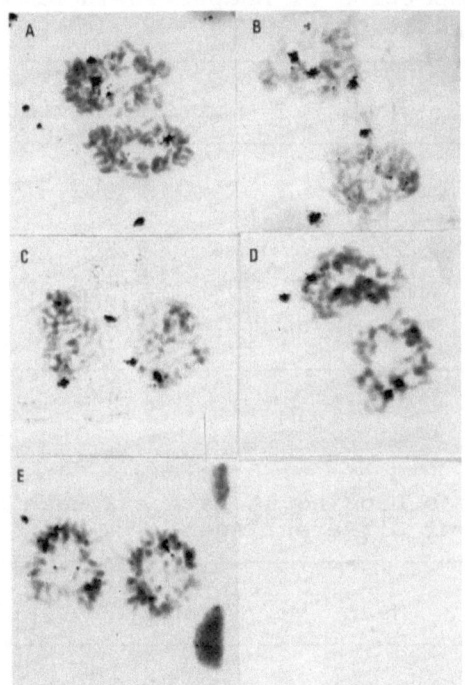

Fig. 3.   A, Nondisjunction. The large group of grains
in the top ring shows the nondisjoined
chromosome. The small group of grains below
this is background. Each of the other group of
grains (one in each ring) shows a chromatid.
B, Chromatid malsegregation. Three disjoined
chromatids have segregated into the upper
ring. C, Lagging chromatid in central position. D, Lagging chromatid in peripheral
position. E, Monopolar division. One of the
chromosomes in the left-hand ring is demonstrating a monopolar division. One chromatid
segregates into the ring whilst its sister
chromatid is expelled into the cytoplasm.

Table 1. Proportion of Arrested Metaphases in Dividing Cell Populations

Age	N	Mean	Standard Deviation	Range
15-25	6	2.7	1.19	0.7-4.2
26-35	8	2.9	2.47	0.7-8.4
36-45	3	5.2	3.25	1.5-7.6
46-55	3	6.2	1.27	5.2-7.6
56-65	3	7.8	0.58	7.5-8.5
66-74	4	9.5	0.54	9.0-10.2

Table 2. Proportion of Anaphases in Dividing Cell Populations

Age	N	Mean	Standard Deviation	Range
15-25	6	24.1	2.99	19.7-28.0
26-35	8	20.7	2.77	18.0-26.5
36-45	3	22.2	3.50	18.2-24.5
46-55	3	16.5	5.19	12.8-22.4
56-65	3	17.8	3.80	13.6-21.0
66-75	4	15.2	2.74	12.3-18.2

Table 3. Chromosomes Lagging at Late Anaphase and Telophase in Women of Different Ages

Age	N	Mean	Standard Deviation	Range
25-35	8	0.74	0.45	0.0-1.5
36-40	6	0.64	0.72	0.0-1.9
41-45	4	0.85	0.44	0.5-1.4
46+	7	1.79	0.94	1.0-3.2

The Behaviour of Individual Chromosomes

In the preparations made to observe the behaviour of individual chromosomes several different types of chromosome error at anaphase were observed. These were defined in the following ways.

Nondisjunction. Nondisunction was recorded when a chromosome had failed to disjoin and the two nondisjoined chromatids had moved into one of the daughter cells (Fig. 3A). In all slides, about 3.0 of the cells appeared to show a 2:1

104

segregation. We consider that the majority of these segregations are normal 2:2 cells in which the two segregating
non-sister chromatids are so close together that they produce
a single image in the photographic emulsion. It is possible,
however, that some of these segregations are nondisjunctions.

Occasionally a chromosome failed to disjoin and also
failed to segregate. This cell was scored in the lagging
category.

Chromatid malsegregation. Chromatid malsegregation was
recorded when one chromatid had moved into the same cell as
its sister chromatid (Fig. 3B). Such sister chromatids were
usually quite separate from one another and it seems likely
that the malsegregation was the consequence of reorientation
on the spindle after initial lagging.

Chromosome and chromatid lagging. Chromosome and
chromatid lagging were recorded when chromatids, disjoined
sister chromatids or nondisjoined chromosomes had failed to
move to the spindle poles with the remainder of the chromatids
and were found either at the metaphase plate or in the
cytoplasm. Lagging was defined as central when the laggard
was in the region of the metaphase plate (Fig. 3C) or peripheral when clearly outside the region of the plate, usually to
the side of the moving chromatids (Fig. 3D).

Monopolar orientation. Monopolar orientation was recorded
when a chromosome had made a connection to only one spindle
pole and had divided such that one chromatid moved to the pole
while the other moved into the cytoplasm (Fig. 3E).

Rates of Error of Chromosomes X, 17 and 18

(i) Total error. The total rates of error were low
(Table 4). For the X chromosome, 56 in 3,504 anaphase cells
showed error, a rate of 1.6%. For chromosome 17, 39 in 3,612
cells showed error, a rate of 1.1%. For chromosome 18, 13 in
2055 cells showed error, a rate of 0.6%.

For chromosomes X and 17, there were differences in the
error rates between individuals (for the X chromosome, $x^2_{(5)}$ or
heterogeneity = 27.5 and p = < .001; for chromosome 17, $x^2_{(5)}$
= 19.78 and .01 > p > .001). The number of errors detected
for chromosome 18 was too small to permit heterogeneity
testing.

The pooled errors in the younger (pre-menopausal)
subjects were compared with those of the older (post-
menopausal) subjects and tested for singificance. No significant difference was found for either of the autosomes but a
difference was found for the X chromosome ($x^2_{(1)}$ = 18.62,
p = < .001).

(ii) Rates of the different types of error. The types of
error found for each of the three chromosomes are shown in
Table 5. Nondisjunction, chromatid malsegregation, central
and peripheral lagging were seen for all chromosomes.

Central lagging was the most common error for all
chromosomes and peripheral lagging the next most frequent

error overall. In every case, lagging accounted for more than half the errors made by each chromosome.

Chromatid malsegregation, a source of 3:1 segregations, occurred at about the same frequency as nondisjuction.

Monopolar segregations were only seen in two of the X chromosome slides. They occurred in the two youngest subjects.

Table 4. Anaphase Error of Chromosomes X, 17 and 18 in Difference Subjects

Subject/ Age	Chromosome X		Chromosome 17		Chromosome 18	
	No. Cells	% Error	No. Cells	% Error	No. Cells	% Error
Error						
SP/23	704	0.4%	423	0.2%	233	0.4%
RS/25	787	0.5%	748	2.0%	421	0.7%
HF/42	778	1.8%	1103	1.3%	429	0.2%
GW/47	473	2.3%	538	1.7%	463	0.4%
TH/62	-	-	369	0.0%	220	1.4%
FD/65	203	4.9%	-	-	-	-
AT/65	559	2.5%	-	-	-	-
FM/82	-	-	435	0.0%	289	1.4%

Note: Dash indicates that no experiment was performed.

Table 5. Rates of Different Error Types in All Subjects (Pooled Data)

Chrom	No. Cells	Nondisjn	Chtid MS	CentLag	PerLag	Monop	"2:1"
X	3504	0.1%	0.2%	0.7%	0.4%	0.06%	3.9%
17	3616	0.2%	0.1%	0.5%	0.3%	0	3.5%
18	3504	0.2%	0.05%	0.3%	0.1%	0	3.5%

Chrom: chromosome; Nondisjn: nondisjunction; Chtid MD: chromatid malsegregation; CentLag: Lagging in the central position in the region of the metaphase plate; PerLag: Lagging peripherally in the cytoplasm; Monop: monopolar spindle; "2:1": segregations which appear to be 2:1, the majority of which are probably normal divisions with chromatids placed so close together that they cannot be resolved.

The rates of nondisjunction, chromatid malsegregation and monopolar segregations were too low to allow statistical analysis between groups. However the total rate of lagging could be analysed for the X chromosome. This analysis gave a $x^2_{(1)}$ of 11.13 and p = < .001. Thus the X chromosome demonstrates age-related differences in the rate of lagging.

(iii) <u>Differences in chromatid and chromosome lagging between subjects</u>. Subjects SP, RS and HF showed only chromatid lagging (2:1 segregations). The three other subjects, however, showed several cells where both sister chromatids were lagging (1:1 segregations). This was especially true for the X chromosome. Gs was the only subject to show lagging of both chromatids of an autosome and this was found for chromosome 17.

For the X chromosome in cells involving error in only one of the X chromosomes, GW showed four cells with lagging of single chromatids and one cell with lagging of both chromatids; FD showed one cell with lagging of a single chromatid and three cells with lagging of both chromatids; AT showed two cells with lagging of one chromatid and three cells with lagging of two chromatids.

The difference between the pre-menopausal and post-menopausal women was tested for significance. Each of the groups showed 14 cells with lagging. The younger group had only single lagging chromatids whilst the older group had seven cells with a single lagging chromatid and seven cells with two lagging chromatids.

The difference was found to be significant at the 1% level ($x^2_{(1)}$ = 6.86; p = < .01).

(iv) <u>Errors of both homologues in a cell</u>. Subjects SP, RS and HF showed only cells with errors in one of the pair of homologous chromosomes. The other three subjects showed complex segregations. In one cell of subject GW, two chromatids of 17 were seen to be lagging at the metaphase plate and two had independently segregated into one of the daughter cells. The resulting daughter cells have thus 2 and 0 chromatids respectively.

In subjects GW, FD and AT several cells showed errors in both X chromosomes. The errors and the resultant cells are respectively:

GW: both chromosomes nondisjoined and lagging (0:0 segregation).
 : all four chromatids segregated to one cell (4:0 segregation).
FD: four disjoined chromatids lagging (0:0 segregation).
 : One nondisjoined and one chromatid lagging (1:0 segregation).
 : One lagging chromatid and three chromatids segregated to one daughter cell, presumably after initial lagging of both homologues (3:0 segregation).
 : Two lagging chromatids and two chromatids segregated into one daughter cell, presumably after initial lagging of both homologues (2:0 segregation).
AT: Both chromatids disjoined and all four lagging (0:0 segregation).

Subjects SP, RS and HF had a total of 22 cells with errors of the X chromosome, each of them only involving a single chromosome. Subjects GW, FD and AT had a total of 30 cells with errors, 23 of which involved one X chromosome and seven of which involved both X chromosomes. The difference was tested and was found to be significant at the 5% level $(X^2_{(1)} = 4.1; .05 > p > .02)$.

DISCUSSION

Disjunction and Segregation of Chromosomes

The disjunction of chromosomes and their segregation are shown here to be independent events. Chromosomes which fail to disjoin may either segregate into one of the daughter cells or they may lag and be eliminated in micronuclei (Ford et al., 1988). Conversely, chromosomes which have disjoined usually segregate normally but they may either lag and be eliminated, or lag and later segregate into one of the daughter cells.

Mechanisms Generating Trisomic and Monosomic Cells

Trisomic cells can be generated by nondisjunction and segregation into only one daughter cell or by the missegregation of a chromatid into the same cell as its sister chromatid following lagging. Monopolar segregations may also generate trisomies if the chromatid which moves into the cytoplasm away from the pole, remains close enough to be incorporated into the nearby nucleus at telophase.

Monosomic cells can be generated as the result of a mechanism which generates a trisomy but can also be generated independently as a result of lagging. Chromatid lag is the most common error and thus 2:1 segregations are the most common. However it has been demonstrated that segregations ranging from 0:0 to 4:0 can be generated as a result of lagging and missegregation.

Differences in Chromosome Behaviour Between Individuals

For chromosome 17, significant heterogeneity in error rates was found between individuals and the error rates ranged from 0% to 2.0%. For each chromosome the data from the younger (pre-menopausal) and older subjects (post-menopausal) subjects was pooled. An age related difference in total error rate was found for the X chromosome but not for either of the autosomes. In older subjects the frequency of cells showing X chromosome error was significantly increased.

Ideally the differences in rates of the different types of error of the X chromosome would have been tested. However, the rates of nondisjunction, chromatid malsegregation and monopolar divisions were too low to allow statistical analysis. In contrast, the numbers of cells showing lagging were high enough to be tested and lagging was shown to increase significantly with age. Within the cells with lagging, lagging of both sister chromatids was relatively common in older subjects but was not observed in younger subjects.

In addition, both of the homologous chromosomes were
sometimes involved in error in older subjects but this was not
observed in the younger subjects.

The ability of a chromatid to reattach to the spindle
after initial lagging, whether it is in the correct orientation or not, would seem to reflect spindle capacity. Perhaps
one of the changes which occurs with ageing is a raduction in
the number of spindle fibrils which are available to be
captured by lagging or malorientated chromosomes. Studies on a
greater number of individuals are needed to test this speculation however the studies on cell division parameters also
demonstrate decreased spindle function. This seems to be a
most dramatic change with age and to have significance for
both mitotic and meiotic aneuploidy.

DATA FROM HUMAN MEIOTIC SEGREGATIONS

It is difficult to obtain objective and unbiased meioicult
data because ethical and experimental considerations preclude
observation of the process of meiosis. Some meiotic segregation data is available, however, and inferences can be drawn
from these:

- Studies of the stages of meiosis can be made from testicu-
  lar biopsies in males.

- Studies of the early stages of meiosis I in females can be
  made from either the ovaries of aborted female fetuses or
  from the ovaries of older women at ovariectomy.

- Oocytes which have failed to fertilize in "in vitro"
  fertilization can be observed in an arrested state at
  meiotic anaphase II.

- Sperm chromosomes, the gametic results of meiosis I and II
  can be observed after the penetration of hamster oocytes.

- The parental source of extra chromosomes and the stage of
  the meiotic error which generated them can often be
  determined from chromosomal and/or DNA studies of parents
  and their abnormal aneuploid conceptions (usually miscarriages).

- The segregation of genes from parents to their trisomic
  offspring can be analysed to determine whether there is an
  altered frequency of chiamata in the cell divisions
  generating trisomies.

RESULTS

Hulten et al. (1985) summarized the data obtained by
herself and others from (i) air-drying and sequential staining
for karyotyping at meiotic metaphase) and (ii) surface-
spreading techniques (for visualization of the synaptonemal
complex at late zygotene and pachytene) of human male and
female meiosis. Table 6 is derived from data reported in the
Hulten paper. The sources of the female data are Sung et al.
(1983), Hulten et al. (1984), Speed (1985), Wallace and Hulten

Table 6. Results of Studies in Human Male and Female
          Meiosis I

Stage	Tissue	No. meioses	% Unpaired	% Super-numery
Pachytene	Fetal ovary	459(21)	4.4%	0.2%
Pachytene	Testicular biopsy	1139(14)	?	0.0%

Key: the number in brackets represents the number of subjects
? means data not given.
% unpaired should give the rate of univalence.
% supernumery should give the rate of premeiotic mitotic
  error.

Table 7. Aneuploid Findings at Metaphase II in Oocytes
          which Fail to Fertilize in "I.V.F." Programmes

Author	No. of oocytes (no. of subjects)	% Aneuploidy
Zenses et al. (1987)	34(28)	20.6
Wramsby et al. (1987)	14(14)	35.7
Our data (unpublished)	47(15)	48.9

The mean age of the subjects in the group studied by Wramsby
et al. was 30.8 (range 25 to 38) and for our group, the mean
was 32.7 (range 28 to 38).

(1985) and the sources of the male data are Solari (1980),
Hulten et al. (1984), Saadallah (1984).

Data Derived from Oocytes at Metaphase II

     Data derived from Oocytes at metaphase II has been
derived from oocytes which failed to achieve fertilization in
in vitro fertilization programmes. This is obviously a
biased sample of subjects since many of the women on these
programmes have unexplained infertility problems and/or are
often aged 35 years or older. Moreover the cells have been
selected by their failure to function normally. Nevertheless
the extent of abnormal findings in this group is remarkable as
shown in Table 7.

     The findings between subjects varied in our study group
as for example, two of the subjects (aged 30 and 31) had 6 and
5 eggs respectively with no hyperploid cells and only one
hypoploid whereas one subject (aged 37) had only four hyperploid cells. Whilst the subjects on IVF programmes are a

highly selected group, Hassold and Chiu (1985) estimated that the age related incidence of trisomies amongst clinically recognized pregnancies might be as high as 6.0% for women aged 30 to 34, 10.5% for women aged 35 to 39 and 29.2% for women aged 40 and older. Thus IVF rates should probably be seen as being elevated but not extreme.

Sperm chromosomes

Sperm chromosomes have been studied after sperm penetration of hamster eggs (Martin et al., 1987). Studies on 1582 sperm from 30 men of proven fertility showed a mean rate of aneuploidy of 4.7% with a standard deviation of 2.9% and range from 0-10%. Sperm have an even higher frequency of structural chromosome abnormalities but that will not be discussed further here.

Meiotic Segregation Data

Meiotic segregation data has been derived for chromosome 21 trisomies using chromosome 21-specific DNA probes and cytogenetic satellite markers. A combination of the two techniques has been found to be more informative than either independently. The meiotic error leading to trisomy 21 has been attributed to paternal meiosis I error in 13% of cases and to paternal II error in 7% of cases. Maternal I error accounts for 67% of cases and maternal II error for 13% of cases.

The number of families who have provided informative data for recombination in an erring meiosis I is still quite small. Warren et al. (1987) used 9 families with maternal I error to derive data which suggested that crossing over is reduced in chromosomes 21 that have undergone error. However in a total of nine families from Stewart et al (1988) and Galt et al (1989), crossing over was demonstrated in three cases. Thus it can only be concluded from the three sources of data that asynapsis may sometimes underlie meiotic I error but that effective synapsis and crossing over often occurs.

CONCLUSION

The results provided by the different types of meiotic analyses lead to the conclusion that meiosis I is more error-prone than meiosis II and that maternal meiosis is more error-prone than paternal meiosis. Furthermore, asynapsis is likely to account for a proportion of meiotic I error but other stages of meiosis I must also be susceptible to error. True nondisjunction, the failure of homologous chromosomes to disjoin, may occur in meiosis as does true nondisjunction, the failure of sister chromatids to disjoin, occur in mitosis. It is also likely that improper segregation of properly disjoined chromosomes occurs in meiosis as it does in mitosis. Both probably occur as a result of inadequate spindle function.

The similarities between the proposed error mechanisms in meiosis and mitosis suggest that in many areas, mitosis is a reasonable model for the study of aneuploid mechanisms. Examination of the data from both types of division stress the necessity to identify the different types of error in different subjects. Only this painstaking approach is likely to reveal the nature of factors initiating error. This may at some stage lead to positive action against aneuploidy.

SUMMARY

Aneuploidy is the most dramatic mutation as it affects large numbers of genes simultaneously. It is the most common cause of congenital abnormalities in humans and probably has a significant role in tumorogenesis and ageing. Aneuploid cells can be generated by errors in meiosis and mitosis, the results of the former affecting all the cells of an organism, the results of the latter only affecting some of the cells.

The errors leading to aneuploidy have often been referred to collectively as nondisjunction. This paper shows that true nondisjunction is a rare cause of mitotic aneuploidy. Chromosome lagging is far more common and lagging of the X chromosome is significantly increased with age. A secondary effect of lagging, reattachment to the spindle with malorientation, is as frequent a cause of 3:1 segregations as mitotic nondisjunction.

The available data on human meiosis is examined. It is concluded that asynopsis accounts for a proportion of meiotic I error but that a significant proportion involves true nondisjunction or segregational errors. In view of the mitotic anaphase data presented, it is suggested that reduced spindle capacity occurs with increasing age and that this is likely to induce both meiotic and mitotic segregational errors.

REFERENCES

Barrett, J.C., Oshimura, M., Tanaka, N., and Tsutsui, T., 1985, Role of aneuploidy in early and late stages of neoplastic progression of Syrian Hamster embryo cells in culture, in: "Aneuploidy, Etiology and Mechanisms", V.L. Dellarco, P.E. Voytek and A. Hollaender, eds., Plenum Press, N.Y.

Bettner, L.G., and Jarvik, L.F., 1971, Stroop color-word test, non-psychotic organic brain syndrome and chromosome loss in aged twins, J. Gerontol., 26:458.

Bond, D.J., and Chandley, A.C., 1983, Aneuploidy, Oxford University Press.

Choo, K.H., Brown, R., Webb, G., Craig, I.W., and Filby, R.G., 1987, Genomic organization of human centromeric alpha satellite DNA: Characterization of a chromosome 17 alpha satellite sequence, DNA, 6:277.

Devilee, P., Cremer, T., Slagboom, P., Bakker, E., Scholl, H.P., Hager, H.D., Stevenson, A.F.G., Cornelisse, C.J., and Pearson, P.L., 1986, Two subsets of human alphoid repetitive DNA show distinct preferential localization in the pericentric regions of chromosomes 13,18 and 21, Cytogenet. Cell Genet., 41:193.

Ford, J.H., and Congedi, M.M., 1987, Rapid induction of anaphase in competent cells by hypotonic treatment, Cytobios, 51:183.

Ford, J.H., and Russell, J.A, 1985, Differences in the error mechanisms affecting sex and autosomal chromosomes in women of different ages within the reproductive age group, Am. J. Human Genet., 37:973.

Ford, J.H., Schultz, C.J., and Correll, A.T., 1988, Chromosome elimination in micronuclei: A common cause of hypoploidy, Am. J. Hum. Genet., 43:733.

Galt, J., Boyd, E., Connor, J.M., and Ferguson-Smith, M.A., 1989, Isolation of chromosome-21-specific DNA probes and their use in the analysis of nondisjunction in Down syndrome, Hum. Genet., 81:113.

Hamerton, J.L., Taylor, A.I., Angell, R., and McGuire, V.M., 1965, Chromosome investigations of a small isolated human population: Chromosome abnormalities and distribution of chromosome counts according to age and sex among the population of Tristan da Cunha, Nature, 206:1232.

Harnden, D.G., and O'Riordan, M.L., 1973, Down's syndrome and leukaemia, Lancet, 1:260.

Hassold, T.J., 1985, The origin of aneuploidy in humans, in: "Aneuploidy, Etiology and Mechanisms", V.L. Dellarco, P.E. Voytek and A. Hollaender, eds., Plenum Press, N.Y.

Hassold, T., and Chiu, D., 1985, Maternal age-specific rates of numerical chromosome abnormalities with special reference to trisomy, Hum. Genet., 70:11.

Hulten, M.A., Laurie, D.A., Martin, R.H., Saadallah, N., and Wallace, B.M.N., 1984, New techniques for detecting chromosome abnormalities in the germ-line in man, in: "Individual Susceptibility to Genotoxic Agents in the Human Population", F.J. de Serres and R.W. Pero, eds., Plenum Press, N.Y.

Hulten, M., Saadallah, N., Wallace, B.M.N., and Cockburn, D.J., 1985, Meiotic investigations of aneuploidy in the human, in: "Aneuploidy, Etiology and Mechanisms", V.L. Dellarco, P.E. Voytek and A. Hollaender, eds., Plenum Press, N.Y.

Jacobs, P.A., and Court Brown, W.M., (1966), Age and chromosomes, Nature, 212:823.

Martin, R.H., Rademaker, A.W., Hildebrand, K., Long-Simpson, L., Peterson, D., and Yamamoto, J., 1987, Hum. Genet., 77:108.

Mattevi, M.S., and Salzano, F.M., 1975, Senescence and human chromosome changes, Humangenetik, 27:1.

Nielsen, J., Jensen, L., Lindhardt, H., Stottrup, L., and Sondergaard, A., 1968, Chromosomes in senile dementia, Br. J. Phsychol., 114:303.

Saadallah, N., 1984, "Cytogenetic aspects of human male meiosis", Ph.D. Thesis, The University of Birmingham.

Solari, A.J., 1980, Synaptonemal complexes and associated structures in microspread human spermatocytes, Chromosoma, 81:315.

Speed, R.M., 1985, The prophase stages in human foetal oocytes studied by light and electron microscopy, Hum. Genet. 69:69.

Stewart, G.D., Hassold, T.J., Berg, A., Watkins, P., Tanzi, R., and Kurnit, D.M., 1988, Trisomy 21 (Down syndrome)): Studying nondisjunction and meiotic recombination by using cytogenetic and molecular polymorphisms that span chromosome 21, Am. J. Human Genet., 42:227.

Sung, W.K., Komatsu, M., and Jagiello, G. (1983), A method for obtaining synaptonemal complexes of human pachytene oocytes, _Caryologia_ 36:315.

Wallace, B.M.N. and Hulten, M.A., 1985, Meiotic chromosome pairing in the human female, _Ann. Hum. Genet._, 49:215.

Warren, A.C., Chakravarti, A.,Wong, C., Slaugenhaupt, S.A., Halloran, S.L., Watkins, P.C., Metaxotou, C., and Antonarakis, S.E., 1987, Evidence for reduced recombination on the nondisjoined chromosomes 21 in Down syndrome, _Science_, 237:652.

Wramsby, H., Fredga, K., and Liedholm, P., 1987, Chromosome analysis of human oocytes recovered from preovulatory follicles in stimulated cycles, _New Eng. J. Med._, 316:121.

Zenzes, M.T., de Geyter, C., Bordt, J., Schneider, H.P.G., and Nieschlag, E., 1987, Cytological investigations of human eggs with unsuccessful fertilization, _Am. J. Hum. Genet._, 41:858.

ADAPTABILITY AND REPAIR MECHANISMS

THE ADAPTIVE RESPONSE TO ALKYLATION DAMAGE IN ESCHERICHIA

COLI

Barbara Sedgwick

Imperial Cancer Research Fund
Clare Hall Laboratories
South Mimms, Potters Bar, Herts, EN6 3LD, U.K.

ABSTRACT

E. coli has both non-inducible and inducible DNA repair activities which remove deleterious lesions from methylated DNA. The non-inducible Tag and Ogt proteins repair toxic 3-methyladenine and mutagenic 06-methylguanine DNA lesions, repsectively. Induction of the adaptive response increases expression of four genes, ada, alkA, alkB, and aidB. The Ada and AlkA proteins are well characterised but the functions of the AlkB and AidB proteins are unknown. The AlkA protein in a DNA glycosylase which excises 3-methyladenine and several other minor lesions from DNA. The Ada protein is a DNA-methyltransfrerase which transfers the methyl groups from 06-methylguanine and methylphosphotriesters on to two of its own cysteine residues. The Ada protein self-methylated by repair of a methylphosphotriester is a strong transcriptional activator of the inducible genes of the adaptive response. The active sites of the multifunctional Ada protein have been identified and studies of the regulatory domains initiated.

INTRODUCTION

Monofunctional methylating agents are widespread environmental mutagens. They alkylate the cellular DNA at many sites generating some base derivatives which miscode and others which block DNA replication. The major miscoding DNA lesion is 0^6-methylguanine which can mispair with thymine, and consequently results in GC to AT transition mutations. The major cytotoxic DNA lesion is 3-methyladenine. The methyl group of this derivative protrudes into the minor groove of DNA and blocks DNA replication. 0^4-methylthymine and 3-methylguanine are minor miscoding and cytotoxic lesions, respectively (reviewed by Saffhill et al., 1985). Alkylphosphotriesters in DNA result from methylation of the phosphate linkages in the DNA backbone and reduce the rate of template-directed DNA synthesis (Miller et al., 1982). Nevertheless, there is evidence to suggest that they are not highly cytotoxic to E. coli (Kataoka et al., 1986).

Mechanisms of Environmental Mutagenesis-Carcinogenesis, Edited by
A. Kappas, Plenum Press, New York, 1990

E. coli has both constitutive and inducible DNA repair activities which specifically remove these deleterious lesions from DNA and, thereby, confer cellular protection against the lethal and mutagenic effects of methylating agents.

THE NON-INDUCIBLE DNA REPAIR ACTIVITIES

Two non-inducible DNA repair functions which specifically remove the major mutagenic and toxic methylated adducts from DNA have been identified in E. coli. These are the tag and ogt gene products (Table 1). Both these genes have been cloned and sequenced (Steinum and Seeberg, 1986; Sakumi et al., 1986; Potter et al., 1987). The Tag protein, 3-methyladenine DNA-glycosylase I, specifically excises toxic 3-methyladenine lesions from DNA (Riazuddin and Lindahl, 1978). Repair of the apurinic sites remaining after action of the DNA glycosylase requires endonuclease, DNA deoxyribophosphodiesterase (Franklin and Lindahl, 1988), DNA polymerase and DNA ligase activities. The recently discovered Ogt protein, 0^6-methylguanine DNA-methyltransferase II, repairs the mutagenic 0^6-methylguanine and 0^4-methylthymine adducts by methyl transfer on to the protein molecule itself (Potter et al., 1987; Rebeck et al., 1988,1989; Shevell et al., 1988). Further details of this type of activity will be described below. These non-inducible DNA repair activities can presumably deal with low levels of DNA methylation produced by either external or endogenous alkylating agents. A possible example of the latter is S-adenosylmethionine (Rydberg and Lindahl, 1982; Barrows and Magee, 1982). More severe DNA methylation damage results in the induction of additional DNA repair functions which constitute the adaptive response to alkylation damage (Samson and Cairns, 1977; Jeggo et al., 1977). As a consequence, the cellular capacity to repair methylated DNA is substantially increased.

THE ADAPTIVE RESPONSE TO ALKYLATION DAMAGE

Several reviews of the adaptive response to alkylation damage have been recently published (Volkert 1988; Demple 1988a; Demple 1988b; Lindahl and Sedgwick, 1988). In this article, I will briefly review the response and enlarge on more recent developments.

The Inducible Genes and their Products

The response is regulated by the ada gene product which controls the expression of four genes, alkA, alkB, aidB and the ada gene itself (Table 1). The ada and alkB genes constitute an operon and their expression is regulated from the ada promoter. The alkA, ada-alkB, and aidB genes are located at 45, 47, and 95 minutes, respectively, on the E. coli genetic map. Three of these genes, ada, alkA, and alkB were first identified by the isolation of mutants sensitive to methylating agents and their products subsequently shown to have roles in DNA repair. All three genes have been cloned and sequenced (reviewed by Lindahl and Sedwick, 1988). The aidB gene was located by isolation of a methylation inducible Mu-dl (Ap[r] lac) fusion. Induction of aidB by methylation damage is ada dependent but it can also be induced in an ada

Table 1. Gene Products which Specifically Repair Alkylated DNA Lesions

Non-inducible genes	Product	Lesions repaired	Detrimental effect of un-repaired lesion
tag	3MeA DNA glycosylase I	3MeA	toxic
ogt	O^6MeG DNA-methyltransferase II	O^6MeG O^4MeT	mutagenic mutagenic

Inducible[1] genes	Product	Lesions repaired	Detrimental effect of un-repaired lesion
alkA	3MeA DNA-glycosylase II	3MeA 3MeG O^2MeT O^2MeC	toxic toxic
ada	O^6MeG DNA-methyltransferase I	O^6MeG O^4MeT MeP	mutagenic mutagenic
alkB	?	?	
alkB	?	?	

[1]Ada regulated inducible genes of the adaptive response to alkylation damage.

independent manner by anaerobiosis. The function of the aidB gene product remains unknown (Volkert 1988; Volkert et al., 1989).

The AlkA protein, 3-methyladenine-DNA glycosylase II, is a DNA glycosylase of broad specificity which excises 3-methyl-purines and O^2-methylpyrimidines from DNA (Evensen and Seeberg, 1982; Karran et al., 1982; McCarthy et al., 1984). Induction of alkA gene expression therefore elevates 3-methyladenine repair and consequently enhances the survival of cells exposed to methylating agents. The non-inducible Tag glycosylase is more substrate specific than the AlkA enzyme and removes only 3-methyladenine from methylated DNA. However, the tag gene on a high copy number plasmid completely complements the alkylation sensitivity of an alkA mutant (Kaasen et al., 1986). It is therefore not apparent that repair of the minor alkylated lesions by the AlkA enzyme contributes significantly to cell survival.

Reduced survival of methylated plasmid DNA in an alkB mutant suggests a role for the AlkB protein in DNA repair

(Kataoka et al., 1983). Although the AlkB protein has been purified, its specific function has not been determined. It does not excise any of the known methylated bases from damaged DNA (Kondo et al., 1986).

The Ada protein is multifunctional. It is a transcriptional activator of the four inducible genes of the adaptive response and is also a DNA repair protein (Teo et al., 1984). It demethylates the highly mutagenic DNA lesions, O^6-methylguanine and O^4-methylthymine, and transfers the methyl groups on to its own cysteine residue, cysteine-321 (Olsson and Lindahl, 1980; Demple et al., 1985). It similarly demethylates the S-diastereoisomer of a methylphosphotriester but transfers the methyl group to a different active cysteine residue, cysteine-69 (McCarthy and Lindahl 1985; Sedgwick et al., 1988; Takano et al., 1988). Self methylation of these active sites of the Ada protein is irreversible and so prevents any further DNA repair activity (reviewed by Lindahl and Sedwick, 1988). Protection against the mutagenicity of alkylating agents conveyed by the adaptive response is therefore saturated on exposure to high doses of these agents (Robbins and Cairns, 1979).

Bacillus subtilis has a non-inducible (Dat-1) and an inducible (Dat-2) O^6-methylguanine-DNA methyltransferase, as well as a separate inducible methylphosphotriester-DNA methyltransferase (Dat-3) (Morohoshi and Munakata, 1987). The Dat-1 gene has been recently cloned and sequenced. The amino acid sequences of the C-terminal domain of the Ada protein, the Ogt protein and the Dat 1 protein, which all demethylate O^6-methylguanine, are highly homologous (47 to 49%) (Potter et al., 1987; Morohoshi et al., 1989). An identical pentapeptide (pro-cys-his-arg-val) to that surrounding the active cysteine-321 residue of Ada occurs in Ogt and Dat-1 and most probably corresponds to their active sites. The functional similarity of these proteins suggests that the conserved amino acid sequences must be important in the mechanism of O^6-methylguanine recognition and/or methyltransfer. For a recent review of suicidal methyltransferases see Demple (1988b). A non-inducible 19 kDa O^6-methylguanine-DNA methyltransferase has also been recently revealed in Salmonella typhimurium (Rebeck et al., 1989). This bacterium, however, does not appear to have a detectable inducible response (Guttenplan and Milstein, 1982). Nevertheless, adaptive responses have been observed in several other bacterial species (reviewed by Lindahl et al., 1988).

The E. coli Ada and Ogt proteins in crude cell extracts can be radioactively labelled at their active sites by incubation with [H^3]-methylnitrosourea-treated DNA and monitored by SDS-polyacrylamide gel electrophoresis and fluorography. Using this method, we have been able to detect the Ada and Ogt proteins in uninduced bacteria and have found that the cells have 10 to 20 times more Ogt than Ada protein (P. Vaughan and B. Sedgwick, unpublished data). In uninduced cells, the Ogt protein therefore mounts the greater defence against mutagenic methylation damage. On induction of the adaptive response, however, the Ada protein is induced approximately 100 fold and then dominates in this defensive role.

Regulation of the Adaptive Response

The inducing signal of the adaptive response, which arises in cells exposed to methylating agents, has been identified as a methylphosphotriester S-diastereoisomer in DNA. This was revealed by demonstrating that the Ada protein, self-methylated at cysteine-69 by repair of a methylphospho-triester, is converted into a strong transcriptional activator of the ada and alkA genes (Figure 1). The Ada protein methylated at cysteine-321 by repair of O^6-methylguanine is not a strong gene activator (Teo et al., 1986). Support for these observations has been recently obtained by the use of mutant Ada protein in which one or other of the active cysteine residues was replaced by alanine (Sakumi and Sekigu-chi, 1989). Methylphosphotriesters in DNA are not considered to be effective cytotoxic lesions because their repair by an amino-fragment of the Ada protein does not significantly increase the survival of cells exposed to a methylating agent (Kataoka et al., 1986). The repair of these adducts by the Ada protein possibly serves primarily as a switch for the adaptive response to alkylation damage. Methylphosphotriester lesions may have been utilised in this role because they are not repaired by the non-inducible Ogt protein. By this reasoning, the response could be up-regulated at the same time as the deleterious O^6-methylguanine and 3-methyladenine lesions are removed by the Ogt and Tag activities.

The Ada protein, self-methylated at cysteine-69, has been demonstrated by DNase 1 footprinting to bind to the promoters of the ada and alkA genes in regions which contain a common nucleotide sequence, AAANNAAAGCGCA (Teo et al., 1986). Deletion analysis from the 5' end of the ada gene has suggested that only the latter 8 nucleotides of this sequence, AAAGCGCA, are essential for ada up-regulation (Nakamura et al., 1988). This eight nucleotide sequence, however, occurs many times in the Genbank database of E.coli DNA sequences (B. Sedgwick and M. Ginsburg, unpublished data). Thus, additional parameters may be involved in defining the specificity of the 'Ada box', the DNA binding site of the methylated Ada protein. Specific binding of methylated Ada protein to the ada promoter has been shown to aid the subsequent binding of RNA polymerase (Sakumi and Sekiguchi, 1989) and presumably, in this way, stimulates transcription (see Figure 1).

Functional Domains of the Ada Protein

Limited digestion with several non-sequence specific reagent proteases established that the Ada protein is composed of two major domains of similar size separated by a central hinge region of approximately ten amino acids. The amino- and carboxy-terminal domains retain their respective DNA repair activities and remove methylphosphotriesters and O^6-methylgua-nine lesions from DNA (Sedgwick et al, 1988). On methylation, a mixture of the two domains, but not the purified C-terminal domain, shows specific binding to the promoter of the ada gene, but is unable to induce transcription. These observa-tions suggest that the 20 kDa amino-terminal domain of the protein contains the DNA binding site, but that the protein must be intact in order to stimulate ada gene transcription (Sedgwick and Hughes, 1988). Yoshikai et al., (1988) have made similar observations on ada gene expression but have found

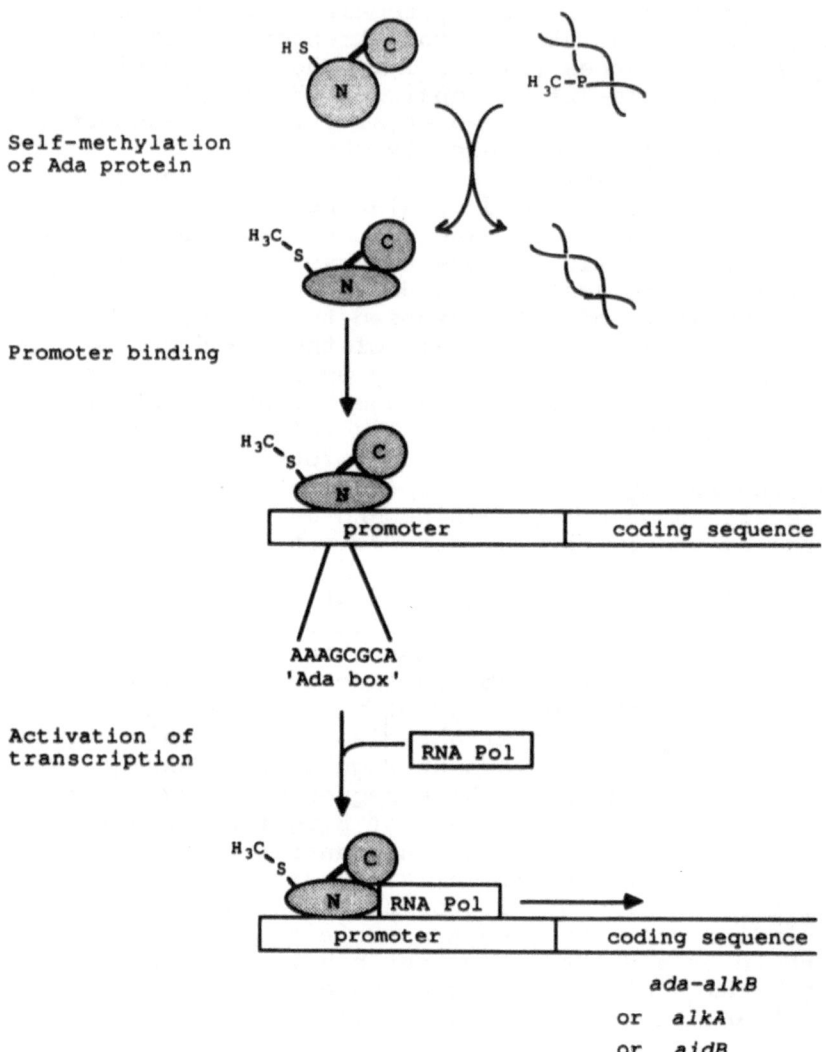

Self-methylation
of Ada protein

Promoter binding

AAAGCGCA
'Ada box'

Activation of
transcription

ada-alkB
or   alkA
or   aidB

Fig. 1.   Diagrammatic representation of transcriptional
activation by the Ada protein. The N-terminal
domain of the Ada protein is self-methylated
by demethylation of a methylphosphotriester in
DNA. This covalent modification, which may
result in a conformation change, enables the
Ada protein to bind the 'Ada box' in the
promoters of the Ada regulated genes. The
bound protein enhances subsequent binding of
RNA polymerase and thus stimulates transcription.

that the methylated 20 kDa amino-terminal domain alone is sufficient to stimulate alkA transcription.

The Ada and AlkA proteins have recently been crystallised (Moody and Demple, 1988; Yamagata et al., 1988). Future X-ray crystallographic studies of these proteins should yield a more detailed understanding of their mechanisms of DNA repair and of transcriptional activation by the Ada protein.

Mutants Ada Proteins

Molecular studies of mutant Ada proteins are also yielding further insights into the functions of the Ada domains. Ada mutants were originally isolated after MNNG mutagenesis as alkylation sensitive strains which cannot be induced for the adaptive response (Jeggo, 1979), or as alkylation resistant strains which express the response constitutively (Sedgwick and Robins, 1980). Other mutant ada genes have been constructed more recently by recombinant DNA techniques. Several mutant Ada proteins with alterations in their C-terminal domains have been shown to be defective in the regulation of ada/alkA gene expression which implies that this domain has a role in gene activation. These studies have included two mutant proteins which have unidentified mutations in their C-terminal domains and were slow to induce ada and alkA gene expression (Demple, 1986); mutant proteins in which the active cysteine-321 residues were substituted by alanine or histidine and had, respectively, an enhanced or decreased ability to induce ada gene expression in vivo (Takano et al., 1988; Tano et al., 1989); and several Ada fusion proteins which derived their amino-termini from ada and their carboxy-termini from vector sequences and either enhanced or inhibited ada gene expression (Shevell et al., 1988).

We have recently cloned and sequenced the ada genes from four mutants which express the adaptive response constitutively. All four mutant ada genes were found to encode Ada proteins with two amino acid substitutions in their amino-terminal domain. One of these amino acid changes, methionine-126 to isoleucine was common to all four mutant proteins, and this mutation alone converted the Ada protein into a strong gene activator (Hughes and Sedgwick, 1989). The Ada[+] protein becomes a strong positive regulator after self-methylation of cysteine-69 as a result of its increased DNA binding affinity for the 'Ada box'. Using the same conditions of DNase 1 footprinting as used for the wild type protein, we have been unable to demonstrate increased specific DNA binding of the mutant protein, Ada-11 (Sedgwick, unpublished data). We are investigating the possibility that binding of Ada-11 is salt or RNA polymerase dependent as observed for a mutant λ repressor and a mutant cAMP receptor protein, respectively (Nelson et al., 1985; Ren et al., 1988).

Proteolysis of the Ada Protein

Proteolysis of the Ada protein has been proposed as a means of down-regulating the adaptive response by inactivation of the methylated Ada protein as a positive regulator. This hypothesis was based on the observation that the Ada protein is sensitive to cleavage by an endogenous protease in crude cell extracts (Teo 1987), and was supported by the finding that the major cleavage products of 20 kDa and 19 kDa were unable to induce <u>ada</u> gene expression <u>in vitro</u> (Sedgwick and Hughes, 1988; Yoshikai, et al., 1988). Proteolysis of the Ada protein, however, has never been observed <u>in vivo</u>. The major proteolytic activity which cleaves the Ada protein in crude extracts has been recently identified as the <u>ompT</u> gene product (Sedgwick 1989), an outer membrane protease (Grodberg and Dunn, 1988). A heat inducible protease which cleaves both the Ada and UvrB proteins (Caron and Grossman, 1988) has also been recognised as the OmpT enzyme (Caron and Grossman, personal communication). The OmpT activity is identical to protease VII, a 33 kDa serine protease (Sugimura and Nishimura, 1988). We have found that deletion of the <u>ompT</u> gene does not delay the decay of the adaptive response after removal of the inducing agent, and thus the OmpT protease cannot be required for down-regulation of the response <u>in vivo</u> (Sedgwick, 1989). A proteolytic activity of approximately 400 kDa and considered to be a thiol protease, which specifically cleaves the Ada protein, has been partially purified from <u>E. coli</u> (Yoshikai et al., 1988). It remains to be ascertained whether proteolysis of the methylated Ada protein by this protease or any other is important in down-regulating the adaptive response.

Alkylating Agents which Induce the Adaptive Response

The ability of various alkylating agents to induce the adaptive response was first examined by a comparison of the sensitivity of wild type and <u>ada</u> mutants to their toxicity and mutagenicity (Sedgwick and Lindahl, 1982). A more recent approach has involved the use of <u>alkA</u>-<u>lacZ</u> fusions from which expression of <u>lacZ</u> is controlled by the <u>alkA</u> promoter. Both chromosomal (Fram et al., 1988; Volkers et al., 1989) and plasmid-borne gene fusions (Otsuka et al., 1985) have been utilised. We have also developed a sensitive immunoblotting technique using monoclonal antibodies raised against the Ada protein. By this method, the Ada protein in uninduced cells can be detected. Any increase in this basal level indicates induction of the adaptive response by the alkylating agent (Vaughan and Sedgwick, unpublished data). Using these tests the ability of a variety of N-alkyl-N'-nitro-N-nitrosogua-nidines, alkyl methanesulphonates, and N-alkyl-N-nitrosoureas to induce the adaptive response has been examined. The overall conclusion is that methylating agents are more efficient inducers than ethylating and also propylating agents (reviewed by Volkert, 1988). The Ada protein is known to be less efficient in repairing ethylated DNA lesions (Sedgwick and Lindahl, 1982). Thus, the slower generation of the self ethylated Ada protein, the transcriptional activator, can at least partly account for poor induction by ethylating agents. It can be concluded that the adaptive response serves primarily to protect cells against DNA methylation damage.

ACKNOWLEDGEMENTS

I thank R. Wood for his expert help with the figure and J. Green for typing the manuscript.

REFERENCES

Barrows, L.R., and Magee, P.N., 1982, Nonenzymatic methylation of DNA by S-adenosylmethionine _in vitro_, _Carcinogen._, 3:349.

Caron, P.R., and Grossman, L., 1988, Potential role of proteolysis in the control of UvrABC incision, _Nuc. Acids Res._, 16:9641.

Demple, B., 1986, Mutant _Escherichia coli_ Ada proteins simultaneously defective in the repair of O^6-methylguanine and in gene activation, _Nuc. Acids Res._, 14:5575.

Demple, B., 1988a, Adaptive responses to genotoxic damage: Bacterial strategies to prevent mutation and cell death, _BioEssays_, 6:157.

Demple, B., 1988b, Self-methylation by Suicide DNA Repair Enzymes, _in_: "Protein Methylation", W.K. Paik and S. Kim, ed., CRC Press, Boca Raton, Florida.

Demple, B., Sedgwick, B., Robins, P., Totty, N., Waterfield, M.D., and Lindahl, T., 1985, Active site and complete sequence of the suicidal methyltransferase that counters alkylation mutagenesis, _Proc. Natl. Acad. Sci. USA_, 82:2688.

Evensen, G., and Seeberg, E., 1982, Adaption to alkylation resistance involves the induction of a DNA glycosylase, _Nature_, 296:773.

Fram, R.J., Marinus, M.G., and Volkert, M.R., 1988, Gene expression in _E. coli_ after treatment with streptozotocin, _Mut. Res._ 198:45.

Franklin, W.A., and Lindahl, T., 1988, DNA deoxyribophosphodiesterase, _EMBO J._, 7:3617.

Grodberg, J., and Dunn, J.J., 1988, ompT encodes the _Escherichia coli_ outer membrane protease that cleaves T7 RNA polymerase during purification, _J. Bacteriol_, 170:1245.

Guttenplan, J.B., and Milstein, S., 1982, Resistance of _Salmonella Typhimurium_ TA 1535 to O^6-guanine methylation and mutagenesis induced by low doses of N-methyl-N'-nitro-N-nitrosoguanidine: an apparent constitutive repair activity, _Carcinogen._, 3:327.

Hughes, S.J., and Sedgwick, B., 1989, The adaptive response to alkylation damage: Constitutive expression through a mutation in the coding region of the _ada_ gene, _J. Biol. Chem._, in press.

Jeggo, P., 1979, Isolation and characterization of _Escherichia coli_ K-12 mutants unable to induce the adaptive response to simple alkylating agents, _J. Bacteriol._, 139:783.

Jeggo, P., Defais, M., Samson, L., and Schendel, P., 1977, An adaptive response of _E. coli_ to low levels of alkylating agent: Comparison with previously characterised DNA repair pathways, _Molec. gen. Genet._ 157:1.

Kaasen, I., Evensen, G., and Seeberg, E., 1986, Amplified expression of the _tag_$^+$ and _alkA_$^+$ genes in _Escherichia coli_: Identification of gene products and effects on alkylation resistance, _J. Bacteriol_. 168:642.

Karran, P., Hjelmgren, T., and Lindahl, T., 1982, Induction of a DNA glycosylase for N-methylated purines is part of the adaptive response to alkylating agents, _Nature_, 296:770.

Kataoka, H., Hall, J., and Karran, P., 1986, Complementation
    of sensitivity to alkylating agents in Escherichia coli
    and Chinese hamster ovary cells by expression of a
    cloned bacterial DNA repair gene, EMBO J., 5:3195.
Kataoka, H., Yamamoto, Y., and Sekiguchi, M., 1983, A new gene
    (alkB) of Escherichia coli that controls sensitivity to
    methyl methane sulfonate, J. Bacteriol., 153:1301.
Kondo, H., Nakabeppu, Y., Kataoka, H., Kuhara, S., Kawabata,
    S., and Sekiguchi, M., 1986, Structure and expression
    of the alkB gene of Escherichia coli related to the
    repair of alkylated DNA, J. Biol. Chem., 261:15772,
Lindahl, T., and Sedgwick, B., 1988, Regulation and expression
    of the adaptive response to alkylating agents,
    Ann.Rev.Biochem., 57:133.
McCarthy, T.V., Karran, P., and Lindahl, T., 1984, Inducible
    repair of O-alkylated DNA pyrimidines in Escherichia
    coli, EMBO J., 3:545.
McCarthy, T.V., and Lindahl, T., 1985, Methyl phosphotriesters
    in alkylated DNA are repaired by the Ada regulatory
    protein of E. coli, Nuc. Acids Res., 13:2683.
Miller, P.S., Chandrasegaran, S., Dow, D.L., Pulford, S.M.,
    Kan, L.S., 1982, Synthesis and template properties of
    an ethyl phosphotriester modified decadeoxyribonucleot-
    ide, Biochem., 21:5468.
Moody, P.C.E., and Demple, B., 1988, Crystallization of
    O^6-methylguanine-DNA methyltransferase from Escherichia
    coli, J. Mol. Biol., 200:751.
Morohoshi, F., Hayashi, K., and Munakata, N., 1989, Bacillus
    subtilis gene encoding for constitutive O^6-methylgua-
    nine-DNA alkyltransferase, Nuc. Acids Res., 17:6531.
Morohoshi, F., and Munakata, N., 1987, Multiple species of
    Bacillus subtilis DNA alkyltransferase involved in the
    adaptive response to simple alkylating agents, J.
    Bacteriol., 169:587.
Nakamura, T., Tokumoto, Y., Sakumi, K., Koike, G., Nakabeppu,
    Y., and Sekiguchi, M., 1988, Expression of the ada gene
    of Escherichia coli in response to alkylating agents:
    Identification of transcriptional regulatory elements,
    J. Mol. Biol., 202:483.
Nelson, H.C.M., and Sauer, R.T., 1985, Lambda repressor
    mutations that increase the affinity and specificity of
    operator binding, Cell, 42:549.
Olsson, M., and Lindahl, T., 1980, Repair of alkylated DNA in
    Escherichia coli, J. Biol. Chem., 255:10569.
Otsuka, M., Nakabeppu, Y., and Sekiguchi, M., 1985, Ability of
    various alkylating agents to induce adaptive and SOS
    responses: A study with lacZ fusion, Mut. Res.,
    146:149.
Potter, P.M., Wilkinson, M.C., Fitton, J., Carr, F.J.,
    Brennand, J., Cooper, D.P., and Margison, G.P., 1987.
    Characterization and nucleotide sequence of ogt, the
    O^6-alkylguanine-DNA alkyltransferase gene of E.coli,
    Nuc. Acids Res., 15:22.
Rebeck, G.W., Coons, S., Carroll, P., and Samson, L., 1988, A
    second DNA methyltransferase repair enzyme in Escheri-
    chia coli, Proc. Natl. Acad. Sci., USA, 85:3039.
Rebeck, G.W., Smith, C.M., Goad, D.L., and Samson, L., 1989,
    Characterisation of the major DNA repair methyltrans-
    ferase activity in unadaptive Escherichia coli and
    identification of a similar activity in Salmonella
    typhimurium, J. Bacteriol., 171:4563.

Ren, Y.L., Garges, S., Adhya, S., and Krakow, J.S., 1988, Cooperative DNA binding of heterologous proteins: Evidence for contact between the cyclic AMP receptor protein and RNA polymerase, <u>Proc. Natl. Acad. Sci.</u>, 85:4138.

Riazuddin, S., and Lindahl, T., 1978, Properties of 3-methyladenine-DNA glycosylase from <u>Escherichia coli</u>, <u>Biochem.</u> 17:2110.

Robbins, R., Cairns, J., 1979, Quantitation of the adaptive response to alkylating agents, <u>Nature</u>, 280:74.

Rydberg, B., and Lindahl, T., 1982, Nonenzymatic methylation of DNA by the intracellular methyl group donor S-adenosyl-L-methionine is a potentially mutagenic reaction, <u>EMBO J.</u>, 1:211.

Saffhill, R., Margison,G.P., and O'Connor, P.J., 1985, Mechanisms of carcinogenesis induced by alkylating agents, <u>Biochimica et Biophysica Acta</u>, 823:111.

Sakumi, K., Nakabeppu, Y., Yamamoto, Y., Kawabata, S., Iwanaga, S., and Sekiguchi, M., 1986, Purification and structure of 3-methyladenine-DNA glycosylase I of <u>Escherichia coli</u>, <u>J. Biol. Chem.</u>, 261:15761.

Sakumi, K., and Sekiguchi, M., 1989, Regulation of expression of the <u>ada</u> gene controlling the adaptive response: Interactions with the <u>ada</u> promoter of the Ada protein and RNA polymerase, <u>J. Mol. Biol.</u>, 205:373.

Samson, L., and Cairns, J., 1977, A new pathway for DNA repair in <u>Escherichia coli</u>, <u>Nature</u>, 267:281.

Sedgwick, B., 1989, In vitro proteolytic cleavage of the <u>Escherichia coli</u> Ada protein by the <u>ompT</u> gene product, <u>J. Bacteriol.</u>, 171:2249.

Sedgwick, B., and Hughes, S., 1988, Multiple functions of the Ada protein in the adaptive response to alkylation damage in <u>Escherichia coli</u>, <u>in</u>: "Mechanisms and Consequences of DNA damage processing", Friedberg, E., and Hanawalt, P., eds., Alan Liss Press, New York

Sedgwick, B., and Lindahl,T., 1982, A common mechanism for repair of 06-methylguanine and 06-ethylaguanine in DNA, <u>J. Mol. Biol.</u>, 154:169.

Sedgwick, B., and Robins, P., 1980, Isolation of mutants of <u>Escherichia coli</u> with increased resistance to alkylating agents: Mutants deficient in thiols and mutants constitutive for the adaptive response, <u>Molec. gen. Genet.</u>, 180:85.

Sedgwick, B., Robins, P., Totty, N., and Lindahl, T., 1988, Functional domains and methyl acceptor sites of the <u>Escherichia coli</u> Ada protein, <u>J. Biol. Chem.</u>, 263:4430.

Shevell, D.E., Abou-Zamzam, A.M., Demple, B., Walker, G.C., 1988, Construction of an <u>Escherichia coli</u> K-12 <u>ada</u> deletion by gene replacement in a <u>recD</u> strain reveals a second methyltransferase that repairs alkylated DNA, <u>J. Bacteriol.</u>, 170:3294.

Shevell, D.E., LeMotte, P.K., and Walker, G.C., 1988, Alteration of the carboxyl-termination domain of Ada protein influences its inducibility, specificity, and strength as a transcriptional activator, <u>J. Bacteriol.</u>, 170:5263.

Steinum, A.L., and Seeberg, E., 1986, Nucleotide sequence of the <u>tag</u> gene from <u>Escherichia coli</u>, <u>Nuc.Acids Res.</u>, 14:3763.

Sugimura, K., and Nishihara, T., 1988, Purification, charac-
    terization, and primary structure of _Escherichia coli_
    protease VII with specificity for paired basic resi-
    dues: Identity of protease VII and OmpT, _J. Bacteriol._,
    170:5625.
Takano, K., Nakabeppu, Y., and Sekiguchi, M., 1988, Functional
    sites of the Ada regulatory protein of _Escherichia
    coli_: Analysis by amino acid substitutions, _J. Mol.
    Biol._, 201:261.
Tano, K., Bhattacharyya, D., Foote, R.S., Mural, R.J., Mitra,
    S., 1989, Site-directed mutation of the _Escherichia
    coli ada_ gene: Effects of substitution of methyl
    acceptor cysteine-321 by Histidine in Ada protein, _J.
    Bacteriol._, 171:1535.
Teo, I.A., 1987, Proteolytic processing of the Ada protein
    that repairs DNA O^6-methylguanine residues in _E. coli_,
    Mut. Res., 183:123.
Teo, I., Sedgwick, B., Demple, B., Li, B., and Lindahl, T.,
    1984, Induction of resistance to alkylating agents in
    E. coli: the _ada_$^+$ gene product serves both as a
    regulatory protein and as an enzyme for repair of
    mutagenic damage, _EMBO J._, 3:2151.
Teo, I., Sedgwick, B., Kilpatrick, M.W., McCarthy, T.V., and
    Lindahl, T., 1986, The intracellular signal for
    induction of resistance to alkylating agents in _E.
    coli_, _Cell_, 45:315.
Volkert, M.R., 1988, Adaptive response of _Escherichia coli_ to
    alkylation damage, _Environ. Mol. Mutagen._, 11:241.
Volkert, M.R., Gately, F.H., and Hajec, L.I., 1989, Expression
    of DNA damage-inducible genes of _Escherichia coli_ upon
    treatment with methylating, ethylating and propylating
    agents, _Mut. Res._, 217:109.
Volkert, M.R., Hajec, L.I., and Nguyen, D.C., 1989, Induction
    of the alkylation-inducible _aidB_ gene of _Escherichia
    coli_ by anaerobiosis, _J. Bacteriol._, 171:1196.
Yamagata, Y., Odawara, K., Tomita, K.I., Nakabeppu, Y., and
    Sekiguchi, M., 1988, Crystallization and preliminary
    X-ray diffraction studies of 3-methyladenine-DNA
    glycosylase II from _Escherichia coli_, _J. Mol. Biol._,
    204:1055.
Yoshikai, T., Nakabeppu, Y., and Sekiguchi, M., 1988, Proteo-
    lytic cleavage of Ada protein that carries methyltrans-
    ferase and transcriptional regulation activities, _J.
    Biol. Chem._, 263:19174.

THE ADAPTIVE RESPONSE OF HUMAN LYMPHOCYTES TO RADIATION

OR CHEMICAL MUTAGENS: CROSS-ADAPTATION AND SYNERGISM

Sheldon Wolff[1], Gregorio Olivieri[1,2] and
Veena Afzal[1]

[1]Laboratory of Radiobiology and Environmental
Health, University of California at
San Francisco, San Francisco, CA 94143, USA

[2]Dipartimento di Genetica e Biologia Molecolare
Universita Degli Studi di Roma, Italia

INTRODUCTION

Exposure of human lymphocytes to tritiated thymidine ($[^3H]$dThd) or low doses of X rays results in the induction of an adaptive response that makes the cells less susceptible to the induction of damage by subsequent high doses of X rays (Olivieri et al., 1984). Several features of this adaptive response have been established in a series of experiments (Wiencke et al., 1986; Shadley and Wolff, 1987; Shadley et al., 1987; Wolff et al., 1989). For instance, it was found that when cells are exposed to low levels of radiation from the incorporated radioisotope (Olivieri et al., 1984) or to X rays (Shadley and Wolff, 1987), which in themselves do not induce chromatid breakage, and were subsequently exposed to 150 cGy (150 rad) of X rays 6 hr before fixation, fewer chromatid aberrations were induced than by the high challenge dose of X rays alone. The low-dose pretreatments did not cause selection against a radiation-sensitive population of lymphocytes and so prevent them from contributing to the yield of aberrations found after the subsequent high challenge doses (Wiencke et al., 1986). This conclusion came from the results of reconstruction experiments in which various proportions of $[^3H]$dThd-labeled cells from a woman were co-cultured with unlabeled cells from a man. The individual types of cells could be distinguished from one another by their XX or XY chromosomal constitution. The labeled and unlabeled cells reached metaphase, where they could be scored for aberrations, in the same proportions in which they were present in the initial cultures. In these co-cultured cells, the adaptive response occurred only in those cells that had been labeled and so had been irradiated. The experiments also showed the adaptive response is not mediated by diffusible factors that could leave the labeled irradiated cells and enter the unlabeled cells. Concentrations of $[^3H]$dThd low enough to give rise to only one disintegration per cell induced this adaptive response (Wiencke et al., 1986). Furthermore, experiments in which initial doses of X rays as low as 0.5-1 cGy also made the cells refractory to subsequent exposures to 150 cGy of X rays (Shadley and Wolff, 1987). It has also been found that the challenge dose can be administered up to three cell cycles after the initial exposure (Shadley et al., 1987).

At present, it is thought that low-dose radiation induces a repair mechanism that, if in place at the time of the high-dose exposure, causes the primary breaks in the chromosomal DNA to restitute more efficiently so that fewer deletions are produced. This working hypothesis came about because the adaptive response can be inhibited by 3-aminobenzamide, an inhibitor of poly(ADP-ribose) polymerase, provided it is added shortly after the challenge dose before any chromosomal repair occurs (Wiencke et al., 1986). It is also known that poly(ADP-ribose) polymerase is induced in response to the production of strand breaks inside the cell, and this enzyme has been suggested to be involved in the repair of such breaks (Benjamin and Gill, 1980). If 3-aminobenzamide is administered between the doses of a fractionated X-ray dose to G_1 lymphocytes, it prevents the repair of the breaks induced by the first dose, so they will be present in the cell at a much later time when the second dose is given (Wiencke et al., 1986). Under such conditions, when the inhibition is released, the breaks from both doses can interact to form dicentrics, rings, and translocations, and the total yield of such two-break aberrations will be proportional to the square of the total dose, rather than being equal to the sum of the yields produced by the two half-doses, as is the case when the dose is fractionated under conditions in which repair can occur between the two doses.

Because this adaptive response is induced by very low initial doses, it bears a resemblance to the adaptive response to very low doses of alkylating agents that make bacteria (Samson and Cairns, 1977; Jeggo et al., 1977), animal cells (Samson and Schwartz, 1980, 1983; Laval and Laval, 1984), and plant cells (Rieger et al., 1982; Heindorff et al., 1987) less responsive to the induction of damage by subsequent exposures to the same or other alkylating agents. The adaptation after exposure to low doses of alkylating agents, however, was related to the induction of an alkyltransferase (Karran et al., 1979; Olsson and Lindahl, 1980) that removes the adducts from the DNA, and that cannot be responsible for the X-ray effect. The reason for this is that although X rays can induce a low level of base damage in DNA, the major DNA lesions responsible for the formation of chromosome aberrations are double-strand breaks (Evans, 1977; Wolff, 1972, 1978; Natarajan et al., 1980). Therefore it is thought that another, hitherto unknown, enzymatic repair mechanism is being induced by the low dose exposure to radiation. With low doses of X rays the response takes some 4-6 hr to become fully operable, i.e., the induction is not complete until 4-6 hr after the initial low dose exposure (Shadley et al., 1987), and it has been found that inhibition of protein synthesis with cycloheximide 4-6 hr after the initial dose prevents the response, indicating that new proteins (enzymes) that could be responsible for the effect might be synthesized at this time (Youngblom et al., 1989; Wolff et al., 1989). Experiments with two-dimensional gel electrophoresis of proteins found in human lymphocytes also show that exposures to 1-cGy doses can reproducibly produce new proteins (Wolff et al., 1989).

In a series of experiments (Wolff et al., 1988) carried out to see if the repair mechanism induced by low doses of radiation from [^{3}H]dThd or from 1 cGy of X rays can affect various types of clastogenic lesions induced in DNA by subsequent exposures to chemical mutagens and carcinogens, pretreated cells were exposed to mitomycin C, which induces cross-links in DNA (Tomasz et al., 1974), bleomycin, an S-independent radiomimetic agent that induces double-strand breaks in DNA (Povirk et al., 1977; Lloyd et al., 1978), and methyl methanesulfonate (MMS), an S-dependent agent that alkylates DNA and then leads to single-strand breaks. The experiments showed that when the cells were challenged with high doses of bleomycin or mitomycin C, approximately half as many chromatid breaks were induced as expected. When, on the other hand, the cells were challenged with the simple alkylating agent MMS, approximately twice as much damage was found as was induced by MMS alone. The results indicated that a prior exposure to low doses of X rays induced some cross-adaptation in that it could reduce the number of chromosome aberrations produced by chemically induced double-strand breaks, and perhaps even by chemically induced cross-links in DNA, but had the opposite effect on breaks induced by the alkylating agent.

Because MMS differs from another methylating agent, N-methyl-N'-nitro-N-nitrosoguanidine (MNNG), in that the two alkylating agents give a different spectrum of methylated bases (Laval and Laval, 1984; Singer and Grunberger, 1983), with MMS producing a far higher ratio of N^7-methylguanine to O^6-methylguanine than MNNG. Experiments have now been carried out with MNNG to see if the synergism obtained between X rays and a subsequent exposure to the simple alkylating agent MMS would be obtained with another methylating agent that produces a different spectrum of lesions. The experiments were also designed to see if chemical agents could also cause cross-adaptation that would reduce the effect of subsequent exposure to X rays as well as other chemicals.

MATERIALS AND METHODS

In all experiments, venous blood from healthy donors was drawn into heparinized (sodium) vacutainers, and 0.5 ml of whole blood was added to 4.5 ml of RPMI 1640 medium containing 10% fetal calf serum, 2 mM glutamine, 100 U/ml of penicillin, 100 µg/ml streptomycin, and 2% phytohemagglutinin (PHA-M, Difco). The blood was cultured at 37°C in 1-oz glass prescription bottles. Two hours before fixation, 0.1 ml of 10^{-5}M Colcemid (final concentration 2 x10^{-7}M) was added to each culture. The cells were then exposed to a hypertonic solution of 0.075 M KCl for about 8 min to spread the chromosomes, and were subsequently fixed in methanol-acetic acid (3:1). Drops of a concentrated suspension of cells in fixative were placed on wet microslides. The cells were then stained with 5% Giemsa in Sorensen's phosphate buffer (pH 6.8). Conditioning doses of X rays were delivered from a Philips RT250 therapeutic unit (250 kVp, 15 mA, HVL 1.06 mm Cu) at a dose rate of 0.2 Gy/min. The challenge doses of X rays were delivered at 1 Gy/min. X-ray pretreatments of PHA-stimulated lymphocytes were

Table 1. The synergistic response of low doses of X rays, MMS, or MNNG with a subsequent high dose of MMS

Conditioning treatment	Challenge treatment (MMS, 0.42 mM)	No. aberrations in 200 cells
1 cGy(1 rad) X rays		
−	−	3
+	−	3
−	+	114
+	+	169*
0.018 mM MMS		
−	−	3
+	−	2
−	+	97
+	+	133*
600 ng/ml MNNG		
−	−	3
+	−	1
−	+	97
+	+	147**

−, No treatment; +, treatment.
Significantly higher than the sum of conditioning and challenge treatments given separately: *$p < 0.01$, **$p < 0.001$ (one-tailed t-test).

administered at 24 hr of culture. The cells were subsequently challenged with MNNG or MMS for various periods of time. MNNG (30 µg/ml) was administered at 40 hr of culture and the chemical was left in until the cells were fixed 26-35 hr later. When MMS (0.42 mM) was used as a challenge, it was added at 43 hr of culture, and also was left in until fixation 23-27 hr later. In those experiments in which the cells were pretreated with low concentrations of MNNG or MMS, they were exposed 26 hr after the cultures were initiated. The cells were then challenged with 30 µg/ml of MNNG at 48 hr, or with 0.42 mM MMS at 51 hr of culture, and were fixed at 77 hr.

RESULTS AND DISCUSSION

The response of human lymphocytes to various challenges after having been exposed to a variety of conditioning doses was dependent upon the agent used for the challenge. When the cells were challenged with 0.42 mM MMS, it was found that a synergistic rather than a protective response occurred. Thus, when the conditioning pretreatment consisted of low doses of X rays (Table 1), a subsequent exposure to MMS resulted in an increased amount of damage. This is similar to the response that had been observed earlier (Wolff et al., 1988). The same

Table 2. Low doses of X rays, MMS, or MNNG cause human
lymphocytes to be less susceptible to the induction
of cytogenetic damage by a high dose of X rays

Conditioning treatment	Challenge treatment (1.5 Gy X rays)	No. aberrations/ no. cells
1 cGy(1 rad) X rays		
-	-	4/100
+	-	3/100
-	+	31/100
+	+	14/100*
0.018 mM MMS		
-	-	4/100
+	-	3/100
-	+	31/100
+	+	16/100*
600 ng/ml MNNG		
-	-	3/100
+	-	2/100
-	+	58/200
+	+	32/200**

-, No treatment; +, treatment.
Significantly higher than the sum of conditioning and challenge treatments given separately: *$p < 0.01$, **$p < 0.005$
(one-tailed t-test).

synergistic response occurred even when the cells had been
pretreated with low doses of MMS or MNNG. Thus, 1 cGy of X
rays, 0.018 mM MMS, or 600 ng/ml MNNG, all of which do not
induce a discernible number of chromatid aberrations in 200
cells, dramatically increased the numbers of chromatid breaks
observed after a subsequent exposure to MMS.

Such a reverse adaptive response, however, was not found
when the cells were challenged with either high doses of X
rays (Table 2) or MNNG (Table 3). In these cases, pretreatment
with low doses of X rays or MNNG, as well as with low doses of
MMS, caused adaptation to occur so that pretreated cells
always contained fewer aberrations than did cells that were
not pretreated.

The synergistic response obtained when MMS, but not other
agents, is used as the challenge treatment, irrespective of
what was used as a pretreatment, highlights the fact that the
alkylating agents, even those that are simple methylating
agents, can differ greatly in their effects (Singer and
Grunberger, 1983). Among the common known differences is that
MNNG methylates guanine at the O^6 position, as well as at the

Table 3. Low doses of X rays, MMS, or MNNG cause human lymphocytes to be less susceptible to the induction of cytogenetic damage by a high dose of MNNG

Conditioning treatment	Challenge treatment (MNNG, 30 µg/ml)	No. aberrations/ no. cells
1 cGy(1 rad) X rays		
−	−	3/100
+	−	1/100
−	+	111/200
+	+	65/200*
−	+	115/100
+	+	81/200*
−	+	120/200
+	+	80/200*
0.018 mM MMS		
−	−	1/100
+	−	3/100
−	+	40/100
+	+	11/100*
600 ng/ml MNNG		
−	−	1/100
+	−	2/100
−	+	40/100
+	+	18/100*

−, No treatment; +, treatment.
*Significantly lower than the sum of conditioning and challenge treatments given separately (p < 0.005, one-tailed t-test).

N^7 position, whereas MMS does not methylate at the O^6 position. Nevertheless, it is as yet unknown whether this or any of the other known effects of MMS are responsible for setting it off from the other agents.

Another remarkable feature of the data in the present experiments is the cross-adaptation that is found between ionizing radiation and alkylating agents. Although some of the clastogens used as a challenge in human lymphocytes produce lesions similar to, or identical to, X rays, e.g., active oxygen radicals or double-strand breaks in DNA, the base damage brought about by alkylating agents is very different from the small amount of base damage found after X rays, which is not a major contributor to chromosomal aberration formation.

Cross-adaptation also has been found in mammalian (Laval and Laval, 1984; Kaina, 1982,1983) and plant cells (Rieger et al., 1985; Veleminsky et al., 1983) for various endpoints. In plants, as in human lymphocytes, the endpoint has been the induction of chromosomal aberrations, whereas in mammalian cells, survival, aberrations, and mutations have been the endpoints observed. The studies with plants have shown that methylnitrosourea and triethylenemelamine can cause cross-adaptation with one another but not with very dissimilar agents such as maleic hydrazide.

In H_4 (rat hepatoma) cells, Laval and Laval (1984) found that MNNG, MMS, and methylnitrosourea could induce cross-adaptation for survival, but not for induction of mutation to thioguanine resistance. Thus, MMS as a pretreatment could decrease the killing effect of higher doses of itself or other agents, but not the mutagenic effect. Since mutations to thioguanine resistance are most likely to be deletions, i.e., the result of clastogenesis, the lack of an effect with MMS pretreatment seems to stand in contrast to the results obtained with human lymphocytes in the present experiments. On the other hand, since killing can be directly related to the formation of chromosome aberrations, the results obtained when survival is used as the endpoint seem to be analogous to those reported here. In Chinese hamster ovary cells, too, a response is obtained that is different from that seen with human lymphocytes. Kaina (1983) found that a single low dose of MNNG reduced the number of chromosomal aberrations induced by a subsequent high dose of MMS, i.e., produced cross-adaptation. The results with human lymphocytes, on the other hand, show a synergistic response whenever MMS is the challenge.

CONCLUSION

Experiments with human lymphocytes show that low dose pretreatments with X rays or with X-ray-like agents such as [^{3}H]dThd, hydrogen peroxide, or bleomycin, as well as alkylating agents such as MMS or MNNG, can induce an adaptive response whereby pretreated cells become somewhat refractory to the induction of cytogenetic damage by high doses administered at a later time. In general, in the human lymphocyte system, all of the agents induced cross-adaptation to themselves. A major exception to this is the synergism obtained when MMS is used as the challenging agent.

ACKNOWLEDGEMENTS

Work supported by the Office of Health and Environmental Research, U.S. Department of Energy, contract no. DE-AC03-76-SF01012.

REFERENCES

Benjamin, R.C., and Gill, D.M., 1980, Poly(ADP-ribose) synthesis _in vitro_ programmed by damaged DNA. A comparison of DNA molecules containing different types of strand breaks, _J. Biol. Chem._, 255:10502.

Evans, H.J., 1977, Molecular mechanisms in the induction of chromosome aberrations, in: "Progress in Genetic Toxicology," D. Scott, B.A. Bridges, and F.H. Sobels, eds., Elsevier/North Holland, Amsterdam.

Heindorff, K., Rieger, R., Michaelis, A., and Takehisa, S., 1987, Clastogenic adaptation triggered by S-phase-independent clastogens in Vicia faba, Mutat. Res., 190:131.

Jeggo, P., Defais, M., Samson, L., and Schendel, P., 1977, An adaptive response of E. coli to low levels of alkylating agents: Comparison with previously characterised DNA repair pathways. Mol. Gen. Genet., 157:1.

Kaina, B., 1982, Enhanced survival and reduced mutation and aberration frequencies induced in V79 Chinese hamster cells pre-exposed to low levels of methylating agents, Mutat. Res., 93:195.

Kaina, B., 1983, Cross-resistance studies with V79 Chinese hamster cells adapted to the mutagenic or clastogenic effect of N-methyl-N'-nitro-N-nitrosoguanidine, Mutat. Res., 111:341.

Karran, P., Lindahl, T., and Griffin, B., 1979, Adaptive response to alkylating agents involves alteration in situ of O^6-methylguanine residues in DNA, Nature, 280:76.

Laval, F., and Laval, J., 1984, Adaptive response in mammalian cells: crossreactivity of different pretreatments on cytotoxicity as contrasted to mutagenicity, Proc. Natl. Acad. Sci., USA, 81:1062.

Lloyd, R.S., Haidle, C.W., Robberson, D.L., and Dodson, M.L., Jr., 1978, A physical map of bleomycin-specific fragmentation sites on PM2 bacteriophage DNA, Curr. Microbiol., 1:45.

Natarajan, A.T., Obe, G., Van Zeeland, A.A., Palitti, F., Meijers, M., and Verdegaal-Immerzeel, E.A.M., 1980, Molecular mechanisms involved in the production of chromosomal aberrations. II. Utilization of Neurospora endonuclease for the study of aberration production by X-rays in G_1 and G_2 stages of the cell cycle, Mutat. Res., 69:293.

Olivieri, G., Bodycote, J., and Wolff, S., 1984, Adaptive response of human lymphocytes to low concentrations of radioactive thymidine, Science, 223:594.

Olsson, M., and Lindahl, T., 1980, Repair of alkylated DNA in Escherichia coli: methyl group transfer from O^6-methylguanine to a protein cysteine residue, J. Biol. Chem., 255:10569.

Povirk, L.F., Wubker, W., Kohnlein, W., and Hutchinson, F., 1977, DNA double-strand breaks and alkali-labile bonds produced by bleomycin, Nucl. Acids Res., 4:3573.

Rieger, R., Michaelis, A., and Nicoloff, H., 1982, Inducible repair processes in plant root tip meristems? "Below-additivity effects" of unequally fractionated clastogen concentrations, Biologisches Zentralblatt, 101:125.

Rieger, R., Michaelis, A., and Takehisa, S., 1985, "Clastogenic cross-adaptation" is dependent on the clastogens used for induction of chromatid aberrations in Vicia faba root-tip meristems, Mutat. Res., 144:171.

Samson, L., and Cairns, J., 1977, A new pathway for DNA repair in Escherichia coli, Nature, 267:281.

Samson, L., and Schwartz, J.L., 1980, Evidence for an adaptive DNA repair pathway in CHO and human skin fibroblasts cell lines, _Nature_, 287-861.

Samson, L., and Schwartz, J.L., 1983, The induction of resistance to alkylating damage in mammalian cells, _in:_ "Induced Mutagenesis: Molecular Mechanisms and Their Implications for Environmental Protection," C.W. Lawrence, ed., Plenum, New York.

Shadley, J.D., Afzal, V., and Wolff, S., 1987, Characterization of the adaptive response to ionizing radiation induced by low doses of X rays to human lymphocytes, _Radiat. Res._, 111:511.

Shadley, J.D., and Wolff, S., 1987, Very low doses of X-rays can cause human lymphocytes to become less susceptible to ionizing radiation, _Mutagenesis_, 2:95.

Singer, B., and Grunberger, D., 1983, "Molecular Biology of Mutagens and Carcinogens," Plenum New York.

Tomasz, M., Mercado, C.M., Olson, J., and Chatterjie, N., 1974, The mode of interaction of mitomycin C with deoxyribonucleic acid and other polynucleotides _in vitro_, _Biochemistry_, 13:4878.

Veleminsky, J., Gichner, T., and Satava, J., 1983, Reduction in the frequency of N-methyl-N-nitrosourea-induced somatic mutations in Tradescantia by pretreatment with low doses of alkylating agents, _Mutat. Res._, 122:229.

Wiencke, J.K., Afzal, V., Olivieri, G., and Wolff, S., 1986, Evidence that the [^{3}H-]thymidine-induced adaptive response of human lymphocytes to subsequent doses of X-rays involves the induction of a chromosomal repair mechanism, _Mutagenesis_, 1:375.

Wolff, S., 1972, The repair of X-ray-induced chromosome aberrations in stimulated and unstimulated human lymphocytes, _Mutat. Res._, 15:435.

Wolff, S., 1978, Relation between DNA repair, chromosome aberrations, and sister chromatid exchanges, _in:_ "DNA Repair Mechanisms", P.C. Hanawalt, E.C. Friedberg, and C.F. Fox, eds., Academic Press, New York.

Wolff, S., Afzal, V., Wiencke, J.K., Olivieri, G., and Michaeli, A., 1988, Human lymphocytes exposed to low doses of ionizing radiations become refractory to high doses of radiation as well as to chemical mutagens that induce double-strand breaks in DNA, _Int. J. Radiat. Biol._, 53:39.

Wolff, S., Wiencke, J.K., Afzal, V., Youngblom, J., Cortes, F., 1989, The adaptive response of human lymphocytes to very low doses of ionizing radiation: a case of induced chromosomal repair with the induction of specific proteins, _in:_ "Low Dose Radiation: Biological Bases of Risk Assessment," Tylor & Francis, London, in press.

Youngblom, J.H., Wiencke, J.K., and Wolff, S., 1989, Inhibition of the adaptive response of human lymphocytes to very low doses of ionizing radiation by the protein synthesis inhibitor cycloheximide, _Mutat. Res._, in press.

EVOLVING MUTATION RATES AND PROSPECTS FOR ANTIMUTAGENESIS

John W. Drake

Laboratory of Genetics
National Institute of Environmental Health
Sciences
Research Triangle Park, North Carolina 27709
U.S.A.

INTRODUCTION

A notion central to biology is that the characteristics of organisms are universally subject to modification by evolution. This has been clear for morphological traits for more than a century and for biochemical traits for a third of a century. It is now also becoming clear that behavioral traits are equally subject to evolutionary molding.

Much theoretical and experimental work during recent decades has established that two modes of evolution operate side by side. One is based on the selective advantages and disadvantages of particular alleles and allele combinations, the other on the selective neutrality of other alleles and allele combinations, allowing them on rare occasions to achieve high frequency by random genetic drift. Both modes of evolution depend ultimately upon the generation of primary diversity by mutation. Is it then possible that the mutation process and the mutation rates it generates are themselves subject to evolution? This article will emphasize that the answer is yes: both the mechanics and the rates of mutation are highly evolved entities.

A consequence of evolved mutation rates is the possibility of artificial evolution by human intervention, much as we have engineered the artificial evolution of economic plants and animals, first by breeding strategies and more recently by direct gene modifications. Towards this ability, we now strive to understand mechanisms of repair and error in DNA metabolism. It would seem that powerful insights into the determinants of mutation rates could be reaped from studies of mutator and antimutator mutations. This article will particularly consider the lessons to be learned from the latter.

SHOULD MUTATION RATES EVOLVE?

There are both theoretical considerations and experimental results that strongly suggest that mutation rates should evolve.

Mechanisms of Environmental Mutagenesis-Carcinogenesis, Edited by
A. Kappas, Plenum Press, New York, 1990

First, natural populations contain readily discovered members displaying substantially higher-than-average mutation rates. (As I will demonstrate later, it is much less clear whether substantially lower than average mutation rates can be so easily found). Thus, mutation rates are polymorphic, indicating the existence of genetic variation upon which selection might act.

When natural selection acts upon newly arisen alleles to increase their frequencies, modifiers that increase mutation rates will also tend to be selected. Several investigators, for instance, have engineered powerful selection for microbial mutator mutations simply by constructing double nutritional mutants, mutagenizing the population to generate new mutators, and then selecting for prototrophs. If the revertant frequency for each single nutritional block in the population was 10^{-9}, for instance, but became 10^{-6} in a mutator background, then the frequency of double revertants would rise from 10^{-18} to 10^{-12}. In a mutagenized population in which all frequencies were increased by 100-fold, the double revertants would rise from 10^{-14} to 10^{-8}. These are all reasonable numbers, and either situation suffices to enrich tremendously for mutator mutants.

There are (at least) three important limits to the efficacy of such selection. First, adaptive variants must arise, which implies not only that the mutation process can generate them in the first place but also that the environment offers opportunity for adaptation. Thus, an inconstant environment will tend to favor increased mutation rates. Second, the frequency and potency of mutator mutants in the population must be sufficiently high so that a substantial proportion of all new adaptive mutations arise in them instead of in the excess of organisms with average mutation rates. Third, recombination between the selected allele and the mutator mutation must be sufficiently infrequent so that the frequency of the mutator rises in parallel with that of the selected allele, that is, the mutator mutation hitchhikes along with the selected allele.

What, then, about selection for smaller mutation rates? The large majority of new mutations will be deleterious or selectively neutral. Thus, a higher proportion of the progeny of a mutator mutant will carry new deleterious mutations than will the progeny of a mutationally average organism. As a result, selection will operate to reduce mutation rates. There are (at least) two forces tending to limit this selection. The first is recombination, which will tend to limit selection against a mutation-rate determinant to the regions recombinationally close to the newly arising deleterious alleles. The second is usually called the physiological cost. Thus, to reduce error frequencies in DNA metabolism, the organism must commit mass and energy, for instance in additional DNA repair systems, increased excision of incorrectly inserted bases (along with substantially increased wastage of correctly inserted bases), or slowed DNA replication in relation to other aspects of cell metabolism. (It should not be forgotten that truly error-free replication will be impossible at any cost for obvious thermodynamic reasons.)

It is thus clear from elementary considerations that

selective forces are likely to act both to reduce and to increase mutation rates, and that an approach to equilibrium may occur in an environment of relatively constant fluctuations. These notions have been addressed in a theoretical framework by a number of workers; see, for example, Kimura (1960, 1967), Levins (1967), Leigh (1970, 1973), Eschel (1973), Painter (1975), Gillespie (1981a,b), Holsinger and Feldman (1983), Ishii and Matsuda (1985), Lieberman and Feldman (1986), Pamilo et al., (1987) and Ishii et al., (1989). Unfortunately, there are two general difficulties with these otherwise often admirable formulations: they contain assumptions either improbable or else difficult or impossible to verify, and/or they contain parameters impossible to measure. It is therefore beyond our reach at present to ascertain within these frameworks whether mutation rates are close to an equilibrium, and what that equilibrium might be. Instead, we must turn to experiment and observation.

CHEMOSTAT RECONSTRUCTION EXPERIMENTS

The exploration of experimental, investigator-prompted evolution has progressed most rapidly with microbial systems because of the short generation times and large populations that they offer. The most interesting experiments bearing on mutation rates are a set of classical studies employing normal and mutator strains of <u>Escherichia coli</u> growing in chemostats. These experiments were designed to reconstruct the expectations for selection for and against high mutation rates outlined above, employing the <u>mutT</u> mutator mutation which increases the rate of the transversion A:T —> C:G by about three orders of magnitude, sufficient even to alter the A:T/G:C ratio noticeably after a few thousand generations (Cox and Yanofsky, 1967).

The key element in the chemostat experiments is the phenomenon of "periodic selection" (see Atwood et al., 1951). Consider bacteria growing in a chemostat. When the frequency of some selectively neutral but infrequent mutant is monitored, it initially rises steadily as expected from the combination of neutrality plus a steady input of new mutations. Periodically, however, this mutant frequency falls sharply, only to rebuilt again. The fall occurs because the chemostat presents an inhospitable environment to <u>E. coli</u> and mutants with increased fitness soon arise. A more fit mutant quickly overgrows the population, sweeping out the accumulated neutral mutations. Later, another even more fit mutant arises to repeat the cycle.

When a chemostat is initiated with a mixture of normal and mutator cells and the frequency of cells bearing the <u>mutT</u> mutation is monitored, it remains steady for a few dozen generations and then rises rapidly and by several orders of magnitude (Gibson et al, 1970; Cox and Gibson, 1974). The reason, of course, is that the new mutant with improved fitness is particularly likely to have arisen in one of the cells with a greatly increased mutation rate. Sex and resulting genetic recombination is rare in chemostats, so that the more fit allele and the modifier of mutation rates remain tightly linked. This experiment was later converted into a more realistic analogue of nature when the ratio of mutator to normal cells was systematically lowered (Chao and Cox, 1983): below a certain <u>mutT</u> frequency, the normally mutating rather than mutator cells won out. This reflects the point made previously, that the (strength X frequency) of mutator mutants in the population must be sufficiently high so that a substantial proportion of all new adaptive mutations arise in them instead of in the excess of organisms with average mutation rates; specifically, the mutator will consistently win only when $N_{wt}\mu_{wt} < N_{mut}\mu_{mut}$.

The converse phenomenon, selection to reduce mutation rates, was reconstructed in chemostats seeded exclusively with <u>mutT</u> cells (Trobner and Piechocki, 1984): after some 2,000 generations, mutation rates had become consistently lower by some 10-fold on average. The cause was the (slow) accumulation of modifier mutations; the <u>mutT</u> allele could be readily recovered intact from such cells in backcrosses.

Thus, the simple expectations of selection to either increase or decrease mutations rates could be well reconstructed in chemostats. What, then, can be learned about neutral populations?

NORMALIZED SPONTANEOUS MUTATION RATES

A large number of measurements have been made of spontaneous mutation rates at diverse sites and loci in diverse organisms. The values range over a dozen orders of magnitude and appear at first glance to be chaotic. They can, however, be rendered sensible by considering the mutational target. When point mutation rates are considered, they are usually rather small and vary tremendously (e.g., hot spots and cold spots) in ways that reflect the effects of local DNA sequence on mechanisms of mutation prevention and enhancement not to be detailed here. When gene mutation rates are considered, they are both considerably larger (typically 100-to 1000-fold) than point mutation rates, and also at least somewhat less variable: dramatic local sequence effects tend to average out to effects of only a few-foldover one or a few kilobases. Nevertheless, gene mutation rates can be easily seen to vary substantially among organisms.

Even this variation can be largely removed, however, when data are available that allow mutation rates per gene to be scaled up to mutation rates per genome. This comparison requires that we have estimated the mutation rate per gene with some confidence (e.g., free of or corrected for selection effects and efficiencies of detection), and that we know the molecular sizes of both the target gene and the genome. When first attempted (Drake, 1969), it was observed that the bacteriophages lambda and T4 and the bacteria <u>E. coli</u> and <u>Salmonella typhimurium</u>, although varying in both genome size and mutation rate per average base pair by about two orders of magnitude, varied in mutation rate per genome by only about three-fold. The significance of this observation was obscured, however, by the fact that all four organisms could be considered to be a kind of family and by the observation that <u>Neurospora crassa</u> offered an apparent mutation rate per genome some three to ten times smaller. Over the years, however, I have added several more organisms to this list, and it is now apparent that the relationship still holds: among DNA bacteriophages, bacteria and simple eukaryotes, the mutation rate per base pair varies by about 10,000-fold but the mutation rate per genome varies only by about three-fold. To the extent that a number of standard laboratory favorites are representative of their respective species in the wild, this observation suggests two attractive and reciprocally related hypotheses: modern mutation rates in natural populations, having reached a consensus value, have therefore also closely approached an equilibrium value; and, since this value is similar for very different organisms, the forces that drive it must be very general in nature. This relationship also has interesting implications for the hypothesis of the molecular clock in evolution as it may apply to microbes (see, for instance, Ochman and Wilson, 1987), and for mechanisms of time-dependent rather than generation-dependent mutation processes as they may contribute to molecular evolution.

THE DOWNWARD ROUTE: ANTIMUTATORS

It seems clear from the above that mutation rates per base pair have evolved sharply downward as genome sizes have increased, at least for microbes. If they could be identified, then, the implied downward modifiers of mutation rates would be antimutator instead of mutator alleles. Not surprisingly, the first antimutator mutations were identified in an organism with a small genome (166 kb) and a concomitantly high mutation rate per average base pair, namely bacteriophage T4 (Drake and Allen, 1968; Drake et al., 1969). These antimutators are alleles of the gene that encodes the viral DNA polymerase. They reduce not only spontaneous but also a variety of chemically induced mutation rates (Drake and Greening, 1970) and, in the case of alleles studied much later, were seen even to reduce ultraviolet mutagenesis mediated by error-prone repair (Drake, 1988).

The two canonical antimutators, tsCB87 and tsCB120, exhibited strong specificities. They generally reduced mutation rates at reversion pathways believed to comprise A:T —> G:C transitions and at others comprising frameshift mutations, but had little effect on pathways believed to comprise G:C —> A:T transitions. Together with Speyer's 1965 report of phage T4 DNA polymerase mutator mutations, these observations spawned a number of biochemical investigations that tended to assign mutation rates to errors of either insertion or proofreading, the latter by the polymerase-associated 3'-exonuclease. One general conclusion of these studies was that the ratio of polymerase to exonuclease activities was a useful index of mutation rates: it was high for mutators and low for antimutators (see, for instance, Bessman et al., 1974 and references therein). Complexity entered, however, when Ripley (1975) demonstrated that A:T —> Py:Pu transversions were increased a few-fold by both of these antimutator mutations. Our original measurements (Drake et al., 1969), together with some recent repetitions (see below), indicate that neither tsCB87 nor tsCB120 particularly affect the average forward mutation rate over a target of several kilobases (e.g., the frequency of r mutants is not affected). Thus, on average, the canonical "antimutator" mutants are either antimutators or mutators when specific pathways (small mutational targets) are examined but are mutationally neutral or nearly so when a larger target is examined. Thus, the analysis of polymerase fidelity must progress (and is, in fact, progressing) from considerations of simple polymerase: exonuclease ratios to those of complex three-dimensional structure and substrate-enzyme contacts.

Dinh Nguyen and I have recently re-examined the question of whether general antimutator mutations can be found among mutations of the T4 DNA polymerase gene. Both published and unpublished records were surveyed to select alleles that exhibited antimutator activity in one or another $rII \rightarrow r^+$ or $r^+ \rightarrow r$ test. These mutants were then carefully assayed for their contents of r mutants, a useful and reliable test for average mutation rates when carefully performed. As shown in Table 1, every "antimutator" turned out to be either indistinguishable from the wild-type control or else more mutable.

This result may perhaps be understood in terms of current structural information about a different DNA polymerase, the E. coli DNA polymerase I large fragment (Ollis et al., 1985; Joyce and Steitz, 1987). This polypeptide exhibits a cleft within which the DNA primer/template is believed to nestle and polymerization to occur, followed by a separate domain for the proofreading 3'-exonuclease. Error discrimination during either polymerization or proofreading is likely to involve diverse deviations from standard DNA dimensions, and the enveloping nature of this enzyme structure clearly offers possibilities for numerous steric contacts between the enzyme and its several substrates. Thus, a change in any single amino acid, although sometimes clearly enhancing discrimination against one particular type of error (such as an A:C mispair, for example), would appear highly unlikely to enhance discrimination against most or all errors, and would in fact be likely to interfere with some other modes of error discrimination at the same time that it improves a single one. Thus, any single missense mutation in the DNA polymerase gene would now appear a highly unlikely candidate for a generalized antimutator mutation; perhaps only a cluster of several or many would produce a general improvement.

A GENERAL SPECULATION ABOUT ANTIMUTATORS AND A SURVEY

The argument advanced above on the improbability of recovering general antimutator mutations in a DNA polymerase gene can be broadened considerably. Consider the situation early in the evolution of the most primitive life forms. The accuracy with which their genomes were replicated was probably far lower than in any organism today, and strong selection probably operated to reduce the error rate. It is likely that a large number of error-generating processes operated, even as they do today. Some of these would be predominant, others less so, and yet others quite moderate in effect. Consider these processes ordered as to importance (that is, as to frequencies of mutations generated), as diagrammed in Figure 1 (Note the log scale: the relative importance of even adjacent processes is imagined to be substantialy). In this situation, selection can operate efficiently only upon the single or few most dominant processes; if a modifier of mutation rate arose that reduced only the magnitude of an already minor process,

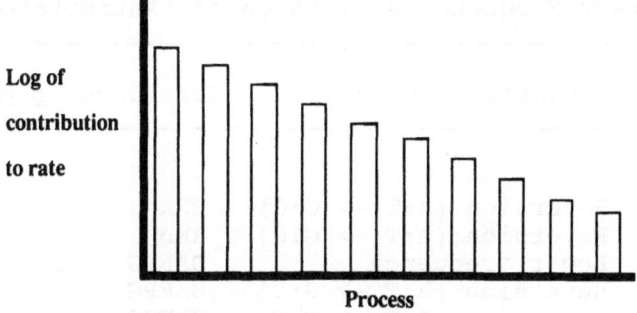

Fig. 1. Hypothetical distribution of the contribution of diverse mutation processes to mutation rates in very early organisms.

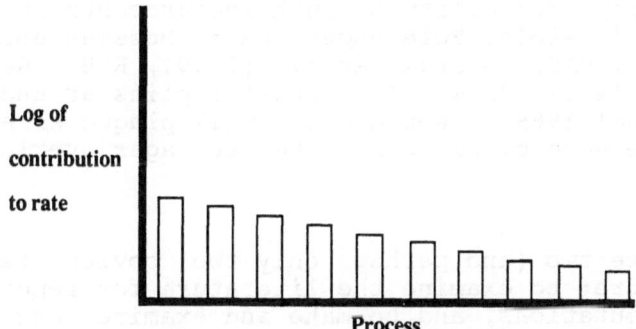

Fig. 2. Hypothetical distribution of the contribution of diverse mutation processes to mutation rates in contemporary organisms.

it would have but a tiny effect upon the overall rate and would be weakly selected, if at all. Thus, at each advance, selection would reduce primarily the predominant mutational processes. As a result, the contemporary distribution would tend to look like that diagrammed in Figure 2: much flatter than at the beginning. (This conjecture is consistent with the general observation - or at least my general impression - that a considerable number of mutational processes operate at similar levels, within a factor of perhaps ten-fold; many other processes no doubt produce mutations at lower frequencies, as implied by Figure 2, but are yet to be discovered.) But Figure 2 implies that a general antimutator mutation would have to reduce the efficacy of at least several processes simultaneously, a probably difficult demand for a single mutation. I therefore predict that general antimutator mutations will be very rare and, if they do occur at all, will either consist of multiple mutations or will be very unfit in some other way, for instance because of very slow DNA replication.

Table 1. Mutant Frequencies of Phage T4 "Antimutator" Mutants

Allele	Antimutator Criterion	Reference	r per 10^4
wild type			6.3
amB22(ser)	Reversion (A:T -> G:C)	BRK	5.8
amC125(ser)	Reversion (A:T -> G:C)	RKB	5.7
tsA58	Low r frequency	DAFPG	21
tsA71	Reversion (A:T -> G:C)	DAFPG	14
tsCB87	Reversion (A:T -> G:C)	DAFPG	12
tsCB120	Reversion (A:T -> G:C)		10
tsL97	Low r frequency	DAFPG	11

Frequencies of r mutants are median values from four to ten
stocks. The ts mutants were grown and assayed on B cells, the
am mutants on BsuI+ cells inserting serine residues at amber
codons. The reproducibility of such measurements is usually
within about 1.5-fold. References: BRK = Bessman and Reha-
Krantz (1977), DAFPG = Drake et al. (1969), RKB = Reha-Krantz
and Bessman (1981). "Low r frequency" implies an underestimate
in the original 1969 screen due to small plaque sizes, which
were overcome here by using a softer top-agar overlay.

There are two (and perhaps only two) obvious tests of
this conjecture: to examine the literature for reports of
antimutator mutations, and to make and examine large collec-
tions of new ones. To date, I have discovered the reports
summarized in Table 2. The following points quickly emerge.
First, most of the assays for antimutator activity employ very
small mutational targets (reversion, and resistance to
streptomycin or rifampicin) which, as we already saw with the
T4 reversion tests, are unreliable. Second, the "NTG^r"
antimutators, of which there were several, have been discarded
and were in any case very sick organisms and therefore of only
limited interest (Zamenhof et al., 1966; Zamenhof, personal
communication). Third, in the case of the E. coli purB
antimutator, the results with large-target mutational assays
are inconsistent: T7^r was unaffected while Val^r was strongly
affected. Although valine resistance can arise at any of
several loci, the predominant mechanism is likely to be via a
particular frameshift mutation (Lawther et al., 1981) and may
thus represent merely another small target. Only one good
candidate remains: Herpes simplex polC7, a mutation of the
viral DNA polymerase (Hall and Almy, 1982; Hall et al., 1984,
1985) that reduces mutation some 60-fold at the thymidine
kinase locus, a gene of unexceptional size. polC7 belongs to
an unusual class of antimutator mutants resistant to the drug
phosphonoacetic acid, which may act as an analogue of the
triphosphate moiety of an incoming dNTP during DNA replication. In any case, this mutation is currently the sole
challenge to my speculation that healthy, generalized antimutator mutations will not arise.

There may, however, be some hope for an additional
exception in a cellular organism. A large set of E. coli
antimutator mutations has been described by Quinones and

Table 2. Reports of Antimumator Mutations

Organism	Antimutator Mutation	Reference	Pathway Monitored	Magnitude of effect
E. coli	NTG^r	ZHZ	Az^s $\rightarrow$ Az^r	13X
	lexA	MLE	Str^s $\rightarrow$ Str^r	13X
	umuC36	SS	lacZ53 $\rightarrow$ LacZ$^+$	4X
	recA56	SS	lacZ53 $\rightarrow$ LacZ$^+$	4X
	recA$^-$	KIIK	argFam $\rightarrow$ Arg$^+$	2X
	purB$^-$	GS	metE$^-$ $\rightarrow$ Met$^+$	36X
			T7^s $\rightarrow$ T7^r	1.2X
			Rif^s $\rightarrow$ Rif^r	1.0X
			Val^s $\rightarrow$ Val^r	500X
S. cerevisiae	spo7-1	EBFE	lys2-1 $\rightarrow$ Lys$^+$	80X
	rev3$^-$	QvBH	lys1-1 $\rightarrow$ Lys1	7X
			his1-7 $\rightarrow$ His$^+$	5X
			arg4-17 $\rightarrow$ ARG4	5X
H. simplex	polC7	HA	tk$^+$ $\rightarrow$ tk$^-$	60X

References: EBFE = Esposito et al., (1975), GS = Geiger and Speyer (1977), HA = Hall and Almy (1982), KIIK = Kondo et al., (1970), MLE = Mount et al., (1972), QvBH = Quah et al., (1980), SS = Sargentini and Smith (1981), ZHZ = Zamenhof et al., (1966).

Piechocki (1985). Although selected on the basis of a reversion test and further characterized by valine and drug resistance and additional reversion tests, a generalized antimutator may yet lurk among these mutants.

REFERENCES

Atwood, K.C., Schneider, L.K., and Ryan, F.J., 1951, Selective mechanisms in bacteria, Cold Spring Harbor Symp., Quant. Biol., 16:345.

Bessman, M.J., Muzyczka, N., Goodman, M.F., and Schnaar, R.L., 1974, Studies on the biochemical basis of spontaneous mutation. II. The incorporation of a base and its analogue into DNA by wild-type, mutator and antimutator DNA polymerases, J. Mol. Biol., 88:409.

Bessman, M.J., and Reha-Krantz, L.J., 1977, Studies on the biochemical basis of spontaneous mutation. V. Effect of temperature on mutation frequency, J. Mol. Biol., 116:115.

Chao, L., and Cox, E.C., 1983, Competition between high and low mutating strains of Escherichia coli, Evolution, 37:125.

Cox, E.C., and Gibson, T.C., 1974, Selection for high mutation rates in chemostats, Genetics, 77:169.

Cox, E.C., and Yanofsky, C., 1967, Altered base ratios in the DNA of an Escherichia coli mutator strain, Proc. Natl. Acad. Sci. U.S.A, 58:1895.

Drake, J.W., 1969, Comparative rates of spontaneous mutation, _Nature_, 221:1132.

Drake, J.W., 1988, Bacteriophage T4 DNA polymerase determines the amount and specificity of ultraviolet mutagenesis, _Mol. Gen. Genet._, 214:547.

Drake, J.W., and Allen, E.F., 1968, Antimutagenic DNA polymerases of bacteriophage T4, Cold Sping Harbor Symp. _Quant. Biol._, 33:339.

Drake, J.W., Allen, E.F., Forsberg, S.A., Preparata, R.M., and Greening, E.O., 1969, The genetic control of mutation rates in bacteriophage T4, _Nature_ 221:1128.

Drake, J.W., and Greening, E.O., 1970, Suppression of chemical mutagenesis in bacteriophage T4 by genetically modified DNA polymerases, _Proc. Natl. Acad. Sci. U.S.A._, 66:823.

Eschel, I., 1973, Clone selection and the evolution of modifying features, _Theor. Pop. Biol._, 4:196.

Esposito, M.S., Bolotin-Fukuhara, M. and Esposito, R.E., 1975, Antimutator activity during mitosis by a meiotic mutant of yeast, _Mol. Gen. Genet._, 139:9.

Geiger, J.R., and Speyer, J.F., 1977, A conditional antimutator in _E. coli_, _Mol. Gen. Genet._, 153:87.

Gibson, T.C., Scheppe, M.L., and Cox, E.C., 1970, Fitness of an _Escherichia coli_ mutator gene, _Science_ 169:686.

Gillespie, J.H., 1981a, Evolution of the mutation rate at a heterotic locus, _Proc. Natl. Acad. Sci. U.S.A_, 78:2452.

Gillespie, J.H., 1981b, Mutation modification in a random environment, _Evolution_ 35:468.

Hall, J.D., and Almy, R.E., 1982, Evidence for control of _Herpes simplex_ virus mutagenesis by the viral DNA polymerase, _Virology_, 116:535.

Hall, J.D., Coen, D.M., Fisher, B.L., Weisslitz, M., Randall, S., Almy, R.E., Gelep, P.T., and Schaffer, P.A., 1984, Generation of genetic diversity in _Herpes simplex_ virus: an antimutator phenotype maps to the DNA polymerase locus, _Virology_, 132:26.

Hall, J.D., Furman, P.A., St. Clair, M.H., and Knopf, C.W., 1985, Reduced _in vivo_ mutagenesis by mutant herpes simplex DNA polymerase involves improved nucleotide selection, _Proc. Natl. Acad. Sci. U.S.A._, 82:3889.

Holsinger, K.E., and Feldman, M.W., 1983, Modifiers of mutation rate: evolutionary optimum with complete selfing, _Proc. Natl. Acad. Sci. U.S.A._, 80:6732.

Ishii, K., and Matsuda, H., 1985, Evolution of the Haldane-Muller principle of mutation load with application for estimating a possible range of relative evolution rates, _Genet. Res._, 46:75.

Ishii, K., Matsuda, H., Iwasa, Y., and Sasaki, A., 1989, Evolutionary stable mutation rate in a periodically changing environment, _Genetics_, 121:163.

Joyce, C.M., and Steitz, T.A., 1987, DNA polymerase I: from crystal structure to function _via_ genetics, _Trends Biochem. Sci._, 12:288.

Kimura, M., 1960, Optimum mutation rate and degree of dominance as determined by the principle of minimum genetic load, _J. Genet._, 57:21.

Kimura, M., 1967, On the evolutionary adjustment of spontaneous mutation rates, _Genet. Res._, 9:25.

Kondo, S., Ichikawa, H., Iwo, K., and Kato, T., 1970, Base-change mutagenesis and prophage induction in strains of Escherichia coli with different DNA repair capacities, Genetics, 66:187.

Lawther, R.P., Calhoun, D.H., Adams, C.W., Hauser, C.A., Gray, J., and Hatfield, G.W., 1981, Molecular basis of valine resistance in Escherichia coli K-12, Proc. Natl. Acad. Sci. U.S.A., 78:922.

Leigh, E.G., 1970, Natural selection and mutability, Am. Naturalist, 104:301.

Leigh, E.G., 1973, The evolution of mutation rates, Genetics Suppl., 73:1.

Levins, R., 1967, Theory of fitness in a heterogeneous environment. VI. The adaptive significance of mutation, Genetics, 56:163.

Lieberman, U., and Feldman, M.W., 1986, Modifiers of mutation rate: a general reduction principle, Theor. Pop. Biol., 30:125.

Mount, D.W., Low, K.B., and Edmiston, S.J., 1972, Dominant mutations (lex) in Escherichia coli K-12 which affect radiation sensitivity and frequency of ultraviolet light-induced mutations, J. Bacteriol., 112:886.

Ochman, H., and Wilson, A.C., 1987, Evolution in bacteria: evidence for a universal substitution rate in cellular genomes, J. Mol. Evol., 26:74.

Ollis, D.L., Brick, P., Hamlin, R., Xuong, N.G., and Steitz, T.A., 1985, Structure of large fragment of Escherichia coli DNA polymerase I complexed with dTMP, Nature, 313:762.

Painter, P.R., 1975, Mutator genes and selection for the mutation rate in bacteria, Genetics, 79:649.

Pamilo, P., Nei, M., and Li, W.H., 1987, Accumulation of mutations in sexual and asexual populations, Genet. Res., 49:135.

Quah, S.K., von Borstel, R.C., and Hasting, P.J., 1980, The origin of spontaneous mutations in Saccharomyces cerevisiae, Genetics, 96:801.

Quinones, A., and Piechocki, R., 1985, Isolation and characterization of Escherichia coli antimutators. A new strategy to study the nature and origin of spontaneous mutations, Mol. Gen. Genet., 201:315.

Reha-Krantz, L.J., and Bessman, M.J., 1981, Studies on the biocchemical basis of mutation. VI. Selection and characterization of a new bacteriophage T4 mutator DNA polymerase, J. Mol. Biol., 145:677.

Ripley, L.S., 1975, Transversion mutagenesis in bacteriophage T4, Mol. Gen. Genet., 141:23.

Sargentini, N.J., and Smith, K.C., 1981, Much of spontaneous mutagenesis in Escherichia coli is due to error-prone DNA repair: implications for spontaneous carcinogenesis, Carcinogenesis, 2:863.

Speyer, J.F., 1965, Mutagenic DNA polymerase, Biochem. Biophys. Res. Commun., 21:6.

Trobner, W., and Piechocki, R., 1984, Selection against hypermutability in Escherichia coli during long term evolution, Mol. Gen. Genet., 198:177.

Zamenhof, S., Heldenmuth, L.H., and Zamenhof, P.J., 1966, Studies on mechanisms for the maintenance of constant mutability: mutability and the resistance to mutagens, Proc. Natl. Acad. Sci. U.S.A., 55:50.

GENETIC ANALYSIS OF DNA REPAIR DEFECT IN XERODERMA PIGMENTOSUM

CELLS: IDENTIFICATION OF COMPLEMENTING GENES

Gursurinder P. Kaur and Raghbir S. Athwal

Department of Microbiology and Molecular
Genetics
UMDNJ-New Jersey Medical School, UMDNJ
185 S. Orange Avenue, Newark, NJ 07103, USA

INTRODUCTION

DNA is subject to damage by a variety of physical and chemical carcinogens. Through evolution, the environment has selected organisms having very efficient repair systems that eliminate much of the damaged DNA. Considerable progress has been made in understanding repair processes in procaryotes and lower eukaryotes and genes involved in several of the various repair pathways have been cloned (See Review in Friedberg, 1985). However, very little is known about the details of biochemical events involved in repair of DNA damage in human cells.

The relationship of defective DNA repair to skin carcinogenesis has been well documented in a heritable recessive disease Xeroderma pigmentosum (XP), (Cleaver, 1968; Robbins et al., 1974; Cleaver, 1983). XP cells are deficient in the removal of pyrimidine dimers (cyclobutane dimers and pyrimidine [6-4] pyrimidone adducts) induced by UV light (Setlow et al., 1969., Mitchell et al., 1985) and in the removal of chemical adducts generated by mutagens such as 4-Nitroquinoline-N-Oxide, psoralen plus near-UV light and benzopyrene (Kaye et al., 1980). Among XP, eight genetically distinct complementation groups designated A through G and a variant (Kraemer et al., 1987; Johnson et al., 1989; Bootsma et al., 1989) have been identified. Although XP provides an excellent model system, genetic analysis of DNA repair has been hampered by experimental difficulties involved in studying recessive phenotypes in human cells. As an alternate approach to delineate the genetic mechanism of DNA repair, mutants belonging to eight different complementation groups have been produced in rodent cell lines (Thompson et al., 1988). However, repair-deficient mutants of rodent cells were found to be defective in functions different from those observed in XP cells (Thompson et al., 1985; Thompson, 1989). In addition, a human repair gene ERCC1 cloned by complementation of the repair defect in Chinese hamster mutants, failed to correct the defect in XP cells (van Duin et al., 1989; Thompson, 1989).

Mechanisms of Environmental Mutagenesis-Carcinogenesis, Edited by
A. Kappas, Plenum Press, New York, 1990

Attempts to rescue human repair genes by transfecting genomic DNA from normal human cells into XP cells has been unsuccessful. Lack of success in these experiments has been attributed to high frequency of reversion to a UV-resistant phenotype in XP cells (Royor-Pokora and Haseltine, 1984; Schultz et al., 1985), low frequency of stably integrated foreign DNA (Hoeijmakers et al., 1987) and possibly large size of repair gene(s).

Considering these limitations, we have explored the possibility of transferring intact human chromosomes to XP cells using the technique of microcell mediated chromosome transfer (MMCT) (Kaur and Athwal, 1989). For this purpose, a panel of mouse/human hybrid cell lines, each containing a single human chromosome, marked with a dominant selectable marker, Ecogpt were used as microcell donors (Athwal et al., 1985). Chromosome transfer clones of XP cells, isolated by selection for Ecogpt were analyzed for increased resistance to UV irradiation. Using this approach, human chromosome 9 has been identified as complementing the repair defect in XP cells of complementation group A. The experimental strategy used here is of general application for the analysis of recessive phenotypes in human cells.

MATERIALS AND METHODS

Cell Lines and Growth Conditions

A hypoxanthine guanine phosphoribosyl transferase deficient (HGPRT⁻) XP-A cell line XPTG-1 was used for complementation studies by MMCT. XPTG-1 was isolated following mutagenesis of XP-A cell line GM4312A (Human Genetic Mutant Cell Repository, Camden, NJ) and selection in 6-thioguanine (6-TG). XPTG-1 cells were also marked with a dominant select-able marker pSV2neo (Southern and Berg, 1982) to facilitate double selection in MMCT experiments.

Mouse/human hybrid cell lines, each containing either individual or multiple human chromosomes, were used as microcell donors for the transfer of human chromosomes to XPTG-1 cells. A single human chromosome present in each hybrid cell line was marked with a dominant selectable marker, Ecogpt (Athwal et al., 1985). Such hybrids are now available in our laboratory for at least 14 different human chromosomes (R.S. Athwal, unpublished results).

The cells were routinely cultured in Dulbecco's Modified Eagles Medium (high glucose) containing 10 percent fetal bovine serum (DMEM) at 37 °C in 10% CO_2 - air atmosphere. The selection medium (MX) for isolation of chromosome transfer clones consisted of DMEM supplemented with mycophenolic acid (25 µg/ml), xanthine (70 µg/ml) and G418 (400 µg/ml) as required. The DMEM containing 6-thioguanine (TG medium) was used to select clones of cells that had lost the Ecogpt marked human chromosome (Sandhu et al., 1988).

Microcell Mediated Chromosome Transfer (MMCT)

Micronucleation was induced in the donor cell lines by treating the cells with colcemid (0.2 µg/ml) for 40 hr.

Micronucleated cells were treated with cytochalasin B for 30 minutes and micronuclei were then isolated by zonal centrifugation in discontinuous ficoll gradients (Athwal et al., 1985). Microcell preparations obtained after centrifugation were size fractionated by sedimentation at unit gravity in ficoll gradients as previously described (Athwal et al., 1985). For fusion, microcells prepared from 1×10^6 cells were added to 1×10^6 recipient cells seeded in 100 mm tissue culture dishes and agglutinated in the presence of phytohemagglutanin-P (PHA-P, 10 µg/ml) at 37°C for 30 minutes. Microcell fusion was facilitated by the addition of a 50% solution of PEG 1500. After fusion, cells were cultured in DMEM for 48 hr expression time and then medium was replaced with MX selective medium containing G418 (400 µg/ml). The addition of G418 selects against surviving donor cells that might have been present among microcells. Independent chromosome transfer colonies isolated individually were propagated in MX medium for further analysis.

Analysis for Complementation

For initial identification of complementation, 1×10^6 cells from each chromosome transfer clone were seeded in the middle of 100mm tissue culture dishes in triplicate, rinsed with phosphate buffered saline (PBS), and UV irradiated with a dose of $6 J/m^2$ given at a rate of $0.5 J/m^2$. The cells were then incubated in fresh medium. After three days, the plates were stained with crystal violet. Cell survival was visually evaluated using an inverted microscope and compared with growth of XPTG-1 and normal human cells (MGH-1) treated in parallel at the same time.

The clones showing increased resistance to UV irradiation at $6 J/m^2$ were further characterized for post UV-irradiation survival. The survival curves were constructed by modification of a method described by Cleaver and Thomas (1988). For this purpose, $5 \times 10^5 - 1 \times 10^6$ cells were plated in triplicate in the middle of 100 mm tissue culture dishes. After 24 hr of growth, the medium was removed, cells were rinsed with PBS, UV-irradiated at doses ranging from $1.5 J/m^2$ to $12.0 J/m^2$ and then refed with MX medium. After 24 hours of cell growth to allow for repair to occur, [^{3}H] thymidine, ([^{3}H]dT, 0.5 µCi/ml; specific activity 40Ci/mMol) was added to the cultures. Following 72 hr of growth in the presence of [^{3}H]dT, cells were harvested by trypsinization, transferred to 15 ml tubes and lysed in Tris-EDTA buffer (pH 7.0) containing 0.1% SDS. Relative incorporation of [^{3}H]dT into cellular DNA determined by scintillation counter, was used as an index of post irradiation cell survival (G. Kaur and R.S. Athwal, unpublished results).

Unscheduled DNA Synthesis (UDS)

In cultured cells, [^{3}H]dT is usually incorporated into DNA during semiconservative replication at the S phase of the cell cycle. However, in cell populations irradiated with UV, [^{3}H]dT is incorporated into DNA at all stages of the cell cycle to replace the regions of DNA containing UV induced pyrimidine dimers. This process of DNA synthesis in response to damage is called "Unscheduled DNA Synthesis" (UDS) which does not occur in repair deficient cells (Cleaver and Thomas,

1981). UDS can be readily detected by autoradiographic methods after incorporation of [^{3}H]dT into cultured cells and provides a visual method to distinguish between repair deficient and repair proficient cells. For UDS assay, 1 x 10^5 cells cultured on microscopic slides were UV-irradiated at 10J/m^2 and then labeled with [^{3}H]dT for 8 hrs. Fixed slides were coated with autoradiographic nuclear emulsion (Kodak NTB2) and developed after 7 days, stained with Giemsa and analyzed for number of grains present in individual nuclei as described by Cleaver and Thomas (1981).

Cytogenetic Analysis

Cells arrested in mitosis with colcemid (0.1 µ/ml) for 2 hrs were harvested, swollen in a hypotonic solution (0.075M KCl) and fixed in three changes of fixative (3:1 Methanol: Acetic acid). Metaphase spreads prepared by air drying were baked overnight at 65°C and then stained with Giemsa to determine the number and identity of human chromosomes present in rodent/human hybrids (Bobrow and Gross, 1974; Seabright, 1971).

Biochemical Analysis

Mouse/human hybrids containing complementing chromosome were analysed for the expression of isozymes of the adenylate kinase (AK$_1$) mapped to human chromosome 9 to confirm the chromosome identification. Human and mouse isozymes were separated by electrophoresis of crude cell extracts in starch gels and detected by histochemical staining using described procedures (Wilson et al., 1976). In addition, the identity of the human chromosome was confirmed by Southern blot analysis (Southern, 1975) using DNA probes D9S12 and D9S11 specific for chromosome 9 (Nakamura et al., 1987). Southern blot analysis was also performed on DNAs isolated from the donor M/H hybrid cell lines RA24-1 and RA9-1, parental human cell line MGT61-1, complemented clones XPM24-1; XPM24-2 and XPM9-1, XPTG-1 cells and back selected clones XPM9-TG1 and XPM24-2TG1, using either pSV2gpt plasmid or Ecogpt fragment DNA as a probe. Southern blot hybridizations were performed using standard described procedures (Maniatis et al., 1982).

RESULTS

An outline of experimental Scheme for the Transfer of human chromosomes to XP-A cells is given in Fig. 1. Human chromosomes, each marked with Ecogpt, present in mouse/human hybrids, were transferred to XPTG-1 cells by MMCT. In these experiments 10 different chromosomes or subchromosomal fragments were transferred to XPTG-1 cells (Table 1). Chromosome transfer clones of XPTG-1 cells each containing a different transferred human chromosome were screened for increased resistance to UV irradiation at 6J/m^2. Increased resistance to UV irradiation was observed only where hybrids RA-24-1 and RA9-1 were used as microcell donors (Table 1). In this case, four independent clones (XPM24-1, XPM24-2, XPM24-3 and XPM9-1) isolated in three different experiments showed increased level of resistance to UV irradiation (Table 1).

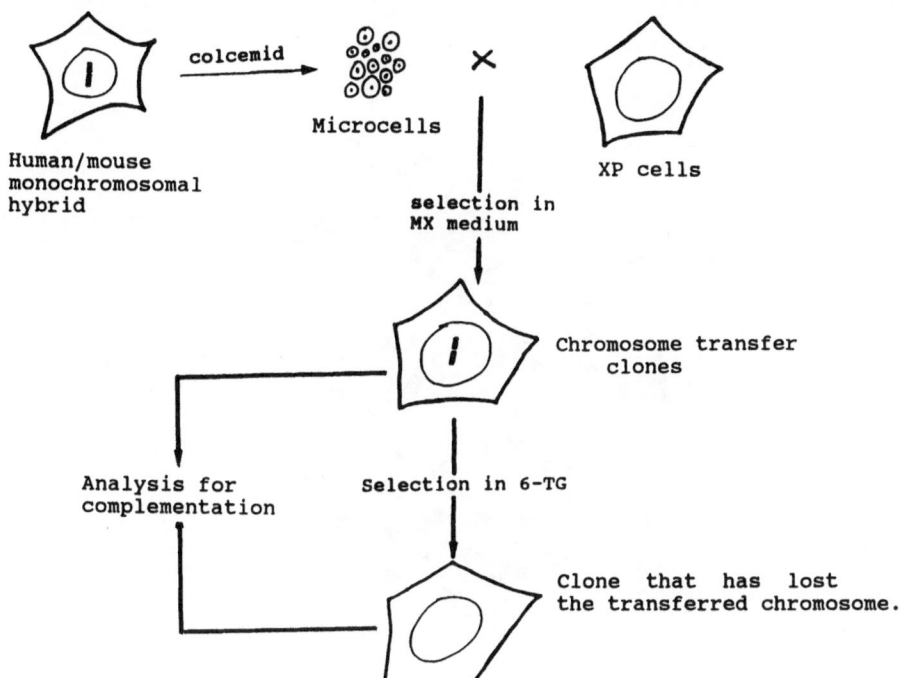

Fig. 1. Experimental scheme representing the transfer
of Ecogpt marked human chromosome to XPTG-1
cells from monochromosomal hybrids and
analysis for complementation. Using this
approach, 10 different human chromosomes were
transferred in XPTG-1 cells to identify the
complementing chromosome.

Table 1. List of Donor Cell Lines and Human Chromosomes
Transferred to XPTG-1 Cells

Exp	Mouse/Human Hybrids	Microcell Hybrids	Human Chromosome	UV Resistance
1	R 3-5	XPM 8	2	−
2	R 12-2	XPM 1-1	5	−
3	RA 5	XPM 16	6	−
4	RA 75	XPM 15-1	22	−
5	RA 12	XPM 23	15	−
6	RA 11-C1	XPM 17-1	NI	−
7	MR 15-45	XPM 20	12	−
8	RA 25	XPM 22	NI	−
9	VR 2-12	XPM 21	NI	−
10	RA 24	XPM 24-1, XPM 24-2, XPM 24-3	9	+
11	RA 9-1	XPM 9-1	9	+

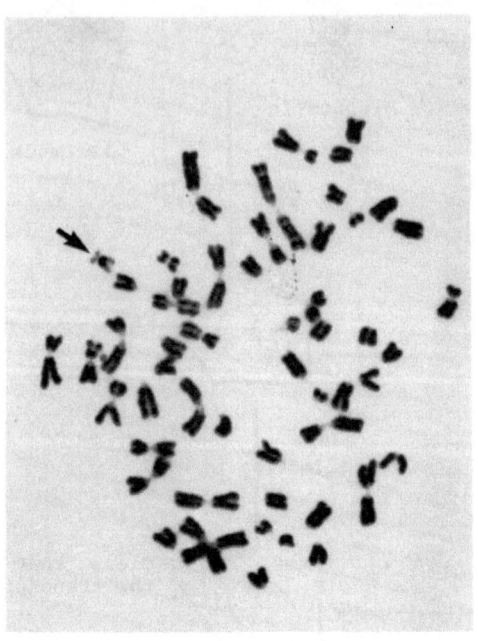

Fig. 2.   Metaphase spread of a monochromosomal hybrid
          RA9-1, stained by G-11 method. A blue stained
          single human chromosome 9 present among
          magenta stained mouse chromosomes is readily
          identified by characteristic staining.

Identification and Lineage of Complementing Human Chromosome

Hybrid RA24-1, a subclone of a segregating cell X cell
hybrid between an hprt⁻ mouse (RAG) cell line and MGT61-1 a
human cell line expressing Ecogpt, contained multiple human
chromosomes. Thus, any of the several human chromosomes
present in RA24-1 cells could have been responsible for the
complementation of the repair defect. In order to determine
which one, marked human chromosomes present in RA24-1 were
transferred individually to mouse and Chinese hamster cells by
MMCT to produce hybrids RA9-1 and CHH9-1 respectively.
Cytogenetic analysis by Giemsa-11 and G-banding revealed the
presence of a single human chromosome 9 in both RA9-1 and
CHH9-1 hybrids (Fig. 2). Chromosomal identity of the human
chromosome in hybrid RA9-1 was further confirmed by biochemical analysis for expression of a human isozyme of adenylate
kinase (AK-1) which maps to chromosome 9 (data not shown).
Southern blot analysis of DNAs from hybrids RA9-1 and CHH9-1
also showed the presence of chromosome 9 specific DNA
sequences D9S11 and D9S12 in these hybrids (Fig. 3A).

Cells from RA9-1 and CHH9-1, cultured in DMEM for 7 days
to facilitate random segregation of the human chromosome, were
"back selected" in TG-medium to generate cells that had lost
the introduced chromosome 9. Cytogenetic analysis of cells
which grew in TG-medium show the absence of any fragment of or

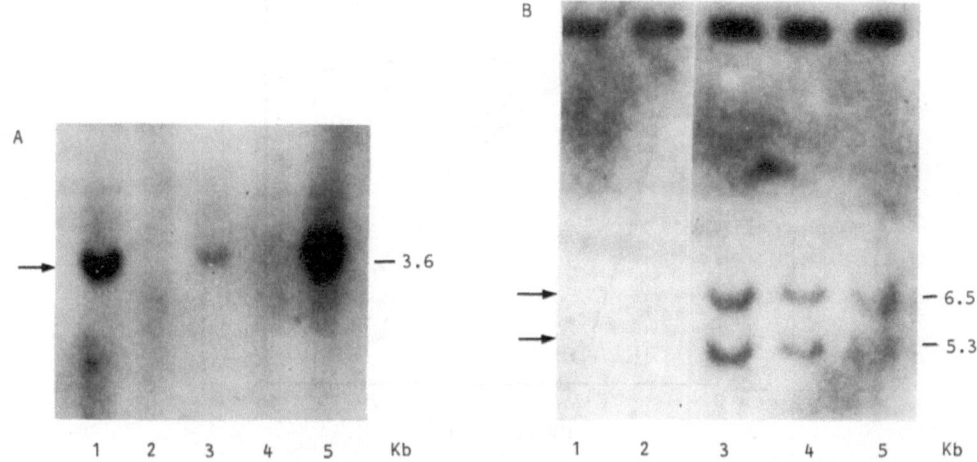

Fig. 3. A. Southern blot analysis of genomic DNA
digested with Pstl and probed with a chromosome 9 specific DNA segment D9S12. Lanes
represent: 1, CHH9-1; 2, CHTG49; 3, RA9-1; 4,
LA9; 5, MGH-1. Identical hybridization bands
present in lanes 1,3 and 5 show the presence
of chromosome 9 specific DNA in human and
hybrid cells. Probe D9S12 did not hybridize
with DNA from Chinese hamster CHTG49 (lane 2)
or mouse LA9 (Lane 4) cells.

B. Southern blot analysis of genomic DNA
digested with restriction enzyme Hind III and
probed with Ecogpt fragment excised from the
vector pSV2gpt. Lanes represent: 1, XPTG-1;
XPM9-TG1; 3, XPM9-1; 4, RA9-1; 5, MGT61-1. Two
identical hybridization bands of 6.5 and 5.3
Kb size, corresponding to integrated pSV2gpt
vector in parental human cells MGT61-1 (lane
5), M/H monochromosomal hybrid RA9-1 (lane 4),
and a complemented clone XPM9-1 (lane 3) show
the presence of the same marked chromosome, in
all cell lines. Lack of hybridization bands
XPTG-1 (lane 1) and a back selected clone
XPM9-TG1 (lane 2), show the absence of the
marked chromosome in these cell lines.

intact human chromosome. Similarly, slot blot analysis with
human DNA as a probe did not detect any human sequences in the
DNA from back selected cell populations. These data show that
a single, marked human chromosome 9 present in hybrids RA9-1
and CHH9-1 was transferred from RA24-1. The identified
chromosome 9 present in RA9-1 was again transferred to XPTG-1
cells by MMCT (Table 1). A single clone (XPM9-1) produced in
this experiment showed elevated level of resistance to UV
irradiation similar in magnitude as observed in other XPM
clones (Table 1).

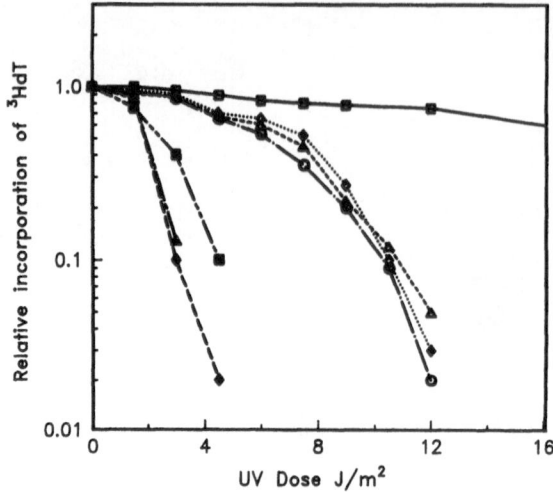

Fig. 4. Cell survival determined by [³H]dT incorporation during a 72 hr labeling period after UV
irradiation at various doses. [³H]dT was added
24 hr after irradiation. <u>Curves represent</u>: UV^S
XPTG-1 cells (--- ---); normal human cells
MGH-1 (- -); Complemented clones XPM24-1
(... ...); XPM24-2 (...Δ...); and (-o-)
XPM9-1; and back selected clones XPM9-TG-1
(...Δ...) and XPM24-2TG1 (... ...) respectively. Each data point represents an average
of three independent experiments. These data
show a partial reversal of UV's phenotype in
clones containing transferred chromosome 9
(XPM24-1; XPM24-2 and XPM9-1), which reverts
to UV's phenotype similar to XPTG-1 in back
selected clones (XPM9-TG1 and XPM24-2TG1).

The fact that our recipient cell line XPTG-1 is hgprt⁻
and can grow in TG medium. This provides a system to select
cells from the complemented clones that have lost the Ecogpt
marked human chromosome. Cells from UV^R clones XPM9-1 and
XPM24-2, cultured in non-selective medium for 10 days to allow
random chromosome segregation, were selected in TG-medium.
Single cell clones XPM9-TG1 and XPM24-2TG1, each originating
from a different complemented clone were analysed for UV^S
phenotype. As anticipated, both of these clones displayed the
UV^S phenotype. As anticipated, both of these clones displayed
the UV^S phenotype characteristic of the recipient cell line
XPTG-1 (Fig. 4). These data show that loss of the marked
human chromosome 9, transferred to XPTG-1 cells to produce
complemented clones, correlates with the loss of UV^R phenotype
(Fig. 4) and therefore gene(s) that complement the repair
defect in XP-A cells are present on this chromosome.

Since a single human chromosome 9 that carried the
integrated Ecogpt gene in MGT61-1 cells was sequentially
transferred to produce hybrids RA24-1, RA9-1 and CHH9-1, it

was possible to follow this chromosome in the various cell lines by assaying for the characteristic Ecogpt integration pattern (Kaur and Athwal, 1989). Southern blot analysis of DNA from all these cell lines, including XPM9TG-1 and XPM24-2TG1, was performed using Ecogpt to probe DNAs digested with restriction enzymes Sacl and Hind III and data on few selected cell lines are presented in Fig. 3B. Identical hybridization bands corresponding to Ecogpt integration were present in the DNAs of MGT61-1, RA24-1, RA9-1, CHH9-1, XPM24-1, XPM24-2 and XPM9-1 (Fig. 3B). As expected, these bands were not present in the DNAs from XPTG-1 and clones XPM9-TG1 and XPM24-2TG1 that were produced by back selection (Fig. 3B).

Identical hybridization bands observed in MGT61-1, RA24-1, RA9-1, XPM24-1, XPM24-2 and XPM9-1 indicate the presence of the same marked chromosome in all cell lines. The absence of the same bands in the DNAs of back selected cell lines XPM9-TG1 and XPM24-2TG1 show the loss of marked chromosome 9 along with reversal of UV^R phenotype.

Characterization of MMCT Clones for Complementation

Quantitative analysis for UV resistance was performed on three of the complemented clones (XPM24-1, XPM24-2and XPM9-1) and two of the back-selected clones (XPM9-TG1 and XPM24-2TG1) as well as on XPTG-1 and normal human MGH-1 cells (Fig. 4). The three clones containing transferred chromosome 9 were significantly more UV resistant than parental XP-A cells but less than MGH-1 cells (Fig. 4). The data show that only a partial complementation of the UV^S phenotype is achieved by transfer of chromosome 9. The clones XPM24-2TG1 and XPM9-TG1 that have lost the normal chromosome 9 were as sensitive to UV irradiation as the parental XPTG-1 cells (Fig. 4). D_{37} (UV dose resulting in 37% cell survival) values observed for MGH-1, complemented clones and repair deficient (XPTG-1, XPM24-2TG1 and XPM9-TG-1) cell lines were >20; 7-8; and 2-3, J/m^2 respectively.

Another measure used for complementation of the repair defect was unscheduled DNA synthesis (UDS) quantified by autoradiography, following the incorporation of $[^3H]dT$ into non S-phase UV irradiated cells (Cleaver and Thomas, 1981). The average number of grains/nucleus was low (0-25 grains/nucleus) in XPTG-1 cells, while the XPM9-1 cells (13-89 grains/nucleus) exhibited a UDS profile closer to MGH-1 cells (22-100 grains/nucleus) (Fig. 5). Considerable variation in number of grains among different nuclei within the same cell line was observed. Heterogeneity in UDS among nuclei in the same cell line has been observed in other studies as well. Such variation may be related to the stage of cell cycle at the time of UV irradiation.

Slot blot analysis of DNA from all complemented clones using total mouse DNA as probe demonstrated the absence of mouse DNA (data not shown). Thus complementation of the repair defect resulting from the transfer of a mouse chromosome or DNA is unlikely.

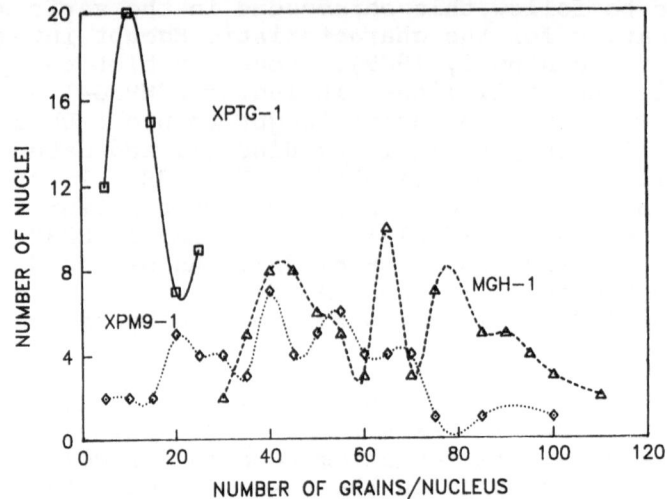

Fig. 5. Unscheduled synthesis as measured by [^{3}H]dt incorporation and autoradiography. Cells UV irradiated at 10 J/m² were labeled for 8 hr: curves show distribution of grains in individual nuclei of: MGH-1 (...Δ...), UV^S cell line XPTG-1 (— —) and a partially complemented clone XPM9-1 (...   ...).

DISCUSSION

We have used MMCT to transfer intact human chromosomes individually to human XP-A cells to identify a chromosome that complements the DNA repair defect. Human chromosome 9 has been determined to carry gene(s) that partially correct the repair defect in XP-A. Our conclusion is based on two lines of evidence: a) chromosome 9 was the only human genomic component present in donor M/H hybrid cell line RA9-1; b) no mouse DNA or chromosome was present in complemented clones. Among 10 different human chromosomes individually transferred to XPTG-1 cells, complementation of the repair defect occurred only with chromosome 9. The UV^R phenotype was eliminated in association with loss of chromosome 9 in back-selected clones XPM9-TG1 and XMP24-2TG1. In previous experiments using a similar approach, complementation of XP-A was achieved, but no specific chromosome was identified (Schultz et al., 1987). In XP-F cells, human chromosome 15 has been shown to complement the repair defect (Saxon et al., 1989 and G. Kaur and R. Athwal, unpublished results). The existence of eight different complementation groups in XP cells point to considerable heterogeneity. It is possible that each complementation group represents an independent gene mutation present on a different chromosome. A precedence for such a phenomenon also exists in repair defective CHO mutants, where complementing human genes have been mapped to several different human chromosomes (see review by Thompson, 1989). Similarly, complementing genes for

XP-A and XP-F also map to chromosomes 9 and 15, respectively (Kaur and Athwal, 1989, Saxon et al., 1989).

Recently, a mouse cDNA that complements the repair defect in XP-A cells has been cloned by a DNA transfection method (Tanaka et al., 1989). It is not certain as yet whether this gene represents the same gene that we have identified on chromosome 9. A hamster DNA segment that complements the repair defect in XP-D cells has also been cloned (Arrand et al., 1989). Partial correction of UV^S phenotype in XP-A and XP-D cells has also been achieved by the transfection of bacteriophage T4, UV specific endonuclease gene denV (Valerie et al., 1987, Arrand et al., 1987). In all reported studies on human XP cells, only a partial correction of UV^S phenotype or repair defect has occurred. Partial complementation of the defect has been attributed to the heterologous nature of the transfected DNA or to the incomplete gene being transferred to XP cells (Arrand et al., 1989). However, in our experiments in spite of the fact that intact human chromosome 9 was transferred, still a partial reversal of XP phenotype occurred in XP-A cells. These results suggest that more than one gene present on different chromosomes may be required for complete complementation. Alternately, it is also possible that the mutant product in XP cells has a harmful effect even on wild type product, for example, by subunit mixing in a multimeric complex. Availability of monochromosomal hybrids and the technology of MMCT offers a promise to sort out the mechanism and identify existence of related genes on various chromosomes.

ACKNOWLEDGMENTS

We thank Dr. Rufus Day (Cross Cancer Institute, Alberta, Canada) for helpful suggestions in the preparation of manuscript. This research was supported in part by New Jersey State Commission on Cancer (Grant No. CCR87-276) and grant No. DE-FG02-89ER-60886 from U.S. Department of Energy. G.P.K was supported by a Postdoctoral Fellowship 668-001 from New Jersey Commission on Cancer Research.

REFERENCES

Arrand, J.E., Bone, N.M., and Johnson, R.T., 1989, Molecular cloning and characterization of a mammalian excision repair gene that partially restores UV resistance to Xeroderma pigmentosum complement group D cells, Proc. Natl. Acad. Sci. USA, 86:6997.

Arrand, J.E., Squires, S., Bone N.M. and Johnson, R.T., 1987, Restoration of UV induced excision repair in Xeroderma D cells transfected with the denV gene of bacteriophage T4, EMBO J., 6:3125.

Athwal, R.S., Smars, M., Searle, B.M., and Deo, S.S., 1985, Integration of a Dominant selectable marker into Human chromosomes and transfer of marked chromosome to Mouse Cells by Microcell fusion, Somatic cell Mol. Genet., 11:177.

Bobrow, M. and Gross, J., 1974, Differential staining of human and mouse chromosomes in interspecific cell hybrids, Nature (London), 251:77.

Bootsma, D.M., Keijzer, Jung, E.G., and Bohnert, E., (1989),
    Xeroderma pigmentosum complementation group XP-1
    withdrawn. Mutat. Res., 218:149.
Cleaver, J.E. 1968, Defective repair replication of DNA in
    Xeroderma pigmentosum, Nature (London), 218:652.
Cleaver, J.E., 1983, Xeroderma pigmentosum, in: "The Metabolic
    Base of Inherited Disease", J.E. Stanburg, J.B.
    Wyngaarden, D.S. Fredrickson, J.L. Goldstein and M.S.
    Brown, eds., 5th ed., McGraw Hill, New York.
Cleaver, J.E., and Thomas, G.H., 1981, Measurement of Unsche-
    duled DNA Synthesis by autoradiography, in: "DNA
    Repair-a Laboratory Manual of Research Procedures",
    E.C. Friedberg and P.C. Hanawalt, eds., vol. 1B, Marcel
    Dekker Inc., New York.
Cleaver, J.E., and Thomas, G.H., 1988, Rapid diagnosis of
    Sensitivity to UV light in fibroblasts from dermatolog-
    ical disorders, with particular reference to Xeroderma
    pigmentosum, J. Invest. Dermatol., 9:467.
Friedberg, E.C. 1985, DNA damage and human disease, in: "DNA
    Repair", Freeman, New York.
Hoeijmakers, J.H.J., Odijk, H., and Westerveld, A., 1987,
    Differences between rodent and human cell lines in the
    amount of integrated DNA after transfection, Exp. Cell
    Res., 169:111.
Johnson, R.T., Elliott, G.C., Squires, S., and Joysey, V.C.,
    1989, Lack of complementation between Xeroderma
    pigmentosum groups D and H, Hum. Genet., 81:203.
Kaur, G.P. and Athwal, R.S., 1989, Complementation of a DNA
    repair defect in Xeroderma pigmentosum cells by
    transfer of human chromosome 9, Proc. Natl. Acad. Sci.
    USA, 86:8872.
Kaye, J., Smith, C.A., and Hanawalt, P.C., 1980, DNA repair in
    human cells containing photo adducts of 8-methoxy-
    psoralen or angelicin, Cancer Res., 40:696.
Kraemer, K.H., Lee, M.M., and Scotto, J., 1987, Xeroderma
    pigmentosum, cutaneous, ocular and neurolgic abnormali-
    ties in 830 published cases, Arch. Dermatol., 123: 241.
Maniatis, T., Fritsch, E.F., and Sambrook, J., 1982, "Molecu-
    lar cloning; a laboratory manual", Cold Spring Harbor
    Laboratory.
Mitchell, D.L., Haipek, C.A., and Clarkson, J.M., 1985, (6-4)
    photoproducts are removed from the DNA of UV-irradiated
    mammalian cells more efficiently than cylcobutane
    pyrimidine dimers, Mutat. Res., 143: 109.
Nakamura, Y., Hoff, M., Gillilan, S, O'Connell, P., Leppert,
    M., Lathrop, G.M., Lalouel J.M., and White, R., 1987,
    Isolation and mapping of a polymorphic DNA sequence
    pMH210 on chromosome 9 (D9S11), Nucleic Acids Res.,
    15:10609.
Robbins, J.H., Kraemer, K.H., Lutzner, M.A., Festoff B.W.,
    and Coon, H.G., 1974, Xeroderma pigmentosum, an
    inherited disease with sun sensitivity, multiple
    cutaneous neoplasms, and abnormal DNA repair, Ann.
    Intern. Med., 80:221.
Royor-Pokora, B., and Haseltine, W.A., 1984, Isolation of
    UV-resistant revertants from a Xeroderma pigmentosum
    complementation group A cell line, Nature (London),
    311:390.
Sandhu, S.S., Gudi, R., and Athwal, R.S., 1988, A genetic
    assay for aneuploidy; quantitation of chromosome loss
    using a mouse/human monochromosomal hybrid cell line,
    Mutat. Res., 201:423.

Saxon, P.J., Schultz, R.A., Stanbridge, E.J., and Friedberg, E.C., 1989, Human chromosome 15 confers partial complementation of phenotypes to Xeroderma pigmentosum group F cell, <u>Am. J. Hum. Genet.</u>, 44:45.

Schultz, R.A., Barbis, D.P., and Friedberg, E.C., 1985, Studies on gene transfer and reversion to UV resistance in Xeroderma pigmentosum cells, <u>Somatic Cell Mol. Genet.</u>, 11:617.

Schultz, R.A., Saxon, P.J., Glover, T.W., and Friedberg E.C., 1987, Microcell-mediated transfer of a single human chromosome complements Xeroderma pigmentosum group A fibroblasts, <u>Proc. Natl. Acad. Sci. USA</u>, 84:4176.

Seabright, M., 1971, A rapid banding technique for human chromosomes, <u>Lancet</u>, 2:971.

Setlow, R.B., Regan, J.D., German, J., and Carrier, W.L., 1969, Evidence that Xeroderma pigmentosum cells do not perform the first step in the repair of ultraviolet damage to their DNA, <u>Proc. Natl. Acad. Sci. USA, 64: 1035.</u>

Southern, E.M., 1975, Detection of specific sequences among DNA fragments separated by gel electrophoresis, <u>J. Mol. Biol.</u>, 98:503.

Southern, P.J. and Berg, L.P., 1982, Transformation of mammalian cells to antibiotic resistance with a bacterial gene under control of the SV40 early region promoter, <u>Jour. Mol. and Appl. Genet.</u>, 1:327.

Tanaka, K., Satokata, I., Ogita, Z., Uchida, T., and Okada, Y., 1989, Molecular cloning of a mouse DNA repair gene that complements the defect of group-A Xeroderma pigmentosum, <u>Proc. Natl. Acad. Sci. USA</u>, 86:5512.

Thompson, L.H., 1989, Somatic Cell Genetics approach to dissecting mammalian DNA repair, <u>Environ. & Molec. Mutagenesis</u>, 14:264.

Thompson, L.H., Carrano, A.V., Sato, K., Salazar E.P., White, B.F., Stewart, S.A., Minkler J.L., and Siciliano M.J., 1987, Identification of nucleotide-excision repair genes on human chromosomes 2 and 13 by functional complementation in Hamster-human hybrids, <u>Somatic cell and Mol. Genet.</u>, 13:539.

Thompson, L.H., Mooney, C.L., and Brookman, K.W., 1985, Genetic complementation between UV-sensitive CHO mutants and Xeroderma pigmentosum fibroblasts, <u>Mutat. Res.</u>, 150:423.

Thompson, L.H., Shiomi, T., Salazar, E.P., and Stewart, S.A., 1988, An eighth complementation group of rodent cells sensitive to Ultraviolet radiation, <u>Somatic Cell Mol. Genet.</u>, 14:605.

Valerie, K., Green, A.P., deRiel, J.K., and Henderson, E.E., 1987, Transient and stable complementation of ultra-violet repair in Xeroderma pigmentosum cells by the denV gene of bacteriophage T4, <u>Cancer Res.</u>, 47: 2967.

van Duin, M.G., Vredeveldt, L.V. Mayne, H. Odijk, W., Vermeulen, B., Klein, G., Weeda, J.H.J., Hoeijmakers, D., Bootsma and Westerveld, A., 1989, The cloned human DNA excision repair gene ERCC-1 fails to correct Xeroderma pigmentosum complementation groups A through I, <u>Mutat. Res.</u>, 217:83.

Wilson, D.E., Povey, Jr. S., and Harris, H., 1976, Adenylate kinases in man: evidence for a third locus, <u>Ann. Hum. Genet.</u>, 39:305.

CHEMICAL CARCINOGENESIS-ONCOGENES

OUTLINE OF A DESCRIPTIVE GENERAL THEORY OF ENVIRONMENTAL

CHEMICAL CANCEROGENESIS - EXPERIMENTAL THRESHOLD DOSES FOR

TUMOR PROMOTERS

Erich Hecker and Friedrich Rippmann

Deutsches Krebsforschungszentrum
Institute of Biochemistry
Im Neuenheimer Feld 280
D-6900 Heidelberg, FRG

INTRODUCTION

For the participants of this meeting on environmental mutagens and cancerogens[*] this paper will provide another challenge in the best possible sense: it adds another new and complex problem to the problems of using mutagenic activity as a surrogate of cancerogenic activity treated extensively during this symposium: that of the non-mutagenic yet perhaps somehow otherwise genotoxic tumor promoters and possible means of establishing appropriate short term assays for their biomonitoring. Before we pinpoint this new problem it might be advisable to provide some background information on cancerogenesis covered essentially by a descriptive general theory of environmental chemical cancerogenesis which we have developed in the last couple of years.

Environmental cancerogenesis may be described in general terms by the simple mathematical expression (Hecker, 1981a; Hecker, 1981b; Schmidt and Hecker, 1989)

$$R_T \sim f(\tau, P_{eprf}, H) \tag{1}$$

It relates R_T, the response by tumors to environmental risk factors as the dependent variable, with all independent variables involved in processes of environmental cancerogenesis: i.e., the host H, the observation time τ, and the

---

[*]The terminology used follows proposals developed by an ad hoc group of toxicologists, pathologists, biochemists and representatives of public health authorities, under the auspices of the Deutsche Forschungsgemeinschaft (DFG), Bonn-Bad Godesberg. In its original language it will be published within the series of DFG documentations, Verlag Chemie, Weinheim/ Bergstrasse. For the English version of it see Appel et al.

Mechanisms of Environmental Mutagenesis-Carcinogenesis, Edited by
A. Kappas, Plenum Press, New York, 1990

typical pattern of exposure (protocol)	effect (tumors)	prototype processes and operationally defined stages
● ● ● ● ●	+	solitary carcinogenesis
●	0	stage » initiation «
○ ○ ○ ○	0	stage » promotion «
● ○ ○ ○ ○	+	initiation / promotion

● = single dose d_i of initiator (usually solitary carcinogen)   ○ = single dose d_p of promotor

Fig. 1. Uni- and multifactorial cancerogenesis as prototype processes of solitary cancerogenesis and conditional cancerogenesis (in terms of initiation/promotion). Prototype experimental pattern show discontinuous chronic exposure of the target tissue to doses d_i and d_o, respectively, of the environmental principal risk factors (eprf) solitary cancerogen (DMBA) and tumorpromoter (3-TI).

pattern of exposure to environmental principal risk factor(s) (eprf) - the protocol P_{eprf}. Eq. (1) was verified theoretically and experimentally for both the principal processes of environmental chemical cancerogenesis i.e., solitary cancerogenesis and conditional cancerogenesis (in terms of intitiation/promotion) recently for the first time (Hecker and Rippmann, 1988; Rippmann and Hecker, 1988).

SOLITARY CANCEROGENESIS

Quantitative Response/Dose Relationships

Using colony inbred female NMRI-mice as the host and a highly standardized protocol providing discontinuous exposures to risk factors (Edler et al., in press, Schmidt and Hecker, 1989) the process of solitary cancerogenesis was modeled on mouse back skin with 7,12-dimethylbenz[a]anthracene (DMBA) as eprf. In the protocol (Fig. 1), one week after initiation of each dose group [48 mice, single dose d_i of DMBA), twice weekly exposures (dose frequency $v = 2$)] to DMBA followed (dose groups with d_i = 100, 10, 2.5, 1.25, 0.625 nMole of DMBA per mouse) up to an exposure time of $t = 60$ weeks. For each dose group the time course of cumulative tumor incidence $F(t)$ was determined using Kaplan-Meier estimates in order to correct the raw incidences for losses of mice (deaths or uncontrolled). A general Weibull model, developed for d_i = const. and $v = 2$, fits well the time course of the cumulative tumor incidences in all five dose groups. To quantify the response parameter R_T[eq.(1)] the median latency time t_{50} [i.e., for $F(t) = 0.5$] including 95% confidence limits were calculated for each dose group (Hecker and Rippmann 1988; Rippmann and Hecker, 1988). By plotting in bilogarithmic coordinates d_i versus t_{50} (with confidence limits), the experimental response/dose relation was obtained. Throughout

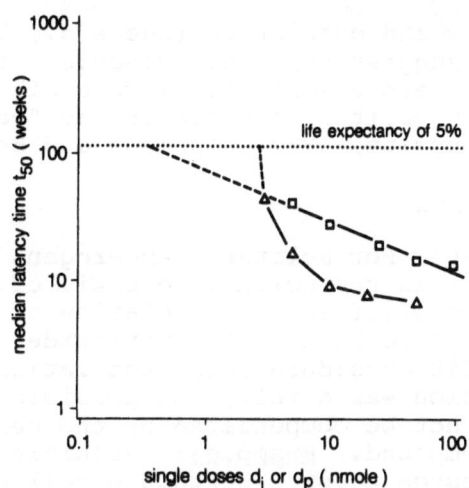

Fig. 2. Response/dose relations in solitary cancer-
         ogenesis by DMBA ( ) and initiation/promotion
         by DMBA/3-TI (Δ) up to a life expectancy of
         124 weeks for the female NMRI mice used; in
         the average 5% of untreated NMRI-mice reach an
         age of 124 weeks (life expectancy of 5% of the
         mice).

the dose range of d_i tested it is strictly linear (Fig. 2).
It can be described quantitatively by the response/dose
function for F = 50% and v = 2 (Hecker and Rippmann, 1988;
Rippmann and Hecker, 1988)

$$d_i \cdot t_{50} = \text{const.}$$

or   (2)

$$\log t_{50} = \frac{1}{n} \log d_i + \text{const.}$$

It is identical in principle with the classical quantitative response/dose relations of solitary cancerogens as
established previously for various tissues, including skin
using mostly protocols with continuous exposure to the
solitary cancerogens involved (e.g., Druckrey, 1967).

For extrapolation of the experimental response/dose
function to reasonable life expectancies (see Fig. 2), e.g.,
to 124 weeks in the present model (life expectancy of 5% of a
group of unexposed NMRI mice, Fig. 2), from the scientific
point of view various hypotheses may be discussed.

Mechanistic hypotheses. Extrapolation in a linear
fashion suggests that a "no effect level" or a "threshold
dose" d_t is non-existent down to very low doses of the
solitary cancerogen DMBA. Usually, from the linearity of the
response/dose function it is concluded that the response of
the target tissue is independent of the size of the total dose
fraction d_i (for constant total dose $D_i = \Sigma d_i$), of the

169

exposure frequency ν and cumulative (see e.g., Druckrey 1967). Such relationships suggest that the molecular effect(s) of solitary cancerogens are essentially irreversible. It is proposed to classify solitary cancerogens as "category 1" of eprf (Hecker and Rippmann, 1988; Rippmann and Hecker, 1988).

Practical Consequences

Risk assessment. For solitary cancerogens - i.e., category 1 of eprf - as the scientific basis of risk assessment, the hypothesis of linear extrapolation of the response/dose function appears to be accepted worldwide. In corresponding risk/benefit considerations -the rationale for preventive legislation -as a rule, the possible risk postulated usually can not be compensated by the benefit expected of an initiating compound. **Examples**: Prohibition of utilization of asbestos despite certain excellent technical properties; for a (rare) exception of the rule: chemotherapy of cancer patients with alkylating cancerostatics only in case of vital indication.

Short term assays. Based upon the mechanistic postulate above, for quick evaluation of environmental compounds _in vitro_ or _in vivo_, irreversible biochemical or biological effects, such as DNA-binding or mutagenicity (in a wide variety of target cells) may be used as a surrogate of initiating and, simultaneously, of solitary cancerogenic activity, although with some scientific reservations, some of which are being discussed in the present meeting.

CONDITIONAL CANCEROGENESIS IN TERMS OF INITIATION/PROMOTION

The standardized protocol using colony inbred female NMRI-mice as the host and discontinuous exposures to eprf (Edler et al., in press; Schmidt and Hecker, 1989) was designed to verify, in a manner comparable to solitary cancerogenesis, the process of initiation/promotion on mouse back skin (Hecker and Rippmann, 1988; Rippmann and Hecker, 1988). To do so DMBA was used as the prototype initiator and the diterpenester (DTE) prototypes 12-O-tetradecanoylphorbol -13-acetate (TPA), 3-O-tetradecanoylingenol (3-TI) and simplexin, respectively, as promoters (Hecker, 1985; Hecker, 1986). In the standardized protocol P_{erf} one week after initiation of the promoter dose groups (48 mice each, single dose d_i od DMBA), twice weekly exposures to TPA as eprf (dose groups with d_p = 10, 5, 2.5, 1.25, 0.625 nMole of TPA per mouse, dose frequency ν = 2) followed, up to an exposure time of t = 90 weeks (for the role of dose frequency ν in initiation/promotion, see also Chouroulinkov et al, 1989). 3-TI and simplexin were used as eprf in an analogous manner. In each promoter dose group tumor incidences were read for individual mice weekly. For each dose group the time course of cumulative tumor incidence F(t) was determined using Kaplan-Meier estimates in order to correct the raw incidences for losses of mice (deaths or uncontrolled). A general Weibull model, developed for d_p = const. and ν = 2, fits well the time course of the cumulative tumor incidences in all five promoter dose groups (Hecker and Rippmann, 1988; Rippmann and Hecker, 1988). To quantify the response parameter R_T [eq.(1)], the median latency time t_{50} [i.e., for F(t) = 0.5 and ν = 2] including 95%

confidence limits was calculated for each dose group. By plotting in bilagarithmic coordinates d_p versus t_{50} (with confidence limits), the experimental response/dose relation for initiation/promotion for initiation/promotion was obtained. Throughout the dose ranges of d_p tested it is non-linear for TPA as well as for 3-TI (Fig. 2) and simplexin. It can be described quantitatively by the newly developed response/dose function for $F = 50\%$ and $v = 2$, whereby

$$(t_{50} - t_{min})\ (d_p - d_t) = \text{const.}$$

t_{min}: extrapolated minimal median latency time   (2)

d_t: threshold dose

[eq. (2)] fits the experimental data satisfactorily (Hecker and Rippmann, 1988; Rippmann and Hecker, 1988).

Mechanistic Hypotheses

Extrapolation of the experimental response/dose relation-ships to reasonable life expectancies (see Fig. 2) e.g., to 124 weeks in the present model (life expectancy of 5% of a group of unexposed NMRI mice), using the hypothetical model of eq. (2), suggests that for the DTE-type promoters of skin "no effect levels" or "threshold doses" d_t definitely exist. From the non-linearity of the response/dose function it may be concluded that the response of the target tissue is dependent of the size of d_p and of the exposure frequency v (at constant total dose $D_p = \Sigma\ d_p$). Furthermore, it is cumulative only above the threshold dose d_t. Such relationships suggest that the molecular effect(s) of conditional cancerogens of the tumor promoter type are essentially reversible. Therefore, it is proposed to classify tumor promoter of the DTE-type as an alternative "category 2" of eprf, in contrast to category 1, i.e., solitary cancerogens (Hecker and Rippmann, 1988; Rippmann and Hecker, 1988).

Practical Consequences

Risk assessment. For tumor promoters - i.e., category 2 of eprf - as the scientific basis of risk assessment, the hypothesis of the existence of finite threshold doses may be used. In corresponding risk/benefit considerations - the rationale of preventive legislation - as a rule, the possible risk postulated may be compensated by the benefit expected of a promoting compound. Example: In the case of the sweeteners saccharin/cyclamate, some time ago these certified promoters of the bladder in animal experiments were taken from the market by some national health authorities, because their benefit had not been considered adequately. After some time, after reactions of the users, the benefit was re-evaluated with the effect that the removal was revoked.

Short term assays. Based upon the mechanistic postulate above, for quick evaluation of environmental compounds in vitro or in vivo, reversible biological or biochemical effects may be used, such as e.g., irritancy on the mouse ear (Hecker, 1985,86; Schmidt and Hecker, 1989), DTE binding to and/or activation of, protein kinase C (PKC), activation of ornithi-nodecarboxylase (ODC) or of Epstein Barr viral genomes in Raji cells, or indirect gene toxicity (in a wide variety of target cells) (e.g., Hecker, 1985; Hecker, 1986; Hecker, 1987). It remains to be seen, if these and other biochemical or biological effects are relevant for promotion of initiated cells and hence appropriate to be used as a surrogate of tumor promoting activity.

SUMMARY

A descriptive general theory of environmental chemical cancerogenesis was developed for a mouse skin model. Solitary and conditional cancerogenesis (initiation/promotion) were tested experimentally by quantititave response/time dose relationships in comparable standardized protocols. The hypotheses that in this system the solitary cancerogen DMBA is without a threshold dose whereas the promoters exhibit a threshold dose d_t were confirmed. Mechanistic criteria are proposed to understand the differences between the two prototype processes of cancerogenesis as pointed out. They may be helpful to draw conclusions as to biological or biochemical effects relevant for the mechanisms of the processes to be used in short term assays as surrogates of solitary and conditional cancerogenic activity, respectively.

REFERENCES

Appel, K.E., Fuerstenberger, G., Hapke, H.J., Hecker, E., Hildebrandt, A.G., Koranksy, W., Marks, F., Neumann, H.G., Ohnesorge, F.K., and Schulte-Hermann, R., Guest Editorial, Chemical Cancerogenesis: Definitions of frequently used terms, J. Cancer Res. Clin. Oncol., in press.

Chouroulinkov, I., Lasne, C., Lowy, R., Wahrendorf, J., Becker, H., Day, N.E., and Yamasaki, H., 1989, Dose and frequency effect in mouse skin tumor promotion, Cancer Res., 49:1964.

Druckrey, H., 1967, Quantitative aspects in chemical carcinogenesis, in: "Potential Carcinogenic Hazards from Drugs", R. Truhaut, ed., UICC Monograph Series, Vol. 7, Springer, Berlin, Heidelberg, New York.

Edler, L., Schmidt, R., Weber, E., Rippmann, F., and Hecker, E., Biological assays for irritant, tumor initiating and promoting activities. III. Revised standardized intitiation/promotion protocol in mouse skin and computerized management of the biodata generated, J. Cancer Res. Clin. Oncol., in press.

Hecker, E., 1981a, Cocarcinogenesis and tumor promoters of the diterpene ester type as possible carcinogenic risk factors, J. Cancer Res. Clin. Oncol., 99:103.

Hecker, E., 1981b, Prototype processes as experimental models in multifactorial carcinogenesis, J. Cancer Res. Clin. Oncol., 99:A29.

Hecker, E., 1985, Multifaktorielle und Mehrstufen-Karzinogenese - aetiologische (epidemiologische) und experimentelle Modelle sowie aktuelle Probleme der Bewertung von Krebsrisikofaktoren der Umwelt, in: "Tumorpromotoren - Erkennung, Wirkungsmechanismen und Bedeutung", K.E. Appel, A.G. Hildebrandt, eds., BGA-Schriften 6/85, MMV Medizin Verlag Muenchen.

Hecker, E., 1986, Multifactorial and multistage carcinogenesis - current problems of assessment of cancer risk of cocarcinogens of the promoter type, Round Table Discussion "Cocarcinogens of the Promoter Type", 14th International Cancer Congress, Budapest, August 21-27, 1986, Book of Abstracts of Lectures, Symposia and Free Communications, Vol. 2:258.

Hecker, E., 1987, Three stage carcinogenesis in mouse skin -recent results and present status of an advanced model system of chemical carcinogenesis, Toxicologic Pathology, 15:245.

Hecker, E., Rippmann, R., 1988, Quantitative determination of experimental threshold doses ("no-effect-levels") for environmental promoters in the initiation/promotion protocol on skin of NMRI-mice, Naunyn-Schmiedeberg's Archives of Pharmacology, 338 (Suppl.):R11.

Rippmann, F., Hecker, E., 1988, General quantitative dose/time (response) relationship for solitary cancerogenesis and initiation/promotion in skin of NMRI-mice and its implications for risk assessment, Naunyn-Schmiedeberg's Archives of Pharmacology 338 (Suppl.):R86.

Schmidt, R., Hecker, E., 1989, Biological assays for irritant, tumor-initiating and tumor-promoting activities. II. Standardized initiation/promotion protocol and semi-quantitative estimation of promoting (or initiating) potencies in skin of NMRI mice, J. Cancer Res. Clin. Oncol., 115:516.

Becker, K., 1980, Multifactorial and multistate relationships - current problems of assessment of ... risk of ... Radiogenic effects of low doses/Very... Wienh Mill... Clas..., .ial coverishapons of the biological Pro... Isinancihat Cancer Congress, Bcdne..., Augut 21... 1980, Book of Abstracts of lectures, Symposia and communications, Vol 2.250.

Bopp, ... 1980, Three Stage Carcinogenesis in a... with ... cancer, studies and present status of an ... Review and ... system of chemical carcinogenesis, Institute of Oncology, Bern, 1979.

Roe... A... Kupahnis K., 1980, Correlation between ... carcinogenic potential in the human ... animal results, ... Carcinogen, annual ... Report of the Programme of ... 1980, ..., ...

Kupahnis K., Roe... A., 1980, Short... report some relationships between the multistate general symbols and partialization and burden of ... epidemiologic application for ... cancer Assessment, Amsterd... IARC Sur... 1980.

Fialkow, P.J., 1974, Classical evidence for ... clonal ... common relationship, New Engl..., Journal ... medicine 291:26.

Scaletta... P... 1...n Instan... Dietary gradient to have found... 1... Literature of prognosis for ... ...

THE INTERACTION OF STEROID HORMONES AND ONCOGENES IN THE

ESTABLISHMENT OF MALIGNANCY

Constantine E. Sekeris

Institute of Biological Research and
Biotechnology
National Hellenic Research Foundation
48, Vas. Constantinou Avenue
116 35-GR Athens, Greece

ABSTRACT

Recent progress in the fields of molecular action of
steroid hormones and of the role of proto-/oncogenes in cell
regulation and transformation can be exploited to delineate
the mechanisms by which steroid hormones act in establishing
malignancy. The alignment of proto-/oncogenes with hormone
responsive elements, either by rearrangement, translocation or
viral integration, will render these genes responsive to
hormones, with all the consequences for carcinogenesis of
their unphysiological expression under the hormonal stimulus.

INTRODUCTION

Effects of steroid hormones on tumor growth have been
demonstrated both by clinical/epidemiological, as well as by
experimental studies. Beatson (1896) first noted that removal
of the ovaries had beneficial effects on patients with
inoperable breast carcinoma. Much later Huggins et al. (1941)
and Huggins and Hodges (1941) demonstrated marked improvement
of patients with disseminated prostatic carcinoma after
castration. Similar effects as castration were observed after
administration of diethylstilbestrol. Many animal (Dunning et
al., 1947; Cutts and Noble 1964; Huggins et al., 1959; Huggins
et al., 1961; Gullino et al., 1975) and cell (Engle and Young,
1978) models have been since introduced to study the effects
of steroids on cancer. In rats prolonged administration of
estrogens induced mammary tumors (Dunning et al., 1947; Cutts
and Noble, 1964), the degree of susceptibility to induction
was strain dependent. Partial regression of the tumors was
observed upon oophorectomy. Huggins et al. (1959, 1961) in
their now classical studies, induced mamma carcinomas in rats
by administering polycyclic aromatic hydrocarbons. Such tumors
appeared also by action of alkylating agents (Dao, 1962).
Breast cancers appeared in female but not in male animals. In
oophorectomized rats, subsequent administration of the

Mechanisms of Environmental Mutagenesis-Carcinogenesis, Edited by
A. Kappas, Plenum Press, New York, 1990

carcinogen did not lead to tumor formation, whereas ovarian
grafting at the time of carcinogen administration restored the
capacity for tumor formation. As an example of a cell model I
refer to the MCF-7 breast cancer cell line (Soule et al.,
1973). Antiestrogens arrest growth of these cells by inducing
a mid G-1 block. Estrogens reverse this effect and restore
growth. In ovariectomized nude mice no growth of transplanted
MCF-7 cells can be observed, but the cells remain viable:
injection of estrogens fully restores the capacity for growth
and tumor formation. In spite of numerous studies, the
mechanism by which steroid hormones affect carcinogenesis is
still unknown. In the following, data supporting the hypothesis, that the effects of steroids in establishing malignancy
steroids are mediated through the modulation of oncogene
expression will be presented. Important in the development of
this concept have been two major advances in molecular
biology, i.e. the deeper understanding of the molecular action
of steroid hormones and the spectacular progress concerning
the role of protooncogenes and v-oncogenes in cell regulation
and cell transformation.

MOLECULAR ACTION OF STEROID HORMONES

The hypothesis that steroid hormones act by way of gene
activation (Karlson, 1961) and induction of proteins, which
are responsible for the phenotypic expression of the hormonal
effects, has been fully vindicated and defined in molecular
terms. A central role has been ascribed to the receptor
protein, as a trans-acting transcription modulator (for review
articles, see Evans, 1988; Green and Chambon, 1988; Beato et
al., 1989). Cloning of many steroid-receptor genes and
experiments with isolated receptor proteins have revealed
common features and structural motifs important for the
interactions of the receptor with the hormone, with the genes
and with other regulatory macromolecules. Interaction of the
receptor with the genetic material is attained through the
DNA-binding domain of the receptor recognizing specific
nucleotide sequences, the hormone responsive elements (HREs).
The DNA-binding domain contains a highly conserved core of
66-68 amino acids rich in cysteines and basic amino acids,
showing similarity with the "zinc-finger" motif of transcription factor TFIIIA and other DNA-binding proteins.

The HREs represent 15 base pair palindrome structures,
six pairs each flanking a central non-conserved trinucleotide.
Mutations within any of the hexanucleotide motifs reduces
hormonal inducibility. As suggested by the dyad symmetry of
the HREs, the receptor indeed seems to bind to HREs as a
dimer. The receptors for glucocorticoids, progesterone,
mineralocorticoids and androgens recognize a common HRE
(5'-AGAACANNNTGTTCT-3'), estrogens a different one (5'-AGGT-
CANNNTGACCT-3'). The HREs can be found at various positions of
the hormone regulated genes, i.e., near the promotor, at
various distances in the 5' upstream region, even within the
gene itself. In some cases a single HRE can confer hormone
inducibility to the gene, in other cases more than one HRE is
needed, each contributing variably to gene activation. The
presence of other regulatory sequences, interacting, among
others, with general or tissue specific transcription factors,
greatly affects the degree of attained hormonal induction.

The importance of protooncogenes in cellular regulation and growth and developmental processes is now well established and the function of the protooncogene encoded products such as growth factors, growth factor receptors, GTP binding proteins and nuclear proteins acting as transcription regulators, among others, has been determined (Varmus, 1984; Weinberg, 1985; Spandidos, 1985; Carrett, 1986). A major finding is that the over-expression, the untimely expression, or the expression of mutated or otherwise impaired protooncogenes is a major causal factor in the cell transformation process.

An increased expression of several protooncogenes has been observed in a variety of human tumors. Very well studied is the relation of the expression of the HER-2 (c-erbB2) protooncogene in relation to human breast and ovarian cancer (Slamon et al., 1989). This gene is the human homologue of the c-neu gene, cloned from a rat neuroblastoma and encodes a transmembrane protein bearing homology to the epidermal growth factor and the other members of the tyrosine kinase class receptors of proto-oncogenes. In one third of the breast and ovarian cancer patients there is an amplification of the gene, negatively correlated with disease free- and overall survival. This correlation is also observed taking as parameters the amounts of the respective RNA and protein products. The expressed protein is identical to the protein expressed in normal tissues with the sole exception of a neutral substitution of isoleucine for valine at position 655 of the transmembrane domain.

A direct implication of the increased expression of HER-2 in cell transformation has been demonstrated by Hudziak et al. (1987) with 3T3 cells. The authors demonstrated in transfection experiments, that amplification of the unaltered gene leads to overexpression of the respective protein, cellular transformation and tumor formation in athymic mice. In experiments with transgenic mice Muller et al. (1988) have shown that the activated c-neu gene linked to a mouse mammary tumor virus (MMTV) promotor induces development of mammary adenocarcinomas, involving the entire epithelium of each gland. These authors conclude that the expression of solely the c-neu oncogene is sufficient for the induction of the malignant transformation process. Experiments of the same nature were performed also by Bouchard et al. (1989) but the conclusion was different, i.e. the expression of the c-neu oncogene was necessary, but not sufficient, to induce malignant transformation. The difference in the experimental set-up was that the construct of Bouchard et al. (1989) used the MMTV-LTR immediately adjacent to the neu c-DNA sequences, in contrast to the constructs of Muller et al. (1988), in which a 600 bp sequence, representing the rat 30 S sequences derived from Harvey MSV genome, separate the MMTV-LTR from the neu c-DNA sequences. It seems the presence of the extra 600 bp sequence and its expression could contribute to the observed differences between the two groups. Independent of these differences, the expression of the c-neu oncogene in breast epithelium, a tissue which is highly influenced by the hormonal status of the organism, and the presence in the LTR flanking the oncogene of several HREs points to the importance of hormones in the expression of the oncogene and its consequences (see below).

HORMONAL CONTROL OF PROTOONCOGENE EXPRESSION

Recent experimental results demonstrate direct effects of steroid hormones on protooncogene expression. Well studied are the effects of estrogens on the expression of c-fos (Loose-Mitchell et al., 1988; Weisz and Bresciani, 1988) and c-myc (Weisz and Bresciani, 1988) in rat uterus. Loose-Mitchell et al. (1988) have demonstrated that administration of estradiol to immature rats causes within thirty minutes increase of c-fos m-RNA, abolished by actinomycin D, reaching a maximum at 3 hours. Other steroid hormones such as dexamethasone, dihydrotestosterone and progesterone, had no such effect. In a study by Weisz and Bresciani (1988), the effects of estrogens on the expression of 20 protooncogenes was followed. Two oncogenes, c-fos and c-myc responded very rapidly within 30 min and 90 min, respectively, to the hormone, whereas the other oncogenes were not affected by hormonal stimulation. It is strongly suggested that the effects of estrogens on c-fos, probably also on c-myc, expression are mediated by direct interaction of the hormone-receptor complex with the DNA. As noted by Loose-Mitchell et al. (1988) a 12 base pair sequence having a sequence 5'-GGTCTAGGAGACC-3' is present, at position -219 — -207 with respect to the start site of transcription, similar to the palindromic estrogen responsive element (ERE) in the Xenopus vitellogenin gene. In addition another similar sequence 5'-GGTCTGCCTAGGC-3' is present at position 2102-2114 of the gene. Both sequences could represent EREs instrumental for c-fos activation. A search for the presence of various HREs in protooncogenes, whose expression is increased in various tumors, has indeed revealed such sequences and has initiated studies on the hormonal inducibility of the respective genes. It is important to note that many retroviruses involved in tumor induction in animals, but also human retrovirus and HIV-1, have HREs in their genome (Miksicek et al., 1986). Markham et al. (1986) has shown that the ability to productively infect human peripheral blood monocytes by HIV-1 was improved by hydrocortisone. That this effect is mediated by the GRE present in the virus genome has been recently demonstrated in transfection assays (Spandidos et al., in preparation). The human papillomavirus type 16 in combination with an activated H-ras gene transforms primary cells only in the presence of the glucocorticoid hormone dexamethasone. It has been found that the HPV-16 genome carries a GRE sequence (Pater et al., 1988). Particularly interesting in this respect is the paper by Stavenhagen and Robins (1988), in which it is shown that the mouse sex limited protein gene has been rendered androgen dependent through endogenous virus insertion, 2 kd upstream of the gene, having an HRE in its 5'-LTR.

POSSIBLE MECHANISMS OF STEROID HORMONE INVOLVEMENT IN CELL TRANSFORMATION

As mentioned above hormone inducible genes contain in some part of their genome, usually in their 5'-upstream region, one or more hormone responsive elements, interacting with the respective hormone-receptor complex. Normal cellular growth, development and regulation is attained by the action of regulatory molecules, including hormones, on specific genes

- important among them being the protooncogenes - in a well defined temporal pattern. Expression of a gene at periods, in which it should under normal conditions be silent, can lead to severe impairment of cellular functions. If the affected gene(s) belongs to the family of protooncogenes, then the link to cellular transformation is obvious. Such gene activation could be attained by placing the respective gene, by gene rearrangement or chromosomal translocation under the control of regulatory elements, which normally do not control it. In this context the role of movable genetic elements should be considered (Georgiev et al., 1981; Chorazy, 1985). If such regulatory elements are HREs then the gene would be subjected to hormonal regulation and the hormone would thus act as a tumor promotor in the establishment of malignancy. As shown recently by de Brekeller (1988) in a study involving chromosomal rearrangement in leukemias and various solid tumors, breakage and deletions occurred preferentially in bands known to contain protooncogenes, growth factor genes, receptor genes and differentiation genes.

In many of the tumors with increased expression of protooncogenes, a parallel amplification of the expressed gene has been also observed (Chorazy, 1985). It is possible that some of the amplified genes are placed near HREs and thus subjected to hormonal control, a hypothesis which is amenable to experimental verification (our work in progress). Other additional mechanisms of hormonal activation of protooncogenes can be envisaged, such as intergration near protooncogenes of viral sequences having in their genome HREs. As it is well known that steroid hormones not only stimulate but also repress gene activity, the possibility that these hormones could repress onco-suppressor genes or tumor inhibitory substances, which are thought to be the products of onco-suppressor genes, should also be considered.

I have refrained from discussing other regulatory elements affecting hormonal inducibility or the effect of changes in the structure of the receptor molecule itself in oncogenesis. These are fields which are now open to experimental study and will surely yield fruitful results in the near future.

REFERENCES

Beato, M., Chalepakis, G., Schauer, M., and Slater, E.P., 1989, DNA regulatory elements for steroid hormones, J. Ster. Biochem., 32:737.
Beatson, G.T., 1986, On the treatment of inoperable cases of carcinoma of the mamma: Suggestions for a new method of treatment with illustrative cases, Lancet, II:104.
Bouchard, L., Lilamarre, L., Tremblay, P.J. and Jolicoeur, P., 1989, Stochastic appearance of mammary tumors in transgenic mice carrying the MMTV/c-neu oncogene, Cell, 57:931.
de Braekeller, M., 1988, Proto-oncogenes, growth factor genes, receptor genes, differentiation genes and structural rearrangements in human cancer, Anticancer Res., 8:1325.
Carrett, C.T., 1986, Oncogenes, Clin. Chim. Acta, 156:1.

Chorazy, M., 1985, Sequence rearrangements and genome insta-
        bility -a possible step in carcinogenesis, J. Cancer
        Res. Clin. Oncol., 109:159.
Cutts, J.H. and Noble, R.L., 1964, Estrone-induced mammary
        tumors in the rat. I. Induction and behavior of tumors,
        Caner Res., 24:1116.
Dao, T.L., 1962, The role of ovarian hormones in initiating
        the induction of mammary cancer in rats by polynuclear
        hydrocarbons, Cancer res., 22:973.
Dunning, W.F., Curtis, M.R. and Segaloff, A., 1947, Strain
        differences in response to diethylstilbestrol and the
        induction of mammary gland and bladder cancer in the
        rat. Cancer Res., 7:511.
Engle, L.W. and Young, N.W., 1978, Human breast carcinoma
        cells in continuous culture: a review, Cancer Res.,
        38:4327.
Evans, R.M., 1988, The steroid and thyroid hormone receptor
        superfamily, Science, 240:889.
Georgiev, G.P., Ilin, I.U.V., Ryskov, A.P., Kramerov, D.A.,
        1981, Mobile dispersed elements in eukaryotes and their
        possible relationship to carcinogenesis (in Russian),
        Genetica, 17:222.
Green, S. and Chambon, P., 1988, Nuclear receptors enhance our
        understanding of transcription regulation, TIG, 4:309.
Gullino, P.M., Pettigrew, H.M. and Grantham, F.H., 1975,
        N-nitrosomethylurea as mammary gland carcinogen in
        rats, J. Nat. Cancer Inst., 54:401.
Hudziak, R.M., Schlessinger, J. and Ullrich, A., 1987,
        Increased expression of the putative growth factor
        receptor p185 HER2 causes transformation and tumo-
        rigenesis of NIH 3T3 cells, Proc. Nat. Acad. Sci.,
        84:7159.
Huggins, C., Briziarelli, G. and Sutton, H., 1959, Rapid
        induction of mammary carcinoma in the rat and the
        influence of hormones on the tumors, J. Exp. Med.,
        109:25.
Huggins, C., Grand, L.C. and Brillantes, F.P., 1961, Mammary
        cancer induced by a single feeding of polynuclear
        hydrocarbons and its suppression, Nature, 189:204.
Huggins, C. and Hodges, C.V., 1941, Studies on prostatic
        cancer. I. The effect of castration of estrogen and
        androgen injection on serum phosphatases in metastatic
        carcinoma of the prostate, Cancer Res., 1:293.
Huggins, C., Stevens, R.E. and Hodges, C.V., 1941, Studies on
        prostate cancer. II. The effect of castration on
        advanced carcinoma of the prostate gland, Arch. Surg.,
        43:209.
Karlson, P., 1961, Biochemische Wirkungsweise der Hormone,
        Dtsch. med. Wschr., 86:668.
Loose-Mitchell, D.S., Chiappetta, C. and Stancel, G.M., 1988,
        Estrogen regulation of c-fos messenger ribonucleic
        acid, Mol. Endocrin., 2:946.
Markham, P.D., Salahuddin, S.Z., Veren, K., Orndorff, S. and
        Gallo, R.C., 1986, Hydrocortisone and some other
        hormones enhance the expression of HTLV-III, Int. J.
        Cancer, 37:67.
Miksicek, R., Heber, A., Schmid W., Danesch,U., Posseckert,
        G., Beato, M. and Schutz, G., 1986, Glucocorticoid
        responsiveness of the transcriptional enhance of
        maloney murine sarcoma virus, Cell, 46:283.

Muller, W.J., Sinn, E., Pattengale, P.K., Wallace, R. and Leder, P., 1988, Single step induction of Mammary Adenocarcinoma in transgenic mice bearing the activated c-neu oncogene, <u>Cell</u>, 54:105.

Pater, M.M., Hughes, G.A., Hyslop, D.E., Nakshatu, H. and Pater, A., 1988, Glucocorticoid dependent oncogenic transformation by type 16 but not type II human papilloma virus DNA, <u>Nature</u>, 335:832.

Slamon, D.J., Godolphin, W., Jones, L.A., Holt, J.A., Wong, S.G., Keith, D.E., Levin, W.J., Stuart, S.G., Udove, J., Ullrich, A. and Press, M.F., 1989, Studies of the HER-2/neu protooncogene in human breast and ovarian cancer, <u>Science</u>, 244:707.

Soule, H.D., Vasquez, J., Long, A., Albert, S. and Brennan,M., 1973, A human cell line from a pleural effusion derived from a breast carcinoma, <u>J. Natl. Canc. Inst.</u>, 51:1409.

Spandidos, D.A., 1985, Mechanisms of Carcinogenesis. The role of oncogenes, transcriptional enhancers and growth factors, <u>Anticancer Res.</u>, 5:485.

Stavenhagen,J.B. and Robins, D.M., 1988, An ancient provirus has imposed androgen regulation on the adjacent mouse sex-limited protein gene, <u>Cell</u>, 55:247.

Varmus, H., 1984, The molecular genetics of cellular onco-genes, <u>Ann. Rev. Genet.</u>, 18:553.

Weinberg, R.A., 1985, The action of oncogenes in the cytoplasm and nucleus, <u>Science</u>, 230:770.

Weisz, A. and Bresciani, F., 1988, Estrogen induces expression of c-fos and c-myc protooncogenes in rat uterus, <u>Mol. Endocrin.</u>, 2:816.

THE RELATIONSHIP BETWEEN DNA-ALKALI-LABILE SITES AND

CARCINOGENESIS IN MAMMALIAN CELLS

Claude Lasne, Luz Orfila and Ivan Chouroulinkov

Institut de Recherches Scientifiques sur le
Cancer
Centre National de la Recherche Scientifique
BP No 8, 94802, Villejuif-Cedex, France

INTRODUCTION

DNA single strand breaks are primary lesions produced in the DNA of cells treated with different chemicals. These alterations to the DNA are mostly represented by alkali labile sites which are converted into DNA single strand breaks after exposure to alkaline solutions (Kohn et al., 1981) and are generally rapidly repaired. The significance of these DNA lesions are not well understood. They might be premutational lesions and converted to gene mutations. The detection of gene mutations and of chromosomal aberrations after treatment of mammalian cells with a chemical is generally predictive of the carcinogenic potential of this chemical. Gene mutations as well as chromosomal aberrations are closely associated with genotoxic events. According to our classification of chemical carcinogens (Chouroulinkov, 1988) they are generally well correlated with the genotoxic stage of carcinogenesis. Three classes of carcinogens have been identified: the carcinogens which exhibit both genotoxic and carcinogenic activities; those which only exhibit the genotoxic activity and the third class of carcinogens which includes the compounds which are carcinogenic but not genotoxic or mutagenic. The carcinogenicity of the last class of chemicals, most of them promotors, is explained by epigenetic mechanisms while the genotoxic carcinogens are associated with initiation.

The purpose of this study was to determine whether the primary damage to DNA expressed as alkali labile sites was able to modify the genome in such a way that malignant transformation will follow or whether no such close relationship may exist between both endpoints (DNA alkali labile sites and chromosomal aberrations) and carcinogenesis. For this purpose, we have studied with different chemicals: 1) the induction of alkali labile sites by means of the alkaline elution method, in the DNA of mammalian cells from various origins such as: rabbit and rat fibroblasts, Syrian hamster embryo (SHE) cells, mouse BALB/c3T3 cells and human lymphocytes; 2) the formation of chromosomal aberrations in human

Mechanisms of Environmental Mutagenesis-Carcinogenesis, Edited by
A. Kappas, Plenum Press, New York, 1990

lymphocytes in resting phase (Go) of growth or after stimulation with phytohaemagglutinin; 3) _in vitro_ cell transformation in the three cell systems using SHE, BALB/c3T3 and C3H10T1/2 cells; and 4) _in vivo_ epidermal hyperplasia and sebaceous gland destruction in the mouse short term skin tests. The last assays were selected for assessing the carcinogenic potential of the substances, since they have been shown to be good indicators of carcinogenic potential (Lazar and Chouroulinkov, 1974; Lasne et al., 1987). All the results were compared to carcinogenicity data from long term mouse skin painting experiments from our laboratory or previously published data.

The chemicals studied included the two aromatic amine oxides: 4-Nitro-quinoline-1-oxide (4NQO) (ICI, Alderley Park, U.K.), which is a potent carcinogen (Purchase et al., 1981) and 3-Methyl-4-Nitroquinoline-1-oxide (3Me4NQO) (ICI, U.K.), generally considered as non carcinogenic (Purchase et al., 1981) or weakly carcinogenic (Nagao and Sugimura, 1976); the two polycyclic hydrocarbons Benzo (a) pyrene (BaP) (ICI, U.K.) and 12-Dimethyl-benz-a-anthracene (DMBA) (Schuchardt, Munich, W. Germany) classically used for mechanistic investigation on skin carcinogenesis; the alkylating agent Mitomycin C (MMC) (ametycine, Choay France), a potent clastogen and presumably carcinogenic (IARC, 1987). In order to discriminate between the genotoxic and epigenetic events, we have studied the effects of Dexamethasone (DXME) (Sigma, St. Louis, USA), an anti-inflammatory steroid known to inhibit some properties of the promotors (Rotstein and Slaga, 1988), on three endpoints: DNA alkali labile sites, cell transformation as well as epidermal hyperplasia and sebaceous gland destruction.

METHODOLOGY

Cells

Rat and rabbit fibroblasts were isolated from the dermis of a 30 day-old male Wistar rat from our breeding house and from a newborn rabbit "Geant des Flandres" respectively, by the method of explants. Syrian hamster embryo (SHE) cells were isolated from 13 day old fetuses by successive trypsinisations. Mouse BALB/c 3T3 cells from the clone A-31-1 as well as V79 Chinese hamster cells were provided by H. Yamasaki (IARC, Lyon, France) and C3H10T1/2 clone 8 from the late Ch. Heidelberger (Los Angeles, USA). All the cells were cryopreserved and an ampoule thawed for each experiment. They were maintained and used in the exponential phase of growth, in the appropriate culture medium supplemented with 10% fetal calf serum (FCS) and 1% mixed antibiotics.

Human lymphocytes: Peripheral blood samples from healthy donors, aged 25-36 years were collected in heparinized tubes. For cytogenetics studies, whole blood samples were used. For alkali labile site analysis, lymphocytes were isolated from whole blood by differential centrifugation on Paque-Ficoll (Pharmacia, Bois d'Arcy, France).

The DNA Alkali-Labile Sites

The DNA alkali-labile sites were determined by means of

the alkaline elution method already described by Kohn et al., (1981), slightly modified by Papadopoulo and Averbeck (1985). In most of the experiments, the cells were labeled with 0.4 μCi/ml^{-1} (^{14}C)-thymidine and reference V79 X-irradiated cells were labeled with 0.2 μCi/ml^{-1} (^{3}H)-thymidine to provide an internal standard. Otherwise, the target cells were labeled with 0.4 μCi/ml^{-1} (^{3}H)-thymidine and each experiment repeated three times. Phtyohaemaglutinnin (PHA) stimulated lymphocytes and rodent cells (rat and rabbit fibroblasts, SHE cells, 3T3 and 10T1/2 mouse cells) were treated for 1 hr without or with FCS and for 22 hr with 10% FCS. The significance of the results was stastistically evaluated by means of variance analysis (Duncan's test or t-test).

Chromosomal Aberrations

Chromosomal aberrations were analysed in human lympho-cytes according to the procedures reviewed by Scott et al.,(1983), slightly modified. Cell divisions were arrested in metaphases by adding 0.8 μg/ml colchicine (Colchineos, Houde, France). Lymphcytes in Go phase were treated in suspension in fresh medium then collected by centrifugation and cultured in the presence of PHA. Stimulated lymphocytes were treated 24 hr after seeding in the presence of PHA. The cultures were recovered 72 hr after the cell seeding. Toxicity was determined by the mitotic index for 1000 cells. Structural chromosomal aberrations were identified.

In Vitro Cell Transformation

In vitro cell transformation was detected in the three cell systems commonly used: the Syrian hamster embryo (SHE) cell system developed by DiPaolo (1980) and reviewed by Tu et al., (1986), the mouse BALB/3T3 cell system developed by Kakunaga (1973) and reviewed by IARC/EPA workshop (1986) and the C3H10T1/2 mouse cell system developed by Reznikoff et al., (1977) reviewed by IARC/EPA (1986).

The in Vivo Mouse Short Term Tests

The in vivo mouse short term tests including the evaluation of epidermal hyperplasia and sebaceous gland destruction were performed according to Lazar and Chouroulin-kov (1974). Briefly, female CD1 mice (Charles River, France) 42±2 day old, were randomized into groups of 15-20 animals and housed individually. Each compound dissolved in acetone or DMSO was applied to the dorsal skin areas which had been closely clipped. Each treatment was repeated on alternate days until a total of three applications (100 μl/application) of the same substance had been given to each animal. A control group received the solvent alone. The mice were killed 8 days after application of the first treatment; the treated areas (4 cm^2) of skin removed, fixed, sectioned and stained for histological examination. The thickness of the epidermis expressed in arbitrary units and the number of sebaceous glands were determined in 12 microscopic fields of each of 6 sections from each specimen. The results were analysed by computer.

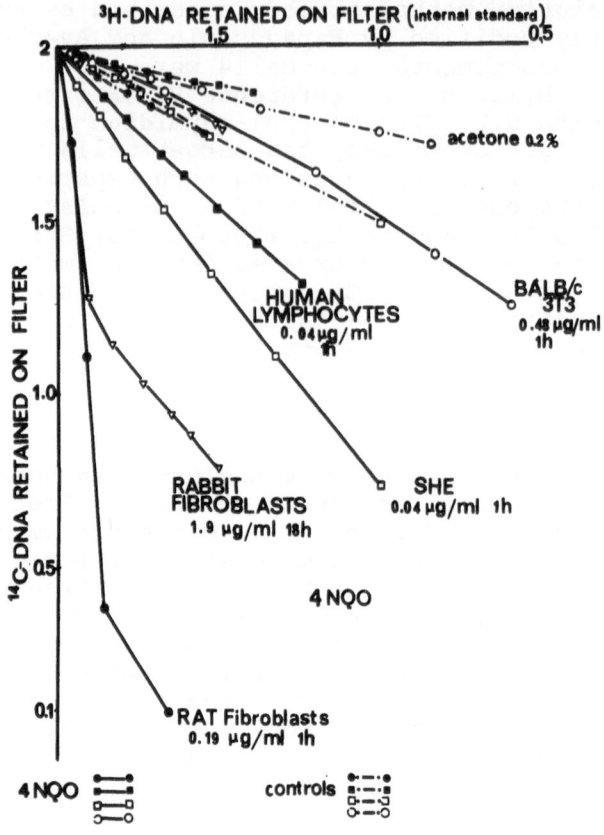

Fig. 1.   Alkaline elution profiles of (14C) DNA from
human lymphocytes, rat and rabbit fibroblasts,
syrian SHE and mouse 3T3 cells treated with
various concentrations of 4NQO. (3H) thymidine
V79 cells irradiated with 3 greys, at 0°, were
used as internal standards. The results
expressed as the per cent of the total
recovered radioactivity are plotted on a
double log scale. The treatment was always
added in the presence of FCS.500,000 target
cells mixed with 300,000 V79 cells were
deposited on each filter.

In Vivo Carcinogenicity Data

In vivo carcinogenicity data concern long term mouse skin
painting experiments (Lazar et al., 1974).

RESULTS

Induction of DNA Alkali-Labile Sites(als)

The results concerning the DNA-als induced by 4NQO in the
different mammalian cell species are presented in Fig. 1. At
the low concentrations tested, the percentage of 14C-DNA

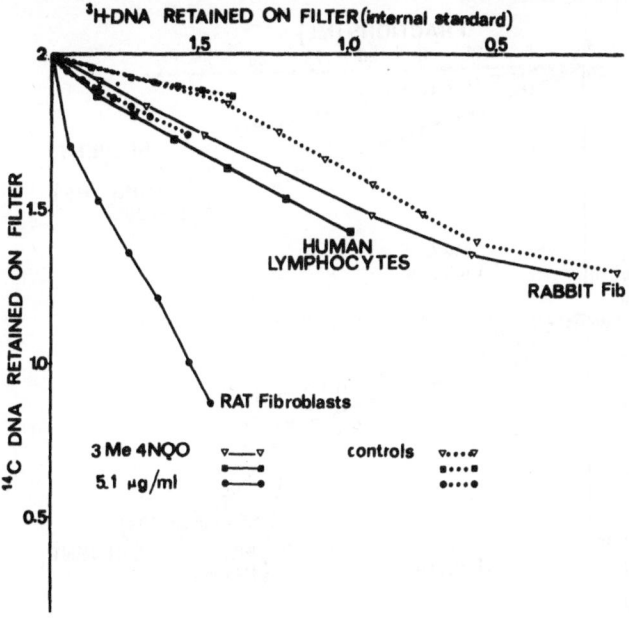

Fig. 2. Alkaline elution profiles of (C14) DNA from
human lymphocytes, rat and rabbit fibroblasts
treated with the same concentration of
3Me4NQO. The experimental conditions and
presentation of the data were the same as
those described in Fig. 1.

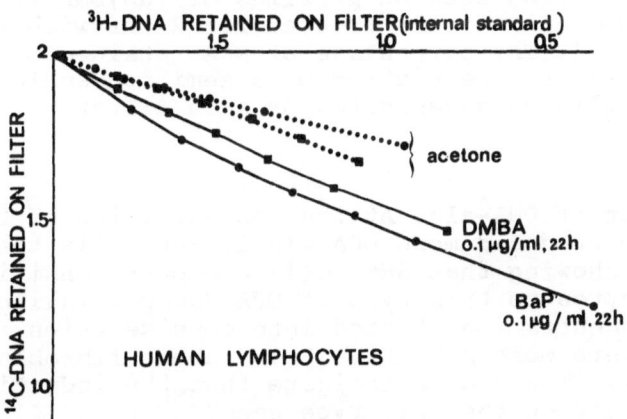

Fig. 3. Alkaline elution profiles of (14C) DNA from
human lymphocytes treated with BaP and DMBA
for 22 hr. The presentation of the data was
the same as that described in Fig. 1.

retained on the filters was highly increased in comparison
with that of the controls. This indicates that 4NQO is a

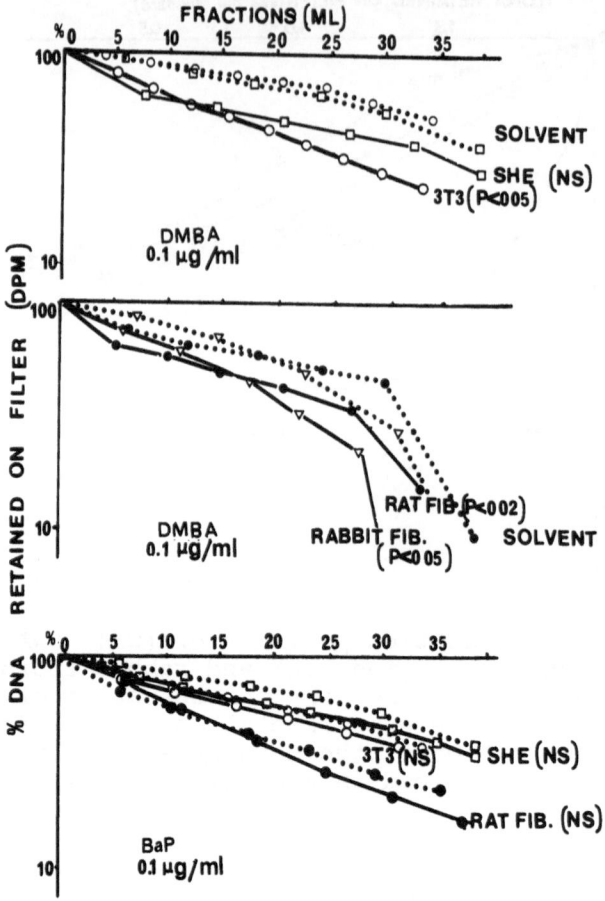

Fig. 4. Alkaline elution profiles of (3H)DNA from
mammalian cells treated for 22 hr with BaP and
DMBA. The percentage of DNA retained on each
filter were plotted on a semi log scale .1x10^6
cells were deposited on each filter.

potent inducer of DNA-als. At the concentration of 0.04 µg/ml
(0.2 µM) 4NQO produced more DNA-als in SHE cells than in human
lymphocytes, showing that SHE cells are more sensitive than
human lymphocytes to this type of DNA damage. Taking the
different concentrations tested into consideration, rat
fibroblasts were more sensitive than rabbit fibroblasts blasts
and 3T3 cells. These data indicate that the induction of DNA
als is dependent on the cell type used.

Figure 2 gives the profiles of the elution curves
obtained after treatment of human lymphocytes, rabbit and rat
fibroblasts with 5.1 µg/ml (25 µM) of 3Me4NQO. The slopes of
the curves indicate that rat fibroblasts are more sensitive
than human lymphocytes, rabbit fibroblasts being most resistant. From the data given in the two figures, it is clear
that 4NQO is a much more potent inducer of DNA-als than
3Me4NQO, since concentrations 200 times higher resulted in
weaker effects. With the two compounds, rat fibroblasts were

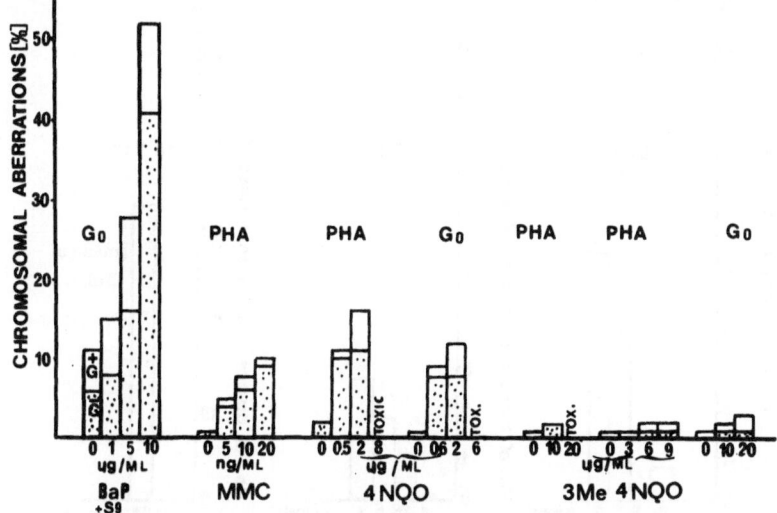

Fig. 5. Induction of chromosomal aberrations in Go and
PHA stimulated human lymphocytes treated with
4NQO, 3Me4NQO, BaP and DMBA. The total number
of chromosomal aberrations was evaluated with
[+G] and without [-G] gaps. Go lymphocytes
were treated for 1 hr without FCS. In the PHA
stimulated lymphocytes, the treatment remained
until the recovery of the metaphase.

more sensitive than rabbit fibroblasts and it is likely that
they are also more sensitive than human lymphocytes.

The next two figures present the alkaline elution curves
obtained with the two polycyclic hydrocarbons BaP and DMBA.
The slopes of the curves in Fig. 3 indicate that 0.1 µg/ml
(0.4 µM) of BaP produced a weak but significant increase in
the DNA als of human lymphocytes ($p < 0.01$) in comparison with
the controls, while the treatment with the same concentration
of DMBA did not result in a significant increase of the
DNA-als.

In contrast, as presented in Fig. 4, BaP did not affect
the DNA elution rate of SHE and 3T3 cells and rat fibroblasts,
as well as rabbit fibroblasts (data not given on the Figure),
while the same concentrations of DMBA induced weakly but
significantly DNA-als in 3T3 cells ($p < 0.05$), rat fibroblasts
($p < 0.02$) and rabbit fibroblasts ($p < 0.05$). An increase to
1 µg/ml in the concentrations of BaP and DMBA did not modify
the results.

In conclusion, BaP was found weakly active in human
lymphocytes and negative in SHE cells, 3T3 cells, rat and
rabbit fibroblasts, while DMBA was found to be weakly active
in 3T3 cells, rat and rabbit fibroblasts and negative in human
lymphocytes and in SHE cells. Negative results were also
observed after treatment of human lymphocytes with concentrations of MMC ranging from 10 µM to 100 µM (3-33 µg/ml). 4NQO

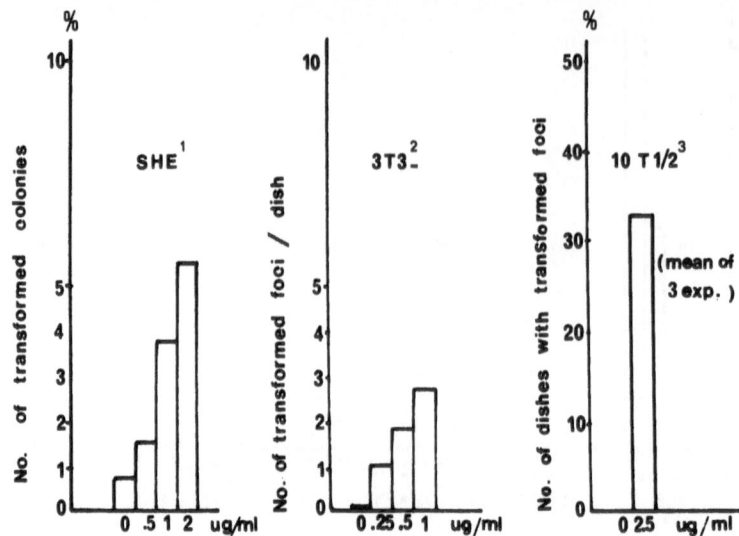

Fig. 6. Transforming activity of BaP in SHE, 3T3 and
10T1/2 cells. In the SHE cell system, the
number of transformed colonies was expressed
per the total number of surviving colonies. In
10T1/2 cell system transformed foci have
included type II and type III foci. 1) 10
dishes per each concentration, with 300 cells
per dish; 2) 15 dishes per concentration, with
10^4 cells per dish and 3) 20 dishes per
experiment with 2,000 cells per dish.

was the only compound found to be very active in all the
cell species studied.

Induction of Chromosomal Aberrations

The results are presented in Fig. 5. The induction of
chromosomal aberrations was expressed as the number of
chromosomal aberrations with and without gaps, per 100
metaphases. From this figure, it is clear that BaP, MMC and
4NQO induced a dose related increase in chromosomal aberrations in human cells treated either in the resting phase of
the cell cycle or after stimulation by PHA. MMC was active at
very low concentrations (5 ng/ml), while 4NQO was active over
a very narrow range of concentrations (0.5-2 μg/ml). No
metaphases were seen after treatment of Go human lymphocytes
for 1 hr with 6 μg/ml 4NQO followed by 72 hr culture. 3Me4NQO
was very weakly active, since a weak but reproducible dose-
related increase in the number of chromosomal aberrations was
observed in three experiments performed using the same
experimental protocol.

Most of the aberrations detected were chromatid breaks
(gaps) and deletions. However, some dicentrics, rings and
symmetric and asymmetric interchanges appeared at high toxic
concentrations. In conclusion, the results show that the four

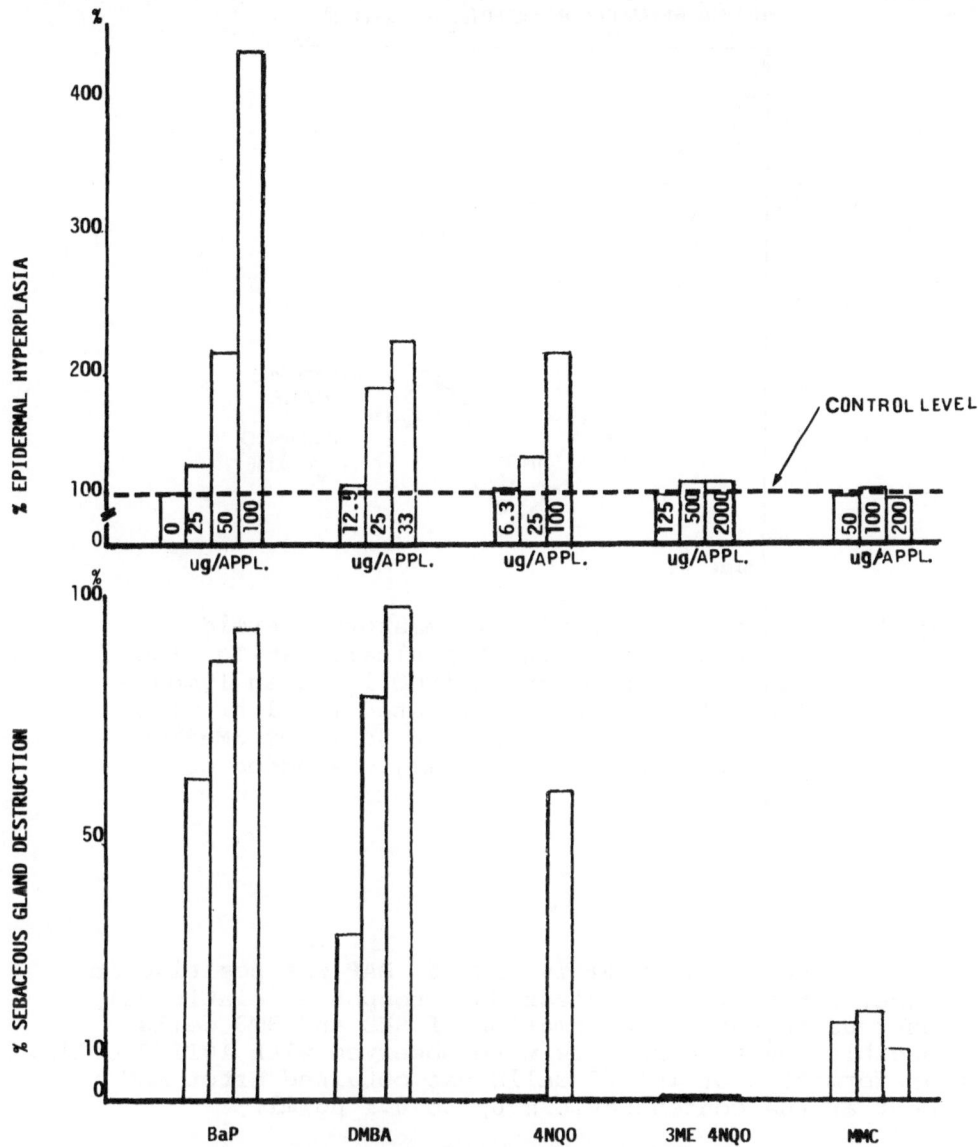

Fig. 7. Epidermal hyperplasia and sebaceous gland destruction after treatment of mouse skin with BaP, DMBA, 4NQO, 3Me4NQO and MMC. BaP, DMBA and MMC were dissolved in acetone; 4NQO and 3Me4NQO, in DMSO. Each graph represents the means of epidermal thickness (in arbitrary units), and of sebaceous glands destroyed, for 20 animals. The dotted line represents the epidermal thickness of the control animals and is equal to 100.

compounds are clastogenic at different orders of magnitude.

Cell Transformation

The transforming activity of BaP is compared in the three

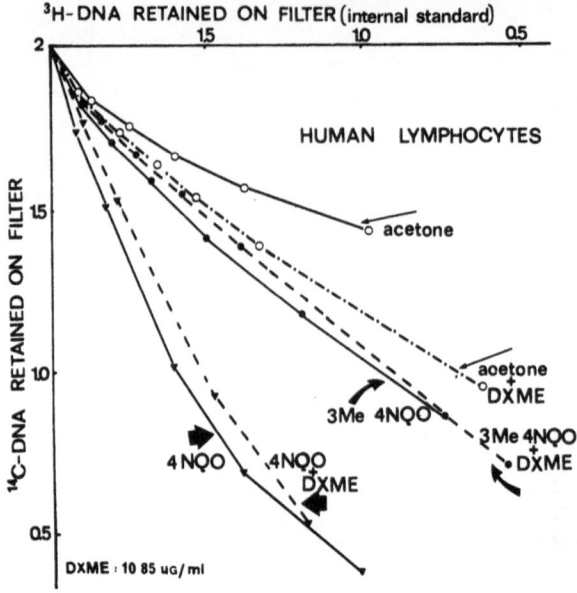

Fig. 8. Effect of the anti inflammatory steroid
Dexamethasone on the DNA-alkali labile sites
induced by 4NQO and 3Me4NQO in human lymphocytes. The cells were treated for 1 hr with
0.06 µg/ml of 4NQO and 5.6 µg/ml of 3Me4NQO.
Dexamethasone (10.85 µg/ml) was added concom-
mitantly.

cell transformation systems in Fig. 6. BAP was positive in
each case. Furthermore, a clear dose response relationship
was observed in the transformation of SHE and 3T3 cells.
Reproducible positive results were observed with 10T1/2 cells.
No transformation of 10T1/2 cells was obtained after MMC
treatment at the concentrations up to 0.1 µg/ml.

Different publications have dealt with transformation of
cultured cells _in vitro_ by DMBA, 4NQO and 3Me4NQO. They have
all pointed out the transforming activity of DMBA in the SHE
cell system (Pienta et al., 1977), in the 3T3 cell system
(Dunkel et al., 1981) and in the 10T1/2 cell system (Reznikoff
et al., 1977; Nesnow et al., 1985). 4NQO was tested as
positive in the 3T3 cell system (Kakunaga, 1973; Dunkel et
al., 1981) and in the 10T1/2 cell system (Nesnow et al.,
1985). In the SHE cell system, 4NQO was found positive in
some laboratories (Pienta et al., 1977; Kuroki and Sato,
1968), negative in the interlaboratory study (Tu et al., 1986)
and slightly positive by Dunkel et al., (1981). On the basis
of these data, we consider that 4NQO has a transforming
potential _in vitro_. 3MeNQO was found to be inefficient in
cell transformation in the single trial reported (Kuroki and
Sato, 1968).

In Vivo Mouse Short Term Tests

Epidermal hyperplasia induction and sebaceous gland

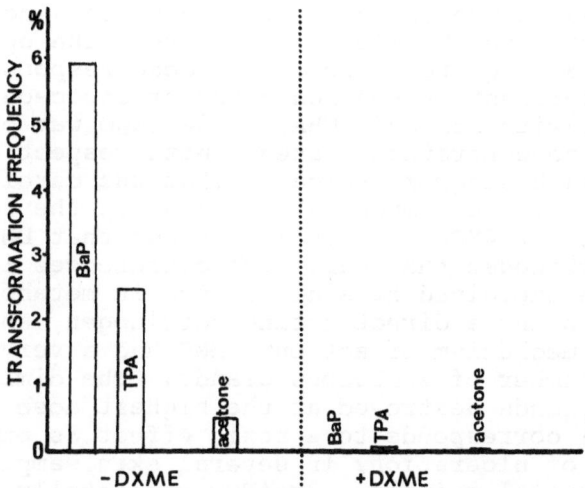

Fig. 9.   Inhibition by Dexamethasone (DXME) of SHE cell
transformation induced by BaP and TPA.BaP (1
µg/ml) or TPA (0.1 µg/ml) were added for four
days in the culture medium, then DXME (10.85
µg/ml) was added for four extra days in half
of the dishes.

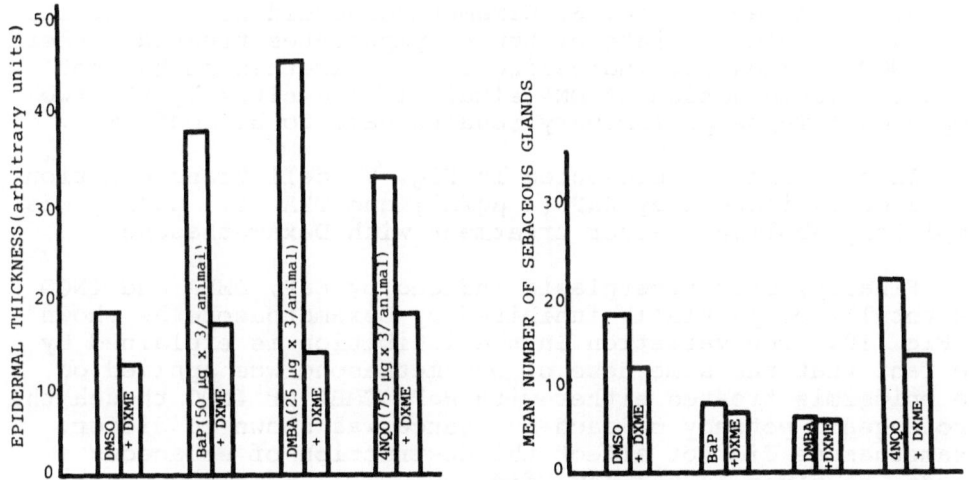

Fig. 10.   Effect of Dexamethasone (DXME) on mouse skin
hyperplasia and sebaceous gland destruction
induced in vivo by BaP, DMBA and 4NQO. DXME
(217 µg per application, in 100 µl acetone)
was applied just after each application of all
the chemical treatments. (3 applications per
treatment). The data were expressed as the
means for 15 animals per each dose.

destruction are presented in Fig. 7. In comparison with the epidermal thickness of the solvent treated animals (as indicated by the dotted line), it is clear that BaP, DMBA and 4NQO have increased the epidermal thickness in a dose-related manner, while MMC and 3Me4NQO did not affect the epidermal thickness even at high toxic doses. A dose response relation ship was also apparent in the reduction of sebaceous glands after treatment with BaP and DMBA, while 4NQO was only active at the highest concentration tested. With respect to the doses used, this histogram indicates that the carcinogenic potential of the three compounds decreases in the following order: DMBA > BaP > 4NQO. It is well known that DMBA is a more potent carcinogen than BaP. The differences in the response can be explained by a difference in metabolism of BaP and DMBA, 4NQO being a direct acting carcinogen, and by differences in mechanism of action. MMC had a very weak effect on the number of sebaceous glands. The diminution of the number of glands destroyed at the highest dose (200 µg/application) corresponds to a toxic effect as expressed by the appearance of ulcerations in several skin samples and by the death of several animals. 3Me4NQO was totally inactive. In conclusion, these results are in good agreement with the data from long term studies which have indicated that BaP (Osborne and Crosby, 1987) and DMBA (Chouroulinkov et al., 1979) and 4NQO are potent carcinogens, while 3Me4NQO is generally considered very weak carcinogen or not carcinogenic and MMC presumably carcinogenic.

Effect of Dexamethasone on DNA Alkali-Labile Sites, Cell Transformation, Epidermal Hyperplasia and Sebaceous Gland Destruction

As presented in Fig. 8, Dexamethasone did not affect the DNA alkaline elution rate of human lymphocytes treated either with 4NQO or 3Me4NQO, indicating that Dexamethasone has not affected the induction of DNA alkali labile sites by the two compounds. These preliminary results have to be confirmed.

In contrast, as presented in Fig. 9, cell transformation of SHE cells induced by BaP (1 µg/ml) and TPA (0.1 µg/ml) was completely abolished after treatment with Dexamethasone.

Finally, skin hyperplasia induced by BaP, DMBA and 4NQO was totally or partially inhibited by Dexamethasone as shown in Fig. 10. The variation in the inhibition is explained by the fact that the same dose of Dexamethasone was applied on the epidermis treated either with BaP, DMBA or 4NQO though the carcinogenic potency of each substance was shown to differ. Dexamethasone did not affect the destruction of sebaceous glands, as given in the same figure.

DISCUSSION

The results presented in the first part of this study show that DNA-alkali labile sites were induced by substances known to be potent carcinogens, such as 4NQO, BaP and DMBA, but also by substances considered as non carcinogens, such as 3Me4NQO. 4NQO was the only substance found positive in all the cell types studied and highly active in all of them. DMBA was weakly but significantly active in rat and rabbit fibroblasts

TABLE 1. Summary of the Results Obtained in the Different Test Systems

Substance	DNA-SSB	Cytogenetics: chromosomal aberrations	In vitro cell transformation	In vivo skin short term tests Hyperpl.	Seb.gland destruction	Carcinogenicity (long term)	Event
4NQO	+ positive	+ positive	+ positive	+ positive	+ positive	complete	Genotoxic
DaP	± positive	+ positive	+ positive	+ positive	+ positive	carcinogens	+
DMBA	± positive	n.d.	+ positive	+ positive	+ positive		Epigenetic
3Me4NQO	+ positive	± positive	- negative	- negative	- negative	non-carcinogen	Genotoxic
MMC	- negative	+ positive	n.d.	- negative	± positive	carcinogen ?	Genotoxic

n.d.: Not determined.
? : Possibly carcinogenic to humans (group 2B, IARC).

195

as well as in mouse 3T3 cells, but negative in human lymphocytes and syrian hamster cells, while BaP was weakly positive in human lymphocytes and negative in the other cell species. These results are rather surprising since SHE cells were shown to metabolize polycyclic hydrocarbons such as BaP (Schechtman, 1985), and a dose related transformation of SHE and 3T3 cells with BaP and DMBA in the absence of a S9 mix has been demonstrated by different groups, as already mentioned in this paper. In addition, we have obtained the cell transformation of rat fibroblasts with the two substances (Lasne et al., 1977). Furthermore, fibrosarcomas are easily produced in rats, syrian hamsters and mice after subcutaneous treatment with polycyclic aromatic hydrocarbons. Finally, BaP and DMBA have been shown to have a strong carcinogenic potential. The clastogenic agent MMC was inactive in human lymphocytes. The absence of DNA alkali labile sites after treatment of these cells with MMC might be a consequence of its high capacity to form DNA-cross-links (Dusre et al., 1989) masking the possible formation of DNA alkali labile sites. This has yet to be demonstrated. Lastly, 3Me4NQO, which is able to induce DNA alkali labile sites, is negative in all the assays except chromosomal aberration. Thus these data seem to indicate that DNA alkali labile sites are not linked to carcinogenesis.

The good correlation between the formation of DNA alkali labile sites and of chromosomal aberrations by BaP, 4NQO and 3Me4NQO in human lymphocytes as well as the absence of alkali labile site inhibition by Dexamethasone, an inhibitor of promotion, suggest that DNA alkali labile sites are associated with a genotoxic event. The inhibition of cell transformation and skin hyperplasia by Dexamethasone suggests that both endpoints are associated with epigenetic events as is already accepted for epidermal hyperplasia (Lasne et al., 1987). With respect to our classification of carcinogens, as proposed in the introduction, these results allow us to classify the three chemicals DMBA, BaP and 4NQO as complete carcinogens, since they result in both genotoxic and epigenetic events, as summarized in Table 1. From this table, a good correlation should also be noted between all the assays, for the carcinogen 4NQO and for the non carcinogen 3Me4NQO; the latter being only genotoxic with respect to our classification. No definitive conclusion can be drawn at present for MMC. More studies have to be done to clarify its effect on mouse skin.

ACKNOWLEDGEMENTS

    This work was partly supported by the Ministere de la Recherche et de la Technologie (Paris).

REFERENCES

Chouroulinkov, I., 1988, Initiation-promotion: Biological and toxicological interpretations of carcinogenesis, in: "Theories of Carcinogenesis", H.O. Iversen, ed., N. Hemisphere Publ. Co., Washington.
Chouroulinkov, I., Gentil, A., Tierney, B., Grover, P.L., and Sims, P., 1979, The initiation of tumors on mouse skin by dihydrodiols derived from 7-12 DMBA and 3-MC, Int. J. Cancer, 24:455.

DiPaolo, J.A., 1980, Quantitative *in vitro* transformation of Syrian golden hamster embryo cells with the use of frozen stored cells, J. Natl. Cancer Inst., 64:1485.

Dunkel, V.C., Pienta, R.J., Sivak, A., and Traul, K., 1981, Comparative neoplastic transformation responses of BALB/3T3 cells, Syrian hamster embryo cells and Rauscher murine leukemia virus infected Fischer 344 rat embryo cells to chemical carcinogens, J. Natl. Cancer Inst., 67:1303.

Dusre, L., Covey, J.M., Collins, C., and Sinha, B.K., 1989, DNA-damage, cytotoxicity and free radical formation by Mitomycin C in human cells, Chem. Biol Inter., 71:63.

IARC, 1987, IARC Monographs on the Evaluation of the Carcinogenic Risk of Chemicals to Humans, Suppl. 7, Lyon, France.

IARC/NCI/EPA working group, 1986, Cellular and molecular mechanisms of cell transformation and standardization of transformation assays of established cell lines for prediction of carcinogenic chemicals: Overview and recommended protocols, Cancer Res., 45:2395.

Kakunaga, T., 1973, A quantitative system for assay of malignant transformation by chemical carcinogens using a clone derived from BALB/3T3, Int. J. Cancer, 12:463.

Kohn, K.W., Ewig, R.A.G., Erickson, L.C. and Zwelling, L.A., 1981, Measurement of strand breaks and cross links by alkaline elution, *in*: "DNA Repair: A Laboratory Manual of Research Procedures", E.C., Friedberg and P.C. Hanawalt, eds., Marcel Dekker, New York.

Kuroki, T., and Sato, H., 1968, Transformation and neoplastic development *in vitro* of hamster embryonic cells by 4-nitroquinoline 1-oxide and its derivatives, J. Natl. Cancer Inst., 41:53.

Lasne, C., Gentil, A., and Chouroulinkov, I., 1977, Two-stage carcinogenesis with rat embryo cells in tissue culture, Br. J. Cancer, 35:722.

Lasne, C., Venegas, W., Royer, R., and Chouroulinkov, I., 1987, Initiating, promoting and carcinogenic activities of three naphthofurans in a mouse skin long term study, Jpn. J. Cancer Res., (Gann), 78:565.

Lazar, P., and Chouroulinkov, I., 1974, Validity of the sebaceous gland test and the hyperplasia test for the prediction of the carcinogenicity of cigarette smoke condensates and their fractions, *in*: "Experimental Lung Cancer", E. Karbe and J.P. Park, eds., Springer Verlag, Berlin.

Lazar, P., Chouroulinkov, I., Izard, C., Moree-Testa, P., and Hemon, D., 1974, Bioassays of carcinogenicity after fractionation of cigarette smoke condensate, Biomedicine, 20:214.

Nagao, M., and Sugimura, T., 1976, Molecular biology of the carcinogen 4-nitroquinoline 1-oxide, *in*: "Advances in Cancer Res.", Vol. 23, G. Klein, S. Weinhouse and A. Haddow, eds, Academic Press, New York.

Nesnow, S., Curtis, G., and Gerland, H., 1985, Tests with the C3H1OT1/2 clone 8 morphological transformation bioassay, *in*: "Progress in Mutation Res.", Vol. 5., J. Ashby, F.J. de Serres et al., eds., Elsevier, Amsterdam.

Osborne, M.R., and Crosby, N.T., 1987, Carcinogenesis, *in*: "Benzopyrenes", M.M. Coombs, J. Ashby and R.F. Newbold, eds., Cambridge University Press, Cambridge.

Papadopoulo, D., and Averbeck, D., 1985, Genotoxic effects and DNA photo-adducts induced in Chinese hamsters V79 cells by 5-Methoxypsoralen and 8-Methoxypsoralen, Mutat. Res., 151:281.

Pienta, R.J., Poiley, J.A., and Lebherz, W.B., 1977, Morphological transformation of early passage golden Syrian hamster embryo cells derived from cryopreserved primary cultures as a reliable in vitro bioassay for identifying diverse carcinogens, Int. J. Cancer, 19:642.

Purchase, I.F.H., Clayson, D.B., Preussmann, R., and Tomatis, L., 1981, Activity of 42 compounds in animal carcinogenicity studies, in: Prog. in Mutation Res., Vol. 1, "Evaluation of Short-term Tests for Carcinogens", F.J. de Serres and J. Ashby, eds., Elsevier, North Holland, New York.

Reznikoff, C.A., Bertram, J.S., Brankow, D.W., and Heidelberger, C., 1977, Quantitative and qualitative studies of chemical transformation of cloned C3H mouse embryo cells sensitive to postconfluence inhibition of cell division, Cancer Res., 33:3229.

Rotstein, J.B., and Slaga, T.J., 1988, Anticarcinogenesis mechanisms as evaluated in the multistage mouse skin model, Mutat. Res., 202:421.

Schechtman, L.M., 1985, Metabolic activation of procarcinogens by subcellular enzyme fractions in the C3H10T1/2 and BALB/c 3T3 cell transformation systems, in: "Transformation Assay of Established Cell Lines: Mechanisms and Application", T. Kakunaga and H. Yamasaki, eds., IARC, Lyon.

Scott, D., Danford, N., Dean, B., Kirkland, D.J., and Richardson, C.R., 1983, in: "Report of the UKEMS Sub-committee on Guidelines for Mutagenicity Testing", B.J. Dean, ed., UKEMS publisher, London.

Tu, A., Hallowell, W., Pallotta, S., Sivak, A., Lubet, R.A., Curren, R.D., Avery, M.D., Jones, C., Sedita, B.A., Huberman, E., Tennant, R., Spalding, J., and Kouri, R.E., 1986, An interlaboratory comparison of transformation in Syrian hamster embryo cells with model and coded chemicals, Env. Mutagenesis, 8:77.

STRUCTURE AND METABOLISM OF MUTAGENS-CARCINOGENS

QUANTITATIVE STRUCTURE-ACTIVITY RELATIONSHIPS, AND MUTAGENS

AND CARCINOGENS

R. Benigni

Laboratory of Toxicology and Ecotoxicology
Istituto Superiore di Sanita
Roma, Italy

INTRODUCTION

The basic philosophy of structure-activity relationship studies is to draw conclusions by analogy. We assume that similar chemical properties of drugs will result in similar biological responses. The problem consists of identifying exactly what these properties are, and how they are connected with the biological activity of interest. Attempts to correlate structure with biological activity date back to the last century. These attempts were essentially qualitative, or semi-quantitative. In the sixties there was a great progress towards the establishment of a Quantitative Structure-Activity Relationship (QSAR) science. This transformation has happened in the field of medicinal chemistry because of the pioneering work of Hansch.

FROM QUALITATIVE TO QUANTITATIVE STRUCTURE-ACTIVITY RELATIONSHIP (QSAR): THE HANSCH PARADYGM

The great dream of the medical chemist has always been the rational design of drugs, most specifically pharmaceutical drugs and pesticides. The Hansch approach (also known as the extra-thermodynamic approach) was created to solve the specific problem of drug modification; it consists of the development of new and more potent drugs through the chemical modification of known ones. The Hansch approach assumes that, when dealing with a series of compounds with a common skeleton and different substituents attached to it, the effect of the substituents on the strength of interaction between a drug and its receptor (or other biomolecules) is an additive combination of the effects of the substituents. Based on physical chemistry theory, the basic interactions are supposed to be electrostatic, steric (repulsion), and hydrophobic; each substituent is assumed to contribute to each of these forces (Hansch and Fujita, 1964; Hansch, 1976). This conceptual framework can be represented with the equation:

$$\log(A) = fh \cdot (xh) + fe \cdot (xe) + fs \cdot (xs) + \text{constant}$$

Mechanisms of Environmental Mutagenesis-Carcinogenesis, Edited by
A. Kappas, Plenum Press, New York, 1990

where the activity A is separated into the contributions of fragment x to hydrophobic (xh), electronic (xe) and steric (xs) forces. If the values of biological activity are measured for a properly selected range of chemicals with different substituents and the parameters xh, xe and xs are known (many of them are actually tabulated), the coefficients of the equation can be routinely estimated by multiple regression analysis. The obtained equation describes the influence of physical chemical forces on the biological activity, and can be used to predict the activity of other chemicals belonging to the same series. This last fact has obvious practical consequences, since it permits the rational selection of the chemicals to be synthesized and tested, and since it avoids a random search, which is expensive and scarcely efficient (Franke, 1984).

The Hansch approach is based largely on the hypothesis of a receptor-substrate-like interaction. A receptor is a molecular structure (usually polypeptide, often located on cell membranes), which specifically recognizes the drug. The drug then binds to the receptor and interacts with it: this interaction produces the stimulus (Lehninger, 1981). The recognition and binding between drug and receptor requires a strict complement of the two structures, including 3-dimensional conformation as well. In practice, a receptor recognizes a family of molecules which have a very similar skeleton, but which are dissimilar in the substituents attached to that skeleton. The Hansch approach, then, is mainly aimed at describing the variation of biological activity due to small variations of the common structure of congeneric chemicals (i.e., belonging to the same chemical class). This conceptual framework also explains why a simple mathematical formula, such as that shown in Equation 1, can work so efficiently in many cases. But one can expect that such linear QSAR's — valid in the range of substituent variation studied — will still function slightly beyond that range; when the variations of the chemical structure become too great, the validity of the equation will decrease in an unpredictable way (Hansch, 1977).

ORIGINALITY OF THE HANSCH APPROACH, AND FURTHER DEVELOPMENTS

In spite of its inherent simplicity, the linear additive form usually taken by the QSAR Hansch solutions has several appealing properties. Probably the most important is simplicity itself. Even though more sophisticated data analyses — in particular the multivariate data analysis methods (Franke, 1984; Lebart et al., 1984; Lewi, 1980) — are often advisable in order to fully explore complex situations, the ability to relate activity and parameters in a simple formula allows a problem to be more easily understood, as well as permitting a more immediate utilization of the results of a study. Sometimes, the derivation of a "simple" Hansch equation is the result of a complex series of analyses and hypotheses on the mechanisms underlying the biological activity; the final equation thus summarizes much preliminary work (Franke, 1984).

The Hansch's contribution to the development of QSAR theory was essential. He demonstrated that the relationship between physical chemical properties and biological activity can be examined statistically. The introduction of statistics allowed more rational studies to be performed: one can either accept or reject apparent relationships, detect outliers, and quantitatively compare different hypotheses. He also introduced multivariate statistics, which evaluate simultaneously the mutual interactions and relative importance of the different properties. A third significant contribution was the discovery that the partition coefficient has additive properties, and can be calculated from the contributions of the individual substituents. In particular, he studied logP, which is the logarithm of the octanol/water partition coefficient: this parameter is a good model for the absorption and transport phenomena between the lipidic and aqueous compartments of cells and organs, and plays an important role in many QSAR's. The ability to calculate relative logP — and by analogy also other physical chemical parameters — enormously increased the practical feasibility of QSAR studies (Martin, 1981).

The research performed by Hansch and coworkers on many different types of biological activity, and by other scientists using his approach has stimulated a series of methodological studies, which have become an important part of the conceptual framework of QSAR field generally speaking, and which represent quality standard criteria for any correct structure-activity study. The first question is how to validate the found QSAR's. The ambition of a structure-activity study is not only the description of the relationships existing in the available data base, but also the derivation of a correlation valid for the entire class of chemicals to which the data base belongs. One possible approach is to split the sample of chemicals into a training and a validation set. The training set consists of one part of the molecules under study, and is used to find a structure-activity relationship. This relationship is then applied to the remaining chemicals (validation set), and the biological activity of each chemical is calculated based on the structural parameters present in the equation. If the calculated and measured activities are significantly similar, then the found relationship is correct, and can be used for any other application. If the gap between calculated and measured activity is too large, the structure-activity study has to be redone based on different hypotheses and parameters. To obtain more soundly based results, it is advisable to repeat the procedure, varying each time the way in which the chemicals are partitioned into the two sets.

In a final Hansch equation, there are a number of parameters representing physical chemical properties. Usually the study is started with a number of parameters greater than that which will be present in the final equation. The statistical analysis excludes the parameters not correlated with the biological activity. If some parameters are highly correlated with each other - i.e. they represent more or less the same fundamental property - the statistical analysis points to the parameter most correlated with the biological activity, and excludes the other ones as redundant. An F- or t-statistics is usually performed. However, a more subtle problem in the choice of parameters has been identified: the possibility of chance correlations. Together with the number and type of parameters to be included in the final equation, also the number of parameters to be screened before formulating the final equation is a crucial aspect. Topliss and coworkers demonstrated that the greater the number of screened parameters, the greater the possibility that the final equation -though statistically significant - includes parameters correlated with the activity merely by chance. In other words, if more and more parameters are screened to construct a statistically significant equation, there exists the possibility that a parameter may be found which correlates only by chance with the activity in that sample of chemicals, but which has no real correlation in other samples nor in the entire class of chemicals. This equation, although statistically significant according to F- or t-values, may give rise to misinterpretations, and - even more serious - may be devoid of any predictive power when applied to other chemicals of the same class. The results of Topliss are very important, and point out the necessity to adequately plan a structure-activity study. This can be done by keeping the number of parameters screened adequately low in proportion to the number of chemicals. As a rule of thumb, Topliss has indicated 5 as the "safe" ratio chemicals to parameters (Topliss and Edwards, 1979).

THE TOPOLOGICAL APPROACHES

In the same period in which Hansch elaborated his model, another method referring directly to the substituents -instead of to the physical chemical forces, was being formulated (Free and Wilson, 1964). This is the so-called Free-Wilson analysis, and can be formulated in the following way:

$$\log(A) = \sum_i x_i z_i + C$$

where the activity A is expressed as a sum of the z_i contributions of the various substituents i. x_i is 1 or 0, according to the presence or absence of the given substituent. It is worth remembering that z_i is the contribution of a specific substituent, in a specific substitution position on the molecule skeleton.

This approach is formally equivalent to that elaborated by Hansch, because the latter is also based on the additive contribution of the substituents to the physical chemical forces, hence to the activity. Moreover, both approaches are aimed at congeneric substances (i.e. belonging to the same chemical class). The Free-Wilson approach, however, permits an analysis to be made even when the physical chemical parameters for some substituents are missing.

The Free-Wilson approach has remarkable scientific importance, also because of the successive scientific developments it provoked. All the so-called topological approaches are "sons" of the Free-Wilson method. The topological approaches are based on the topology, or 2-dimensional representation of the molecular structure. At present, several topological approaches exist. They differ from each other for the kind of molecular descriptor that they generate, and for the kinds of statistical mathematical methods used. These approaches define the descriptors within a more general conceptual framework that the Free-Wilson method. The types of descriptors are: a) simple counts of the different types of atoms and bonds; b) substructures, i.e. any ensemble of atoms interconnected. These can be classical functional groups (such as $-NH_2$, $-CO$, etc) or any other atom subgroup identified according to any criterion; c) environmental descriptors. These descriptors are subgroups of atoms characterized by the types of atoms to which they are connected in the molecule (Franke, 1984).

These methods arise from 2 primary requirements: a) simplicity, because the molecular descriptors are obtained directly from the structural formula of the molecule. The generation of the descriptors is easily implemented in computer programs which can perform this operation automatically; b) the necessity to deal with non congeneric groups of chemicals, which however share the same kind of biological activity. Non congeneric chemicals are those which do not possess a common skeleton, and belong to different chemical classes. This is a common occurrence in mutagenicity and carcinogenicity. Both the Hansch and Free-Wilson approaches are not applicable in these non congeneric situations.

Among the different topological approaches applied in mutagenicity and carcinogenicity, it is worth considering ADAPT and CASE. They have developed a more precise configuration, and have been implemented in computer program packages that perform the various steps of the structure-activity analysis in sequence.

ADAPT (Automated Data Analysis by Pattern Recognition Techniques) (Jurs et al., 1983) reads the structure formula of the molecule - which is coded in the computer - and generates a wide range of descriptors of three different classes: a) topological (number of the different types of atoms, bonds and rings; presence or absence of substructures; electronic charges on the substructures; b) geometrical, calculated on the basis of the 3-dimensional structure; c) whole-molecule, such as logP. Once the descriptors have been calculated, ADAPT applies different mathematical methods (pattern recognition) to point out those correlated with the biological activity.

The CASE (Computer Automated Structure Evaluation) method has an underlying rational different than ADAPT and other topological approaches (Klopman, 1984). The latter methods generate the molecular descriptors based on a library, or pre-defined list. CASE generates all the possible substructures consisting of 3 to 10 non-hydrogen atoms interconnected for each molecule. For a structure-activity study all the molecules are fragmented in this way, and the correlation of each substructure with the biological activity is screened. The substructures not correlated with the activity are eliminated, and a function - usually a linear one similar to the Hansch approach - including the relevant features is constructed. The coefficients of the equation define the relative weights of the substructures for the activity. It is evident that the CASE method, which places no restrictions on the types of substructures generated, has the advantage of being able to identify important substructures never considered relevant before. A problem with this approach seems to be - at least in principle - the possibility of chance correlations. When analyzing a set of molecules with many atoms, the number of combinations of interconnected atoms increases exponentially: thus, the number of substructures screened may become very large, with a consequent increase in the probability of chance correlations. It is important in such cases to plan the study in such a way as to eliminate this danger. For example, with a data base of 1000 molecules no more than 200 substructures should be generated for the screening.

An important question concerns the difference, with relative advantages and disadvantages, between the approaches using physical chemical properties and topological descriptors. A first consideration is that the Hansch approach, in its original formulation, cannot be used with non congeneric chemicals, while some topological methods can. In general, the topological methods have the advantage of simplicity in the generation of descriptors. The advantage of the Hansch approach - when applicable - is in its ability to express solutions in terms of physical chemical forces, which are a consistent scientific theory. In this sense, these solutions are more general, and can be used for a clearer understanding of the activity mechanisms: thus, they also permit a better control of the validity and reasonability of the results of the analysis. To extend the results of a topological analysis beyond the mere empirical correlation, it is necessary to interpret and "translate" them into more general concepts.

QSAR IN MUTAGENICITY AND CARCINOGENICITY

The concepts outlined so far deal with the general questions of structure-activity studies. Let us see how these apply in the specific case of chemical mutagenicity and carcinogenicity. A large part of the genotoxic effects, and a somewhat important fraction of the carcinogenic effects due to chemicals, have been rationalized in terms of DNA modifications. In principle, the same physical chemical forces correlated with the biological activity in the Hansch approach can be postulated to rule mutagenic and carcinogenic activity also. Electronic effects will rule both covalent and non covalent binding to DNA. Steric factors will determine the accessibility to structures such as the minor and major DNA grooves, and linker and core DNA. Because of the 3-dimensional interaction between enzyme and substrate, steric factors will also be important in the metabolic processes, which often lead to the production of the reactive species. Moreover, hydrophobicity will rule the absorption and transport of the molecules. On this basis, the Hansch method can be extended to mutagenic and carcinogenic effects. By analogy, we can postulate that the topological approaches generated for drug design will also be valid. Actually, many successful applications of the various methods to mutagenicity and carcinogenicity exist in scientific literature (Frierson et al., 1986; Benigni et al., 1989).

There are, however, some specificities which make the application of QSAR in mutagenicity and carcinogenicity different than that in classical drug design. A primary difference is the strong linking of classical drug design with the concept of receptor. This interaction is specific to a high degree, and limits the range of chemical structures that produce a given effect. In genotoxicity, very different molecular structures can give rise to the same effect. This is probably because the DNA is not a receptor. The DNA is rather an acceptor for different molecules at different sites. It is worth mentioning that one of the most ambitious, and still today rarely achieved goal in the search for new antitumoral drugs are compounds, which are able to interact in a specific way with DNA sequences. If a QSAR study focuses only on one class of agents, which act via the same mechanism, the QSAR models for drug design are likely to work efficiently also in mutagenicity and carcinogenicity. But in mutagenicity and carcinogenicity there can be several possible objectives for a QSAR study. One important field of interest is mutagen and carcinogen risk assessment. The goal here is the invention of a tool for easily and quickly predicting the toxicological properties of other agents, for which little or nothing is known in terms of their possible risks. This can be useful for making decisions in specific situations, or for setting priorities for further experimentation. In practicing risk assessment, one frequently deals with lists of non congeneric chemicals. These situations are determined by the social and economical history, and cannot be planned in advance as in drug design. For non congeneric chemicals generating the same biological effect -e.g. mutation - there is no common skeleton that remains fixed in the correlation analysis. Therefore, the goal of a QSAR study becomes two-fold; a) to describe the basic determinants of the activity; b) to describe how the effects of these determinants are modulated by the substituents. This usually implies a more complex description of molecules than in drug design, which analyzes only how the activity is modulated by the substituents. In spite of the inherent difficulties, the importance of risk assessment stimulates such studies.

Another problem is that a pharmacological activity concerns only one very specific activity. Actually, carcinogens act via very different mechanisms and produce different spectra of induced tumors; they can be classified all together in the same category only for a very rough approximation. The same is true for mutagens. For instance, Salmonella mutagens can have different strain specifities. Thus, putting together in the same class different carcinogens or mutagens - independently of their mechanism of action - is pretending to identify a unifying activity which actually does not exist.

Other difficulties in QSAR studies can arise because of the transformation of compounds inactive "per se", but which are metabolized into reactive species. Molecules which belong to the same chemical class are not very different from each other to our eye; however, they can be very different when viewed by the cellular machinery. Relatively small chemical differences can lead the compounds to completely different metabolic pathways. Consequently, no common QSAR holds for all molecules in the class. In these cases, a valid QSAR can only be obtained if we already know enough, or we make a good enough hypothesis, about the metabolic fate of the compounds.

This problem is more general and does not apply only to

the metabolism. The biological systems recognize the molecules according to their own rules, and discriminate very accurately between them. This fact is source of complexity when one wants to classify the chemicals, because there is no single definition of a chemical class. We cannot describe the compounds by deriving as many chemical descriptors as possible, and then classifying them in an automatic way. A class of compounds may be such for a certain biological effect, but may not exist for another biological effect. The construction of classes of compounds for SAR studies necessarily implies a knowledge of, or a hypothesis for, which chemical parameters, among the almost infinite possibilities, are relevant for a specific activity.

This problem has, together with theoretical aspects, immediate practical consequences. The definition of a chemical class is important in several instances: a) to design efficient QSAR studies. An application area is risk assessment. In front of so many existing chemicals never tested for toxicity, one strategy is to sample a number of them, in such a way that they are representative of "chemical classes". These representative chemicals should be tested experimentally. The results of the experiments would be used for deriving a QSAR for the entire class of compounds; b) to define the validity limits of established QSAR's; c) to give a more solid ground to the relationships between different biological actions. For example, the concordance between mutagenicity and carcinogenicity varies very much depending on the specific sample of chemicals tested.

In all the cases examined before, a wrong definition of the chemical class undermines the results of the study. An approach exclusively based on chemical data - how sophisticated are chemical parameters and mathematical classification methods used - has not much hope for success, if also biological aspects are not previously introduced into the model. This is probably one of the most exciting areas of research in QSAR studies. Together with the difficulty of using chemical information which consists of many parameters, that may have complex relationships to each other, there is also the problem of treating, in a coordinate way, biological information. Due to its complexity, biological information is rarely easily defined by standard mathematical procedures. This is an area for soft modeling techniques, such as the multivariate data analysis methods and the expert systems of artificial intelligence. These approaches maintain part of the formal rigor of mathematics, but also have much greater flexibility.

CONCLUSIONS AND PERSPECTIVES

It is important to stress that existing QSAR methods are effective in many situations. Several examples of successful studies exist in mutagenicity and carcinogenicity. On the other hand, these methods are not advanced enough to be able to solve all kinds of problems; further investment and study is necessary. As we have seen before, one area of research is the rational definition of chemical classes. A related question is the definition of criteria of chemical similarity. Together with these more basic questions, it is also important to continue the QSAR analysis of groups of congeneric com-

pounds, using and comparing different approaches. The accumulation of knowledge from individual chemical class analyses
can give new insights into mutagenicity and carcinogenicity
phenomena, and can lead to more general and unifying theories.
Furthermore, the QSAR studies have several appealing and
fascinating qualities, for those involved in it. I should cite
the challenge posed to the researcher by the fact that a QSAR
study is not only a piece of knowledge which is added to a
list of other information, but is something whose validity is
immediately controlled. No QSAR can survive the predictivity
test, if it has not a solid ground. Another reason for which
QSAR research is particularly interesting is that it straddles
many different areas: chemistry, biochemistry, statistics,
mathematical modeling, informatics. Several areas of expertise
are required for what is really an interdisciplinary science.

REFERENCES

Benigni, R., Andreoli, C., and Giuliani, A., 1989, Quantita-
        tive structure-activity relationships: principles, and
        applications to mutagenicity and carcinogenicity,
        Mutat. Res., in press.
Franke, R., 1984, "Theoretical Drug Design Methods", Elsevier,
        Amsterdam.
Free, S.M., and Wilson, J.W., 1964, A mathematical contribu-
        tion to structure-activity studies, J. Med. Chem.,
        7:395.
Frierson, M.R., Klopman, G. and Rosenkranz, H.S., 1986,
        Structure-activity relationships (SAR's) among mutagens
        and carcinogens: a review, Environ. Mutagenesis, 8:283.
Hansch, C., 1976, On the structure of medicinal chemistry, J.
        Med. Chem., 19:1
Hansch, C., 1977, On the predictive value of QSAR, in:
        "Biological Activity and Chemical Structure", J.A.
        Keverling Buisman, ed., Elsevier, Amsterdam.
Hansch, C., and Fujita, T., 1964, Analysis. A method for the
        correlation of biological activity and chemical
        structure, J. Am. Chem. Soc., 86:1616.
Jurs, P.C., Hasan, M.N., Henry, D.R., Stouch, T.R. and
        Walen-Pederson, E.K., 1983, Computer-assisted studies
        of molecular structure and carcinogenic activity, Fund.
        Appl. Toxicol., 3:343.
Klopman, G., 1984, Artificial intelligence approach to
        structure-activity studies. Computer automated struc-
        ture evaluation of biological activity of organic
        molecules, J. Am. Chem. Soc., 106:7315.
Lebart, L., Morineau, A. and Warwick, K.M., 1984, "Multivari-
        ate descriptive statistical analysis", Wiley, New York.
Lehninger, A.L., 1981, "Biochemistry: The Molecular Basis of
        Cell Structure and Function", Worth, New York.
Lewi, P.J., 1980, Multivariate data analysis in structure-
        activity relationships, in: "Drug Design", E.J. Ariens
        ed., Vol. X, Academic Press, New York.
Martin, Y.C., 1981, A practicioner's perspective of the role
        of quantitative structure-activity analysis in medici-
        nal chemistry, J. Med. Chem., 24:229.
Topliss, J.G., and Edwards, R.P., 1979, Chance factors in
        studies of quantitative structure-activity relation-
        ships, J. Med. Chem., 22:1238.

ENZYMIC ASPECTS ON THE METABOLIC ACTIVATION OF AROMATIC AND HETEROCYCLIC AMINE MUTAGENS IN MAMMALIAN AND BACTERIAL CELLS

Ryuichi Kato and Yasushi Yamazoe

Department of Pharmacology
School of Medicine
Keio University
Shinjuku-ku, Tokyo, Japan 160

INTRODUCTION

Aromatic amines and heterocyclic amines, which are included in food pyrolysates, show their mutagenicities to Salmonella and mammalian cells in the presence of activating system, S9-mix, added externally (Sugimura and Sato, 1983; Kato, 1986; Kato and Yamazoe, 1987). These aromatic and heterocyclic amines are N-hydroxylated at the amino moiety by cytochrome P-450 species, especially P-448-H (P-450d) type enzymes in the liver microsomes of experimental animals (Sugimura and Sato, 1983; Kato, 1986; Kato and Yamazoe, 1987; Yamazoe et al., 1988). The resulting N-hydroxyaryla-mines are capable of forming covalent bonds with biological macromolecules, such as nucleic acids and proteins, but further metabolic activations are needed to produce the maximal mutagenic activity (Kato and Yamazoe, 1987). These metabolic activations consist of enzymic esterifications, such as amino acid O-acylation, O-acetylation, N,O-acetyltransfer and O-sulfation, which generate reactive ultimate forms. In this article, we describe the enzymic basis of the metabolic activation of aromatic and heterocyclic amine mutagens, N-hydroxylation and esterification, in comparison with mammalian and bacterial cells.

ABBREVIATIONS: Glu-P-1, 2-amino-6-methyldiphenyl-dipyrido [1,2-a: 3',2'-d]imidazole; Trp-P-2, 3-amino-1-methyl-5H-pyrido [4,3-b]indole; IQ, 2-amino-methylimidazo[4,5-f]quinoline; MeIQx, 2-amino-3,8-dimethyl-3H-imidazo[4,5-f]quinoxaline; MC, 3-methylcholanthrene; PCB, polychlorinated biphenyl; AHH, arylhydrocarbon hydroxylase; AF, 2-aminofluorene; AAF, 2-acetylaminofluorene; PABA, p-aminobenzoic acid; AAB, 4-aminoazobenzene.

Mechanisms of Environmental Mutagenesis-Carcinogenesis, Edited by
A. Kappas, Plenum Press, New York, 1990

N-Hydroxylation of Aromatic and Heterocyclic Amines in Mammalian and Bacterial Cells

Since P-448-type cytochrome P-450s, such as P-448-H (P-450d) and P-448-L (P-450c) are responsible for N-hydroxylation of aromatic and heterocyclic amines, cultured mammalian cells show very low activity or deficiency in N-hydroxylations by means of low contents or deficiency in P-448-H and P-448-L in cultured mammalian cells. Moreover, N-hydroxylation and cytochrome P-450 are lacking in <u>Salmonella typhimurium</u>. Among the tissue preparations from experimental animals, the liver shows the highest activity and other tissues show low or very limited activities. Although N-hydroxylation of aromatic and heterocyclic amines in livers of intact experimental animals is not so high, the activity is induced by treatment with methylcholanthrene (MC*[1]) or PCB. However, there are marked species differences in the extent of induction of N-hydroxylation and mutagenic activating capacity (Fig. 1) (Yamazoe et al., 1981; Yamazoe et al., 1988).

P-448-H shows activity over 5 ~ 30 times higher than P-448-L in the formation of N-hydroxyarylamines and mutagenesis in Salmonella test system (Kato et al., 1983). Recently, we have shown that the hepatic content of P-448-H was increased in diabetic rats, and that the mutagenic activation and N-hydroxylation of Glu-P-1 by hepatic microsomes from diabetic rats was increased by 100% in parallel with the increase in P-448-H content (Fig. 2) (Yamazoe et al., 1988). Mutagenic activation of IQ and MeIQx by hepatic microsomes from diabetic rats was also increased by 111 and 133%, respectively. On the other hand, N-hydroxylation and

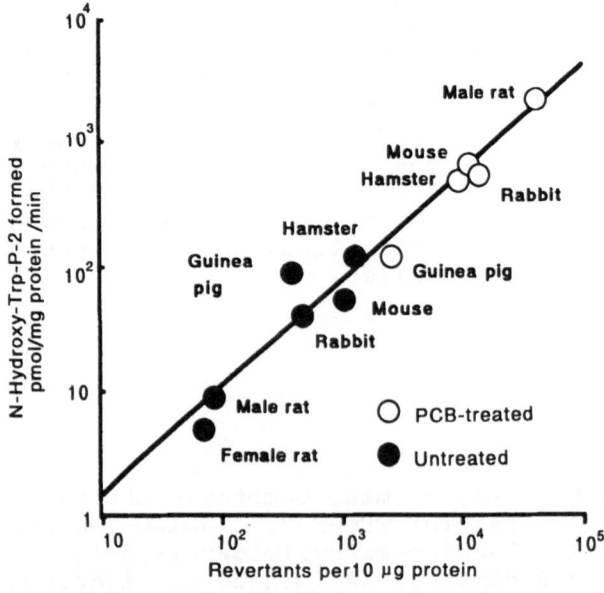

Fig. 1. Relationship between the formation of N-OH-Trp-P-2 and the induction of revertants by liver microsomes from rats, mice, hamsters, guinea pigs, and rabbits.

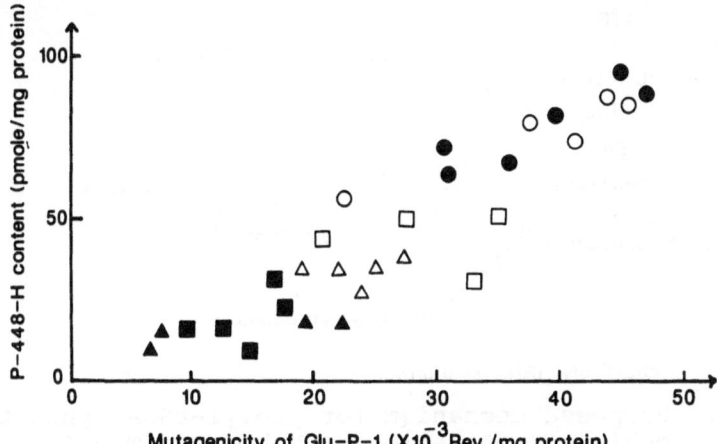

Fig. 2. Relation between P-488-H content and the
mutagenicity of Glu-P-1 in rat hepatic
microsomes. Key: male control (▲), female
control (Δ), male treated with alloxan plus
insulin (■), female treated with alloxan plus
insulin (□), male treated with alloxan alone
(●), and female treated with alloxan alone
(○).

mutagenic activation of these heterocyclic amines by other
forms of cytochrome P-450s, such as phenobarbital-inducible
P-450b and P-450e, and P-450-male and P-450$_{6\beta-1}$, are low
(Sugimura and Sato, 1983; Kato et al., 1983; Yamazoe et al.,
1984).

Moreover, we have shown that the number of revertants
induced by Glu-P-1, IQ and MeIQx were nearly the same or
rather higher in the presence of hepatic microsomes from
humans than those from rats (Yamazoe et al., 1988). In
accordance with these results, the level of Glu-P-1 N-hydroxylation by human liver microsomes was similar to that of rat
liver microsomes. The content of immunoreactive P-448-H in
human liver is as high as in rat liver and is well correlated
with the extent of mutagenic activation of these heterocyclic
amines. Moreover, the anti-P-448-H IgG markedly inhibited the
mutagenic activation of Glu-P-1 without inhibition of AHH
activity. All those results indicate that mutagenic activation of Glu-P-1, IQ and MeIQx by human liver microsomes is
mainly carried out by P-448-H type cytochrome P-450 through
the N-hydroxylation (Yamazoe et al., 1988).

O-Prolylation of N-Hydroxyarylamines in Mammalian and
Bacterial Cells

Using rat hepatic cytosol, we found that the covalent
binding of N-hydroxy-Trp-P-2 to DNA and tRNA is markedly
stimulated in the presence of proline and ATP. The enzyme
responsible for the activation of N-hydroxy-Trp-P-2 was
copurified with the activity of prolyl-tRNA synthetase
(Yamazoe et al., 1982; Yamazoe et al., 1985). The mechanism
of covalent binding of N-hydroxy-Trp-P-2 is considered to

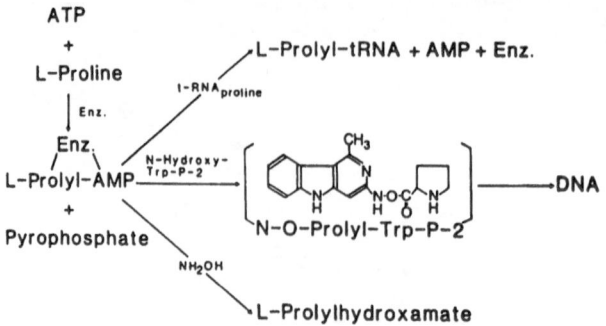

ATP
+
L-Proline

L-Prolyl-tRNA + AMP + Enz.

t-RNA proline

Enz.

Enz.

N-Hydroxy-
Trp-P-2

L-Prolyl-AMP

N-O-Prolyl-Trp-P-2

DNA

+
Pyrophosphate

NH₂OH

L-Prolylhydroxamate

Enz.; Prolyl-tRNA synthetase

Fig. 3.   Proposed mechanism for prolyl-tRNA synthetase-
mediated activation of N-hydroxy-Trp-P-2.
Enz., enzyme.

proceed through the formation of L-prolyl-AMP-enzyme complex
and then this complex reacts with N-hydroxy-Trp-P-2 to form a
reactive intermediate, probably N-O-prolyl-Trp-P-2 (Fig. 3).

In addition, clear species differences were observed in
the activities of hepatic cytosols to catalyze the covalent
binding of N-hydroxy-Trp-P-2 to DNA in the presence of ATP and
proline: Highest in Sprague-Dawley rats, followed by CDF_1
mice, Syrian golden hamsters, New Zealand white rabbits in
that order (Yamazoe et al., 1985). With regard to other
N-hydroxyarylamines, prolyl-tRNA synthetase-dependent covalent
binding in rat liver cytosol is also found with N-hydroxy-IQ,
but it is not detectable with N-hydroxy-AF. On the other
hand, the clear activity of prolyl-tRNA synthetase-dependent
covalent binding is not detected in the Salmonella tester
strain.

O-Acetylation of N-Hydroxyarylamines in Mammalian and Bacterial Cells

Yamazoe et al (1982) firstly found that the covalent
binding of N-hydroxy-Trp-P-2 to DNA by rat liver cytosol is
markedly stimulated in the presence of acetyl CoA. Furthermore, acetyl CoA dependent covalent binding of N-hydroxyarylamines, such as N-hydroxy-Glu-P-1, N-hydroxy-IQ, N-hydroxy-AF
and N-hydroxy-3', 2-dimethyl-4-aminobiphenyl was demonstrated
(Yamazoe et al., 1982; Shinohara et al., 1985; Flammang et
al., 1985; Saito et al., 1986; Shinohara et al., 1986;
Flammang and Kadlubar, 1986). Marked species differences are
observed in the acetyl CoA dependent binding of N-hydroxy-Trp-
P-2, N-hydroxy-Glu-P-1 and N-hydroxy-AF. The rank order of
these species differences is dependent on the substrates
(Shinohara et al., 1986) The tissues other than the liver
showed activity for acetyl-CoA dependent binding to DNA of
N-hydroxyarylamines but there was a marked substrate-dependent
species difference (Shinohara et al., 1986).

Furthermore, we purified N-hydroxyarylamine O-acetyltransferase from hamster livers to electrophoretical homoge-

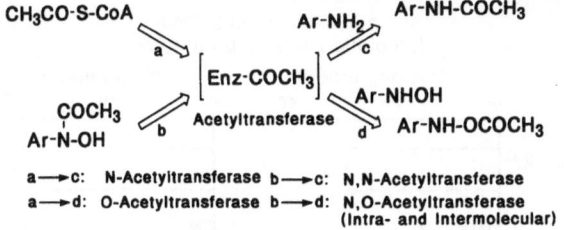

a→c: N-Acetyltransferase   b→c: N,N-Acetyltransferase
a→d: O-Acetyltransferase   b→d: N,O-Acetyltransferase
                              (Intra- and Intermolecular)

Fig. 4. Mechanisms and relationships of N-acetylation,
         O-acetylation, N,N-acetyltransferase and
         N,O-acetyltransfer by acetyltransferase.

neity. The activity of N-hydroxyarylamine O-acetyltranferase
was copurified with those of arylhydroxamic acid N,O-acetyl-
transferase and arylamine N-acetyltransferase. The purified
enzyme was capable of mediating the covalent binding of
various N-hydroxyarylamines to DNA, tRNA and poly(G) (Shinohara et al., 1986). Thus, the mechanism and relationship of
O-acetylation, N-acetylation, N,N-acetyltransfer and
N,O-acetyl transfer by acetyltransferase may be understood by
the scheme presented in Fig. 4.

     Recently, we purified two O-acetyltransferase designated
AT-I (30KD) and AT-II (28KD) from hamster livers of rapid
acetylators to electrophoretical homogeneity (Kato and
Yamazoe, 1987; Kato and Yamazoe, 1989). Using antibodies
against two O-acetyltransferase, we have found that both AT-I
and AT-II are detected in liver and other tissues of the rapid
acetylator, but in the slow acetylator only AT-I is detected
(Kato and Yamazoe, 1989). Moreover, we have found a remarkable acetyl donor and substrate specificity between AT-I and
AT-II (Tables 1 and 2). It is of special interest that AT-I is
capable of using acetyl CoA and arylhydroxamic acid as acetyl
donors, whereas AT-II is not capable of using arylhydroxamic
acid. Moreover, acetyl CoA dependent N-hydroxy-AF O-acetylation is catalyzed by both AT-I and AT-II, but AT-II is not
able to catalyze acetyl CoA dependent N-hydroxy-Glu-P-1
O-acetylation. In contrast, acetyl CoA dependent N-acetylation
of PABA is not catalyzed by AT-I. The inhibition by anti-AT-I
serum of acetylation in cytosol from hamster livers of rapid
acetylator, therefore, shows clear differences depending on
the contribution of AT-I to the total acetylation activity
(Kato and Yamazoe, 1989). AAB N-acetylation and N-hydroxy-
Glu-P-1 O-acetylation are clearly inhibited, whereas PABA
N-acetylation is not inhibited and AF N-acetylation is
partially inhibited.

     The cytosol from S. typhimurium TA98 shows high activity
in the O-acetylations of N-hydroxyarylamines, N-hydroxy-AF,
N-hydroxy-Trp-P-2, N-hydroxy-Glu-P-1 and N-hydroxy-IQ (Fig. 5)
(Saito et al., 1983; Saito et al., 1985; Yamazoe et al.,
1989). It is of special interest that the purified acetyltransferase can catalyse both activities, which were copurified and inseparable (Saito et al., 1985). In accordance with
very low mutagenic activities of N-hydroxy-Glu-P-1, N-hydroxy-
IQ and N-hydroxy-AF in S. typhimurium TA98/1,8-DNP$_6$, a

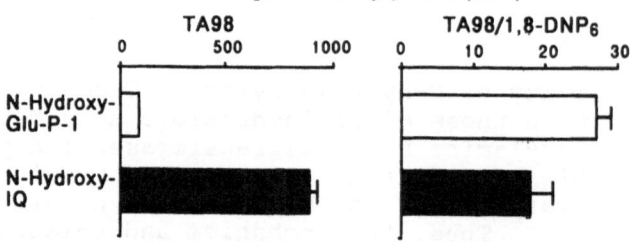

Fig. 5. The rate of the covalent binding of N-hydroxy-
Glu-P-1 and N-hydroxy-IQ to calf thymus DNA
byrat hepatic cytosols and mutagenicities to
Salmonella.

Table 1. Acetyl Donor and Substrate Specificities of AT-I and
AT-II and Acetylation Polymorphism in Hamster Liver

		AT-I	AT-II	Polymorphism
1)	Acetyl CoA-dependent N-OH-Glu-P-1 O-acetylation	++	-	-
2)	Acetyl CoA-dependent N-OH-AF O-acetylation	+	+	+
3)	Acetyl CoA-dependent AF N-acetylation	+	+++	++
4)	Acetyl CoA-dependent PABA N-acetylation	-	+++	+++
5)	N-OH-AAF-dependent PABA N-acetylation	+	-	-
6)	Arylhydroxamic acid dependent AAB N-acetylation	+++	-	-
7)	Arylhydroxamic acid dependent N, O-acetyltransfer	+++	-	-

Table 2. Characteristics of AT-I and AT-II

		AT-I	AT-II
1)	Molecular weight	30KD	28KD
2)	Acetyl CoA-dependent		
	a) N-OH-Glu-P-1 O-acetylation	++	-
	b) PABA N-acetylation	-	+++
3)	Utilization of cofactor		
	a) Butyryl-and hexanoyl CoA	++	-
	b) Arylhydroxamic acid	++	-
	c) p-Nitrophenyl	+	-

Table 3. Acetyl CoA-Dependent DNA Binding of N-Hydroxy-Glu-P-1 and IQ

Enzyme source	Amounts bound to DNA (pmol/mg DNA/mg protein/min)	
	plus cofactor	minus cofactor
N-Hydroxy-Glu-P-1		
Rat hepatic cytosol(male)	28.4±3.8	0.8±0.1
(female)	22.3±5.1	0.8±0.2
TA98 cytosol	14.6±3.1	1.9±0.5
TA98/1.8-DNP$_6$ cytosol	ND[*1]	ND
N-Hydroxy-IQ		
Rat hepatic cytosol(male)	7.3±1.1	1.1±1.0
(female)	4.9±1.8	ND
TA98 cytosol	41.6±6.9	1.5±1.1
TA98/1.8-DNP$_6$ cytosol	ND	ND

[*1]ND, Not detected (less than 0.5 pmol/mg DNA/mg protein/min).

mutant deficient in N-hydroxyarylamine O-acetyltransferase (Saito et al., 1983), acetyl CoA-dependent DNA bindings of these N-hydroxyarylamines by cytosols of this mutant strain are not detected (Table 3) (Saito et al., 1983; Saito et al., 1985; Kato and Yamazoe, 1989; Yamazoe et al., 1989).

O-Sulfation of N-Hydroxyarylamines in Mammalian and Bacterial Cells

N-Hydroxyarylamines are mutagenically activated by a phenol-sulfotransferase in mammalian tissues through O-sulfation (Yamazoe et al., 1987; Yamazoe et al., 1989). The activity of N-hydroxyarylamine O-sulfotransferase was highest

Table 4. O-Acylations of N-Hydroxyarylamines by Rat Liver Cytosol

	O-Prolylation	O-Acetylation	O-Sulfation
N-Hydroxy-Trp-P-2	+++	++	±
N-Hydroxy-Glu-P-1	-	+++	++
N-Hydroxy-IQ	++	+	+
N-Hydroxy-AF	-	++	+

in the livers. However, there are marked species differences in O-sulfation of N-hydroxyarylamines. The O-sulfation of N-hydroxy-Glu-P-1 is highest in rat liver, but relatively low in livers of other animals and humans. In dog liver, no measurable activity was detected. Rank order of the activity is as follows: Guinea pig = human, rabbit = hamster > mouse. On the other hand, the O-sulfation of N-hydroxy-IQ was detected only in rat livers, and no activity was detected in the cytosols from other experimental animals. In rat liver, the activity is higher in male than in female, but the extent of sex difference was dependent on the N-hydroxyarylamine used (Yamazoe et al., 1987; Yamazoe et al., 1989) indicating that multiple forms of sulfotransferases may be involved in O-sulfation of N-hydroxyarylamines. The activity of O-sulfotransferase in rat liver is regulated by the mode of secretion of growth hormone (Yamazoe et al., 1987). On the other hand, no O-sulfotransferase activity is found in bacterial cytosols, even though p-nitrophenol is used as a substrate, indicating that sulfotransferase is deficient in bacteria (Yamazoe et al., 1989).

Finally, the substrate specificity of O-acylation of N-hydroxyarylamines by rat liver cytosol is summarized in Table 4.

Formation of N-Hydroxyarylamines and N-Hydroxyarylamides, their Esterifications and Covalent Binding to DNA in Salmonella Ames Assay System

In Salmonella Ames assay system, mutagenic arylamines are N-hydroxylated by cytochrome P-450 and resulting N-hydroxyarylamines are transported into bacterial cells to cause mutation. Moreover, the esterifications of N-hydroxyarylamines in the Salmonella are considered to stimulate markedly covalent binding to DNA and mutation. However, the esterifications outside bacterial cells usually decrease the extent of mutation owing to formations of reactive intermediates (N-acetoxyarylamines) which react immediately into biological nucleophiles before being transported into bacterial cells (Kato and Yamazoe, 1987). On the other hand, N-hydroxyarylamides, such as N-hydroxy-AAF, undergo N,O-acetyltransfer to form a reactive intermediate, N-acetoxy-AF, in mammalian cells or in S-9 Mix. N-Hydroxy-AAF is relatively stable and transported into bacterial cells where no N,O-acetyltransfer activity is detected. Moreover, in O-acetyltransferase deficient Salmonella TA98/1,8-DNP$_6$, no mutation is observed

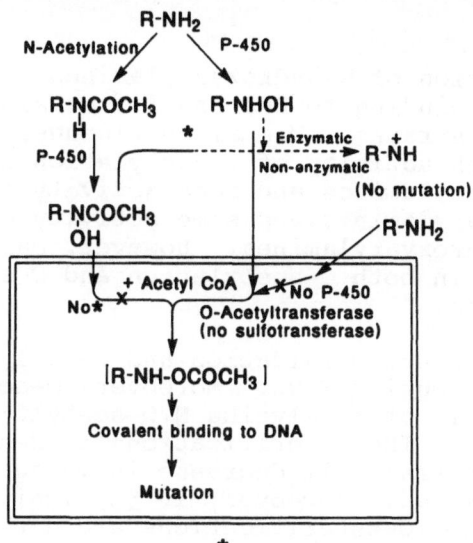

Fig. 6.   The mechanism and factors affecting th
emutagenic activation of mutagenic arylamines
in the Salmonella Ames test system.
*N,O-acetyltransferase

with N-hydroxy-AAF owing to a lack of O-sulfotransferase. The
comparison between mammalian and bacterial cells for metabolisms of mutagenic arylamines, N-hydroxyarylamines and N-hy-
droxy-arylamides is given in Table 5. In conclusion, it is
important to know the activation and inactivation pathways in
mammalian and bacterial cells of environmental mutagens for
the evaluation of their mutagenic potency.

Table 5.   Comparison between Bacterial and Mammalian
Cells for Metabolisms of Mutagenic Arylamines,
N-Hydroxyarylamine and N-Hydroxyarylamides

	Mammalian liver	Bacteria	
		TA98	TA98/1,8-DNP$_6$
Arylamine			
N-Oxidation	+ ~ +++	−	−
N-Acetylation	+ ~ +++	+++	−
N-Hydroxyarylamine			
O-Prolylation	+ ~ ++	−	−
O-Acetylation	++ ~ +++	++ ~ +++	−
O-Sulfation	+ ~ ++	−	−
N-Hydroxyarylamide			
N,O-Acetyltransfer	++ ~ +++	−	−
O-Sulfation	+ ~ ++	−	−

SUMMARY

O-Esterification of N-hydroxyarylamines is an important step for covalent binding to DNA and for causing mutation of Salmonella tester strains. Mammalian tissues, especially livers, have a high activity of O-acetylation and O-sulfation for all N-hydroxyarylamines and some activity of O-sulfation for all N-hydroxyarylamines and some activity of O-prolylation for specific N-hydroxyarylamines. However, marked species differences exist in both O-acetylation and O-sulfation of N-hydroxyarylamines.

On the other hand, O-sulfation and O-prolylation are lacking in S. typhimurium TA98. Moreover, O-acetyltrasferase in TA98 is incapable of catalyzing N,O-acetyltransfer of N-hydroxyarylamide. The esterifications of N-hydroxyarylamines outside bacterial cell decrease in mutagenic activities of N-hydroxyarylamines. Moreover, in S. typhimurium TA98/ 1,8-DNP$_6$ these three O-esterifications are not detectable. The deficiency in O-acetylation is a factor responsible for the low mutagenic activity of arylamine in TA98/1,8-DNP$_6$ in Ames test.

REFERENCES

Flammang, T.J., and Kadlubar, F.F., 1986, Acetyl coenzyme A-dependent metabolic activation of N-hydroxy-3', 2'-dimethyl-4-aminobiphenyl and several carcinogenic N-hydroxyarylamines in relation to tissue and species differences, other acyl donors, and arylhyroxyamic acid-dependent acyltransferases, Carcinogenesis 7:919.

Flammang, T.J., Westra, J.G., Kadlubar, F.F., and Beland, F.A., 1985, DNA adducts formed from the probable proximate carcinogen, N-hydroxy-3', 2'-dimethyl-4-aminobiphenyl, by acid catalysis or S-acetyl coenzyme A-dependent enzymatic esterification, Carcinogenesis, 6:251.

Kato, R., 1986, Metabolic activation of mutagenic heterocyclic aromatic amines from protein pyrolysates, CRC Crit. Rev. Toxicol., 16:307.

Kato, R., Kamataki, T., and Yamazoe, Y., 1983, N-Hydroxyulation of carcinogenic and mutagenic aromatic amines, Environ. Health Perspect., 49:21.

Kato, R., and Yamazoe, Y., 1987, Metabolic activation and covalent binding to nucleic acids of carcinogenic heterocyclic amines from cooked foods and amino acid pyrolysates, Gann 78:297.

Kato, R., and Yamazoe, Y., 1987, N-hydroxyarylamine O-acetyltransferase in mammalian livers and Salmonella, in: "Carcinogenic and Mutagenic Responses to Aromatic Amines and Nitroarenes", C.M. King, L.J. Romano, and D. Schuetzle, ed., Elsevier Science Publishing Co., Amsterdam.

Kato, R., and Yamazoe, Y., 1989, Further metabolic activations of mutagenic and carcinogenic N-hydroxyarylamine by conjugating enzymes, in: "Xenobiotic Metabolism and Disposition", R. Kato, R.W. Estabrook and M.N. Cayen, ed., Taylor and Francis Ltd., London.

Saito, K., Shinohara, A., Kamataki, T., and Kato, R., 1985, Metabolic activation of mutagenic N-hydroxyarylamines by O-acetyltranferase in <u>Salmonella typhimurium</u> TA98, <u>Arch. Biochem. Biophys</u>., 239:286.

Saito, K., Shinohara, A., Kamataki, T., and Kato, R., 1986, N-Hydroxyarylamine O-acetyltransferase in hamster liver: identity with arylhydrozamic acid N,O-acetyltransferase and arylamine N-acetyltranferase, <u>J. Biochem</u>., 99:1689.

Saito, K., Yamazoe, Y., Kamataki, T., and Kato, R., 1983, Mechanism of activation of proximate mutagens in Ames'tester strains: the acetyl-CoA dependent enzyme in <u>Salmonella typhimurium</u> TA98 deficient in TA98/1, 8-DNP$_6$ catalyzes DNA-binding as the cause of mutagenicity, <u>Biochem. Biophys. Res. Commun</u>., 116:141.

Shinohara, A., Saito, K., Yamazoe, Y., Kamataki T., and Kato, R., 1985, DNA binding of N-hydroxy-Trp-P-2 and N-hydroxy-Glu-P-1 by acetyl-CoA dependent enzyme in mammalian liver cytosol, <u>Carcinogenesis</u>, 6:305.

Shinohara, A., Saito, K., Yamazoe, Y., Kamataki, T., and Kato, R., 1986, Acetyl coenzyme A-dependent activation of N-hydroxy derivatives of carcinogenic arylamines: mechanism of activation, species difference, tissue distribution, and acetyl donor specificity, <u>Cancer Res</u>., 45:2495.

Sugimura, T. and Sato, S., 1983, Mutagens-carcinogens in foods, <u>CRC Crit. Rev. Toxicol</u>., 8:189.

Yamazoe, Y., Abu-Zeid, M., Yamauchi, K., and Kato, R., 1988, Metabolic activation of pyrolasate arylamines by human liver microsomes: possible involvement of a P-448-H type cytochrome P-450, <u>Gann</u>, 79:1159.

Yamazoe, Y., Abu-Zeid, M., Yamauchi, K., Murayama, N., Shimada, M., and Kato, R., 1988, Enhancement by alloxaninduced diabetes of the rate of metabolic activation of three pyrolysate carcinogens via increase in the P-448-H content in rat liver, <u>Biochem. Pharmacol</u>., 37:2503.

Yamazoe, Y., Abu-Zeid, M., Gong, D., Staiano, N., and Kato, R., 1989, Enzymatic acetylation and sulfation of N-hydroxyarylamines in bacteria and rat livers, <u>Carcinogenesis</u>, 10:1675.

Yamazoe, Y., Kamataki, T., and Kato, R., 1981, Species difference in N-hydroxylation of a tryptophan pyrolysis product in relation to mutagenic activation, <u>Cancer Res</u>., 41:4518.

Yamazoe, Y., Manabe, S., Murayama, N., and Kato, R., 1987, Regulation of hepatic sulfotranferase catalyzing the activation of N-hydroxyarylamide and N-hydroxyarylamine by growth hormone, <u>Mol. Pharmacol</u>., 32:536.

Yamazoe, Y., Shimada, M., Kamataki, T., and Kato, R., 1982, Covalent binding of N-hydroxy-Trp-P-2 to DNA by cytosolic proline-dependent system, <u>Biochem. Biophys. Res. Commun</u>., 107:165.

Yamazoe, Y., Shimada, M., Maeda, K., Kamataki, T., and Kato, R., 1984, Specificity of four forms of cytochrome P-450 in the metabolic activation of several aromatic amines and enzo[a]pyrene, <u>Xenobiotica</u> 14:545.

Yamazoe, Y., Shimada, M., Shinohara, A., Saito, K., Kamataki, T., and Kato, R., 1985, Catalysis of the covalent binding of 3-hydroxyamino-1-methyl-5H-pyrido[4,3-b] indole to DNA by a L-proline- and adenosine triphosphate-dependent enzyme in rat hepatic cytosol, <u>Cancer Res</u>., 45:2495.

BIOMONITORING AND EPIDEMIOLOGY OF HUMANS EXPOSED TO ENVIRONMENTAL MUTAGENS-CARCINOGENS

CURRENT TECHNIQUES FOR HUMAN POPULATION MONITORING FOR GENETIC EFFECTS

A.T. Natarajan and A.D. Tates

Dept. Radiation Genetics and Chemical
Mutagenesis, Sylvius Laboratory
Medical Faculty, State University of Leiden
Wassenaarseweg 72, AL 2333 Leiden
The Netherlands

Exposure of human populations to genotoxic agents can lead to genetic effects which can be monitored in somatic cells. Lymphocytes and erythrocytes from human peripheral blood are the target cells which can be easily obtained and studied. The consequences of exposure of these cells during circulation as well as effects to their stem cells in the bone marrow can be estimated by either taking blood samples shortly or longer times after the event. The end points that are currently studied are (1) chromosomal aberrations and micronuclei in lymphocytes, (2) Sister chromatid exchanges in lymphocytes (3) aneuploidy in lymphocytes, (4) HPRT-mutations in lymphocytes, (5) glycophorin mutations in erythrocytes and (6) haemoglobin mutations in erythrocytes. The current techniques to estimate these end points will be described briefly and the relative merits of these techniques will be discussed.

The sensitivity of a given technique to detect an effect due to exposure to genotoxic agent(s) will depend on the target size as well as the spontaneous frequency of the end point in question. For example, when measuring chromosomal aberrations or micronuclei, which occur with a frequency in the order of 0.5 to 4%, one is dealing with a target which involves the whole genome. HPRT-mutations which include single base pair changes as well as gross rearrangements and deletions, represent events that occur in a single gene comprising of 41.5 kilo bases. The spontaneous frequency of such events is about 1 to 5 mutants per million clonable or stimulatable T-lymphocytes. Glycophorin mutations which, like HPRT mutations, represent loss of function of the gene, also occur at a frequency of 1 to 5 per million erythrocytes. Haemoglobin mutations which involves single base pair changes in a single codon, occur at a frequency of 1 in 2-3 hundred million erythrocytes. Thus, the smaller the size of the target, the rarer the occurrence of changes. Therefore techniques to detect an increase in these rare events should aim at screening large numbers of cells with ease. Consequently automation has played an important role in achieving this aim and some of the techniques developed in our laboratory in this direction will be presented in this paper.

CHROMOSOMAL ABERRATIONS

The types and frequencies of induced chromosomal aberrations depend on the type of genotoxic exposure and the cell cycle stage at the time of exposure. In peripheral blood lymphocytes, most of which are long lived, ionizing radiation induces directly DNA strand breaks some of which are converted into chromosome type aberrations, namely dicentrics, reciprocal translocations, rings and acentric fragments which show up in metaphases when the lymphocytes are stimulated to divide in vitro. The frequency of chromosomal aberrations is dose dependent and therefore can be used to estimate absorbed radiation dose in case of accidents (IAEA Technical Report, 1986). This technique can also be used to discriminate cases such as partial body irradiation, mixture of high and low LET radiations, chronic versus acute exposure etc. This method was proved very useful in a recent radiation accident in Goiania, Brazil for estimating the absorbed radiation doses in the victims (Ramalho et al., 1989). The frequency of radiation induced aberrations due to external exposure diminishes with time whereas, the frequency due to internal contamination increase initially and then decline (Ramalho et al., 1990). Due to the extremely long life span of a subpopulation of lymphocytes, unstable chromosomal aberrations induced in lymphocytes of atom bomb survivors of Hiroshima and Nagasaki can be detected even after 45 years (Awa, 1983).

However, exposure to most of the chemical mutagens induces lesions in the DNA (covalent adducts, cross links etc.) which can undergo cellular repair. The non-repaired lesions give rise to chromatid type of aberrations when lymphocytes are stimulated in vitro. (Natarajan and Obe, 1980, Carrano and Natarajan, 1988). An increase in the frequency of chromosomal aberrations has been found following occupational exposure to carcinogens (such as ethylene oxide, vinyl chloride) as well as clinical treatment with cytostatics (Natarajan and Obe, 1980). There are several confounding factors which can interfere with the effective usage of this technique and these have been discussed elsewhere (Carrano and Natarajan, 1988).

Recently improved staining methods have become available which can increase the resolving power of conventional chromosomal aberration detection techniques. Chromosome specific libraries have been generated which can be used for specifically staining individual human chromosomes by in situ hybridization. When one chromosome pair is exclusively stained, it is easy to detect translocations involving this chromosome accurately (Pinkel et al., 1986). Using this "chromosome painting" technique, translocations involving chromosome 2 in lymphocytes of individuals involved in the radiation accident in Goiania, Brazil have been accurately detected (Vyas and Natarajan, 1990).

226

MICRONUCLEI

Micronuclei are formed by chromosomal fragments or lagging chromosomes in anaphase which are not included in the nuclei of the daughter cells. In contrast to chromosomal aberrations, micronuclei can be detected easily and quickly. However, to quantify and make comparative evaluations bet between individuals or populations, it is necessary to identify the lymphocytes which have undergone a division, because for the formation of micronuclei a cell division is necessary. This identification of divided cells has become possible either by incorporating bromodeoxyuridine in prolif-erating cells and differential staining (Pincu et al, 1984) or by growing the lymphocytes in the presence of cytochalasin B which arrests cytokinesis thereby leading to binucleated cells (Fenech and Morley, 1985). The latter method is preferred in view of the ease with which the cells once divided can be recognized and its sensitivity appears to be high (Ramalho et al., 1988). We have studied, both prospectively and retros-pectively, the frequency of micronuclei in lymphocytes of cancer patients receiving chemotherapy. In a cohort of testis carcinoma patients treated with bleomycin in combination with other cytostatic drugs, an increase in the frequency of micronuclei could be detected even eight years following therapy (Osanto et al., 1990). Similarly in a prospective study with different types of cancer patients, exposure to different types of cytostatic drugs during therapy with Cis-platina, cyclophosphamide etc. also resulted in an increase in the frequency of micronuclei (Tates et al., 1990).

Attempts have been made to automate scoring of micronu-clei in binucleated lymphocytes (Fenech et al., 1988). Collaborative studies in our department and the Department of Cytometry and Cytochemistry have demonstrated the feasibility of automated detection of micronuclei in binucleated lympho-cytes. In these studies, use was made of two different image analysis systems (the expensive and sophisticated LEYTAS-MIAC system and a cheaper AT-PC based system). The specific merits of these systems are being evaluated at present (Tates et al., 1990). In view of the fact that more cells (in thousands) can be scored for the presence of micronuclei in comparison to the scoring of chromosomal aberrations (in hundreds), it is likely, that in the future automated detection of micronuclei may replace manual scoring for chromosomal aberrations in population monitoring studies.

SISTER CHROMATID EXCHANGES

Sister chromatid exchanges (SCEs) are cytological manifestation of DNA double strand breakage and rejoining at homologous sites between two chromatids of a chromosome. These exchanges can be detected easily by growing lymphocytes in a medium containing tritiated thymidine or 5 bromodeoxyuridine for two cell cycles followed by autoradiography or differen-tial staining respectively. This technique has been exten-sively used in human population monitoring. In view of the confounding factors which can lead to a small increase in the frequencies of SCEs, it is difficult to interpret the results obtained in population monitoring studies, using this tech-nique. The advantages and disadvantages of using SCEs for monitoring of human populations have been discussed in detail elsewhere (Carrano and Natarajan, 1988).

HPRT - MUTATIONS IN LYMPHOCYTES

Mutants at the HPRT (hypoxanthine-guanine phosphoribosyl-transferase) locus can be easily detected in human lympho-cytes. Lymphocytes with a functional HPRT gene will phospho-rylate toxic purine analogues such as 6 thioguanine or 8 azaguanine and will incorporate these analogues in their DNA and RNA leading to their death. Mutants which have lost a functional HPRT gene will grow in medium containing 6-thiogua-nine or 8-azaguanine. Strauss and Albertini (1979) used tritiated thymidine in the culture medium to detect HPRT-mutants in human lymphocytes. In two parallel cultures, one containing tritiated thymidine and 6 thioguanine and the other containing only tritiated thymidine (control), human lympho-cytes are grown, fixed and processed for autoradiography. The frequency of labelled nuclei in the control culture will give the value of the labelling index. Few labelled cells growing in the medium containing 6 thioguanine represent the HPRT-mutants (variants). Variant frequency can be estimated easily using the values obtained from the two cultures. This technique is fast and has proved to be efficient in detecting mutants in people exposed to radiation and cytostatics. It is also possible to replace tritiated thymidine with bromodeoxy-uridine in the culture medium and the cells incorporating this analogue (mutant cells) can be detected by staining with Hoechst 33258, instead of autoradiography.

It is now feasible to clone T- and B- lymphocytes in vitro by adding growth factors such as interleukin 2 to the medium. The lymphocytes are grown in microwells over a heavily irradiated feeder layer of appropriate HPRT-human lymphoblastoid cells. To determine the cloning efficiency of lymphocytes of an individual, one, two or more cells are seeded per well and for selecting mutants in the presence of 6-thioguanine, 20,000 cells are seeded per well. Non-mutant clones in cloning efficiency plates and mutant clones in selection plates are usually large in size and therefore can be easily detected visually under an inverted microscope. Using this technique, increase in the frequency of mutations in the lymphocytes of cancer patients treated with cyclophos-phamide, isophosphamide, adriamcycin etc., (Tates et al., 1990).

It has been argued that the use of the HPRT mutation assay (and similar types of mutation detection assays as well) for monitoring human population exposed to potential or proven mutagens is likely to be disappointing as the exposure of each individual to the mutagen is likely to be small compared to the exposure levels involved in the therapy of cancer patients with cytostatic drugs or in radiation accidents (Morley, 1990). Whether this scepticism is justified remains to be tested.

Although the cloning technique is rather expensive and time consuming, it has the added advantage that the mutants can be mutliplied and characterized with regard to the exact nature of the mutation. The HPRT gene is 41.5 kb long comprising of 9 exons. A procedure to sequence the coding region of the gene using mRNA (1.6 kb) has recently been developed (Vrieling et al., 1988). The mRNA is used to make cDNA by means of reverse transcriptase and appropriate primers. The cDNA is further amplified by using appropriate primers and the polymerse chain reaction (PCR) with TAQ polymerase. The DNA is then cloned in M13 vector and sequenced to detect exactly the nature of the change at the DNA level leading to HPRT mutations (Rossi et al., 1990). About 30 spontaneously occurring mutations have been sequenced and the results indicate that they occur at all regions of the gene and comprise of single base pair changes (transitions, transversions), frame shifts, small deletions and large deletions with one or two exons missing (splice mutations). This technique can also be used effectively to check whether an exposure to a particular chemical exposure results in significant changes in the mutation spectrum and whether perhaps, chemical specific mutational hot spots in the gene can be identified.

ERYTHROCYTES

Mutations of the Glycophorin A locus

This assay detects loss of gene expression at the glycophorin A (GPA) locus. This cell surface glycoprotein occurs in two allelic forms, namely M and N and is codominantly expressed. Monoclonal antibodies for individual allelic forms are conjugated with a different fluorescent dye and used to label fixed erythrocytes from heterozygous MN donors. Flow cytometry and sorting are used to estimate the frequency of cells that lack the expression of one of the GPA alleles. Using this system, an increase in the frequency of GPA variants has been found in chemotherapeutically treated cancer patients, atom bomb survivors etc., (Langlois et al., 1987). One of the short comings of the technique is that only MN heterozygotes (about 50% of the population) can be screened. A further disadvantage is that the mutants cannot be characterized at the DNA level.

HAEMOGLOBIN MUTATIONS

There are many haemoglobin mutations such as sickle cell anaemia, which occur in nature. These mutations involve changes in single base pair of a codon (transition, transversion or deletion) in the alpha or beta chain of the haemoglobin molecule. In erythroid precursor cells of normal individuals, these mutants occur very rarely and they are expressed in the matured erythrocytes. Monospecific polyclonal antibodies have been raised against three mutations, namely, haemoglobins, haemoglobin San Jose and haemoglobin Leiden (Tates et al., 1989, Bernini et al., 1990). Since millions of cells have to be screened to detect these rare events an automated image analysis system has been developed (Verwoerd, et al., 1987). This system uses large sized microscopic slides (8 x 8 cm) on which purified erythrocytes are attached as a monolayer, automated stage movement and focusing, and two television cameras. There are two light sources and two sets of filters to detect haemoglobin absorption and to detect fluorescence of the mutated erythrocytes. The equipment scans through the slide, counts the number of erythrocytes, detects the fluorescent erythrocytes and stores them in the memory. The spontaneous occurrence of these mutations is about 8 in 100 million cells. If all the three antibodies are used, the frequency of mutants increases by a factor of three (Bernini et al.,1990). Using this technique, a big increase of the frequency of haemoglobin mutations has been found in the radiation accident victims from Goiania, Brazil (Natarajan et al., 1990). The sensitivity of this system can be increased if more antibodies for other haemoglobin mutations can be generated and used for screening.

CONCLUSIONS

At present several techniques are available for monitoring chromosomal changes and mutations in blood cells of human populations. Their sensitivites vary. Though an increase in the frequency of any of the biological end points discussed above indicate an exposure to a potential mutagenic/ carcinogenic agent, it cannot predict any specific health risk to the exposed human population. More quantitative studies involving animal models using different short term end points and long term end point (occurrence of tumours) are needed to make relevant correlations and extrapolations.

ACKNOWLEDGEMENTS

These studies were financially supported in part, by an EEC Environment Programme grant (contract 0187) and by Shell International, The Hague.

REFERENCES

Awa, A.A., 1983, Chromosome damage in atomic bomb survivors and their offspring-Hiroshima and Nagasaki, in: "Radiation Induced Chromosome Damage in Man", T. Ishihara and M.S. Sasaki, eds., Alan R. Liss Inc., New York.

Bernini, L.F., Natarajan, A.T. Shreuder-Rotteveel, Giardono, P.G., Ploem, J.S. and Tates, A.D., 1990, Assay for somatic mutation of human hemoglobin, in: Proceedings of ICEM-5, Alan R. Liss. Inc., New York, in press.

Carrano, A.V. and Natarajan, A.T., 1988, Considerations for population monitoring using cytogenetic techniques. ICPEMC Publication no. 14, Mutation Res., 204:379.

Fenech, M., Jarvis, L.R. and Morley, 1988, Preliminary studies on scoring micronuclei by computerised image analysis, Mutation Res., 203:33.

Fenech, M. and Morley, A.A., 1985, Measurement of micronuclei in lymphocytes, Mutation Res., 147:29.

International Atomic Energy Agency; 1986, Biological Dosimetry: Chromosome Aberration Analysis for Dose Assessment, IAEA Technical Reports Series, 260:69.

Langlois, R.G., Bigbee, W.L., Kyoizumi, S., Nakamura, N.,
     Bean, M.A., Akiyama, M. and Jensen, R.H., 1987,
     Evidence for increased somatic cell mutations at the
     Glycophorin A locus in atomic bomb survivors, <u>Science</u>,
     236:445.
Morley, A.A., 1990, Human somatic mutation - where are we
     going? <u>in</u>: Proc. ICEM-5, Alan R. Liss Inc. New York,
     in press.
Natarajan, A.T., and Obe, G. 1980, Screening human popula-
     tions for mutations induced by environmental pollu-
     tants; use of human lymphocyte system, <u>Ecotoxicol.
     Enivron. Safety</u>, 4:468.
Natarajan, A.T., Vyas, R., Ramalho, A. (1990), in preparation.
Osanto, S., Tates, A.D., Thijssen, J.C.P., Woldering, V.M. van
     Rijn, J.L.S., and Natarajan, A.T., 1990, Increased
     frequency of chromosomal damage in peripheral blood
     lymphocytes up to nine years following curative
     chemotherapy of patients with testicular carcinoma,
     <u>Cancer Res.</u> (submitted).
Pincu, M., Bass, D. and Norman, A., 1984, An improved
     micronucleus assay in lymphocytes, <u>Mutation Res.</u>,
     139:61.
Pinkel, D., Straume, T. and Gray, J.W., 1986, Quantitative
     analysis using high sensitivity fluorescence hybridiza-
     tion, <u>Proc. Natl. Acad. Sci. U.S.A.</u>, 83:2934.
Ramalho, A.T., Nascimento, A.C.H., and Natarajan, A.T., 1989,
     Dose assessments by cytogenetic analysis in the Goiania
     (Brazil) radiation accident, <u>Radiation Protection
     Dosimetry</u>, 25:97.
Ramalho, A.T., Nasimento, A.C.H. and Bellido, P., 1990, Dose
     estimates and the fate of chromosomal aberrations in
     Cesium-137 exposed individuals in the Goiania radiation
     accident, <u>in</u>: "Chromosome Aberrations - Basic and
     Applied Aspects", G. Obe, and A.T. Natarajan, eds.,
     Springer Verlag, Heidelberg (in press).
Ramalho, A.T., Sunjevaric, I. and Natarajan, A.T., 1988, Use
     of the frequencies of micronuclei as quantitative
     indicators of x-ray induced chromosomal aberrations in
     human peripheral blood lymphocytes: Comparison of two
     methods, <u>Mutation Res.</u>, 207:141.
Rossi, A. et al., 1990, in preparation.
Strauss, G.H. and Albertini, R.J. 1979, Enumeration of
     6-thioguanine resistant peripheral blood lymphocytes in
     man as a potential test for somatic mutations arising
     <u>in vivo</u>, <u>Mutation Res.</u>, 61:353.
Tates, A.D., Bernini, L.F., Natarajan, A.T., Ploem, J.S.,
     Verwoerd, N.P., Cole, J., Green, M.L.A., Arlett, C.F.
     and Norris, P.N., 1989, Detection of somatic mutation
     in man: HPRT mutations in lymphocytes and hemoglobin
     mutations in erythrocytes, <u>Mutation Res.</u>, 213:73.
Tates, A.D., Poll van de, M.L.M., Wells van M. and Ploem,
     S.J., 1990, Use of <u>in vivo</u> micronucleus tests with
     mammalian cells for clastogenicity and carcinogenicity
     studies, <u>in</u>: "Chromosome Aberrations - Basic and
     Applied Aspects", G. Obe, and A.T. Natarajan, eds.,
     Springer Verlag, Heidelberg, in press.
Tates, A.D., et al., 1990, (in preparation).

Verwoerd, N.P., Bernini, L.F., Bonnet, J., Tanke, H.J., Natarajan, A.T., Tates, A.D., Sobels, F.H., and Ploem, J.S., 1987, Somatic cell mutations in humans detected by image analysis of immunofluorescently stained erythrocytes, in: "Clinical Cytometry and Histometry", G. Burger, J.S. Ploem, and K. Gorttler, eds., Acad. Press, New York.

Vrieling, H., Simmons, J.W.I.M., and van Zeeland, A.A., 1988, Nucleotide sequence determination of point mutations at the mouse HPRT locus using in vitro amplification of HPRT mRNA sequences, Mutation Res., 198:107.

Vyas, R.C. and Natarajan, A.T., 1990, Follow up studies of the victims of radiation accident at Goiania (Brazil). Detection of translocations using chromosome specific probes. (in preparation).

USE OF AQUATIC ANIMALS FOR MONITORING GENOTOXICITY IN

UNCONCENTRATED WATER SAMPLES

Catherine Zoll, Vincent Ferrier
and Laury Gauthier

Centre de Biologie du Developement
Universite Paul-Sabatier
118, Route de Narbonne
31062 Toulouse Cedex - FRANCE

INTRODUCTION

The increasing quantity of pollutants in water is getting more and more concerning. Their origins are divers: industry, agriculture, domestic activities, but their ultimate destination is aquatic environment. The general toxicity of many of these pollutants, being usually spectacular and immediate, is easily detected. On the other hand, because the risk related to genotoxicity is less apparent and often delayed, it is consequently more insidious. The presence of genotoxins, even in low doses, does concern non aquatic species by the mean of the food chain and drinking water as well as aquatic ones. Therefore, their detection becomes an issue of prime importance for public health.

Several genotoxicity tests, which can be applied to hydric pollutants, are now available. The most commonly used ones are the _in vitro_ tests. They are executed on prokaryotes or on cell cultures. The Ames test, using the bacteria _Salmonella typhimurium_, is by far the mostly known one, even for the studies of hydric pollutants (Ames et al, 1975; Dutka et al, 1981; Kool et al, 1981; Maruoka and Yamanaka, 1982). Cheap and rapidly performed, its use is subject to some drawbacks. In the first place,its sensibility forbids the test of raw waters and requires the recourse to concentration methods or even to extraction ones. These techniques can provoke either quantitative or qualitative modifications of the genotoxicity of the tested waters. Secondly the Ames test, as well as the other _in vitro_ tests, does not take into account neither the integrated physiological functioning of the organism, nor its interactions with the environment. As a matter of fact, the phenomena of metabolic activation, of detoxification and of excretion are essential data of the animal response to xenobiotics. The objections raised above are removed when the _in vivo_ tests involving organisms evolving in raw waters are being used.

Mechanisms of Environmental Mutagenesis-Carcinogenesis, Edited by
A. Kappas, Plenum Press, New York, 1990

In this paper, we present the tests using lower aquatic vertebrates (fish and amphibians): chromosome aberrations tests, sister-chromatid exchanges (SCE) tests and micronucleus tests.

CHROMOSOME ABERRATIONS TESTS

Used mostly in fish, these tests rely on the existence of chromosome anomalies (usually breaks) induced by some clastogenic substances. The anomalies are observed in dividing cells (metaphase figures). The level of abnormal metaphases which are present in treated animals is compared with the one in controls. This kind of test can only be performed on a few species having a small number of large metacentric or submetacentric chromosomes with centromeric regions well defined.

Since 1975, Kligerman et al, working on _Umbra limi_ (2n = 22), observed that subjects which have been submitted to X-rays possess a rate of abnormal metaphases three hundred times higher than those which were not. Therefore this species seems suited to the study of chromosome aberrations induced by physical or chemical agents. Due to the increasing pollution of the Rhine, Prein et al (1978) and Hooftman and Vink (1981) have been testing the clastogenic capacity of this water on the species _Umbra pygmaea_. After eleven days of exposure to the Rhine water, fish present chromosome breaks in 30% of the metaphases, compared to 8% on reference animals reared in a non polluted medium. A relation between the time of exposure to polluted water and the rate of chromosome aberrations has also been shown. However, _Umbra pygmaea_ can only be found in a geographically limited area. It is becoming scarce, and it is not easy to rear in the laboratory. Its mitotic index is rather low, and it takes time to select enough metaphases for statistically significant results. Hooftman (1981) has proposed the use of the tropical fish _Nothobranchius rachowi_. This species has 16 large chromosomes, and it has a higher mitotic index than that of _Umbra pygmaea_. It is readily reared in the laboratory. Hooftman has shown that chromosome aberrations are as easily induced in this species as in _Umbra pygmaea_ after exposure to known mutagens such as benzo(a)pyrene (B(a)P) or ethyl methanesulfonate (EMS).

Yet, the study of chromosome aberrations cannot be sufficient to detect the presence of mutagenic agents. Many products can cause important genetic damages without producing any visible structural aberration (Kligerman 1982). As far as studies and detection of mutagens in waters are concerned, the use of SCE is preferred to that of chromosome aberrations, because it is known to be more sensible (Perry and Evans 1975).

SISTER-CHROMATID-EXCHANGES TESTS

This techique is based on the differential staining of the two chromatids of the same chromosome.

5-bromodeoxyuridine (BrdU), an analogue of thymidine, is readily incorporated into DNA during replication. If carried

out over two cell cycles it leads to an unequal distribution
of this base in the sister chromatids. After bringing
together the cells and the BrdU for a single replication
cycle, only one strand of DNA in each daughter chromatid is
substituted. After the next replication cycle, one chromatid
contains a single substituted strand while both strands of the
sister chromatid are substituted. The substituted and
unsubstituted chromatids can be distinguished by suitable
stains. Sister chromatid exchanges due to the interchange of
replicating DNA between chromatids at apparently homologous
loci can thus be determined. After SCE there is an alterna-
tion of doubly substituted and singly substituted fragments.
The level of SCE is proportional to the mutagenic effect of
the substance to which the cells are exposed.

The SCE system was first used _in vitro_ on cell cultures
(Barker and Rackham, 1979; Quin et al, 1985).

Kligerman and Bloom (1976) and Kligerman (1977) were the
first to apply _in vivo_ the SCE analysis on an aquatic animal:
the fish _Umbra limi_. They observed a singificant increase of
the SCE level on that species after the treatment by injection
of a direct mutagen, the methyl-methanesulfonate (MMS), and of
an indirect one, the cyclophosphamide (CP); they also obtained
a positive result with subjects reared in water supplemented
with neutral red dye.

Alink et al (1980), encouraged by Kligerman's results,
exposed some subjects of the _Umbra limi_ species to Rhine water
continuously renewed. The rate of their SCE, compared to
those of the reference animals reared in non polluted water,
doubled and tripled in three and eleven days. Hooftman and
Vink (1981) obtained similar results on the same species
reared under the same conditions.

The tropical species _Nothobranchius rachowi_ is also used
for the SCE tests. Van der Hoeven et al (1982) carried out a
comparative study with the two species _Umbra pygmaea_ and
Nothobranchius rachowi, by using EMS, a direct mutagen. The
same sensitivity was established for both species. _Nothobran-
chius rachowi_ also gave significant results with the CP. For
some reasons (high mitotic activity, fast and easy reproduction in laboratory), _Nothobranchius rachowi_ constitutes a more
favourable material than _Umbra sp_ for the SCE tests. Its
adequacy to _in situ_ studies is confirmed by Alink's team
(1982): the subjects exposed to the Rhine water during seven
days presented a rate of SCE almost as high as those treated
with EMS.

One of the great difficulties related to the SCE analysis
is to obtain a good differential staining of the two sister
chromatids. This is the reason why the experimental procedures are continually refined as far as treatment modalities
and cytologic techniques are concerned. In 1985, Van der Gaag
and Van de Kerkoff, and Van de Kerkoff and Van der Gaag
describe some modifications applied to the staining method.
They hope that an eventual standardization of this test on
Nothobranchius rachowi will permit a routine use.

Therefore, some laboratories, which had been giving up
this test because of the difficulty to obtain differential

staining of sister chromatids, could come back to it. Van der Gaag and his team are continuing their studies on the SCE test on <u>Nothobranchius rachowi</u>; Thus in 1989, the genotoxicity of an industrial effluent has been estimated by the simultaneous use of two bacterial assays (the Ames test and the SOS-chromotest), the SCE test in the fish <u>Nothobranchius rachowi</u>, and the micronucleus tests in the newt <u>Pleurodeles waltl</u> and the mussle <u>Mytilus edulis</u>.

For the detection of mutagens in water (Perry and Evans, 1975) SCE have been found to be more sensitive than the observation of chromosome aberrations. Yet, the SCE test (as well as the chromosome aberrations test) still requires a rather long training period as long as experimental techniques, reading and interpreting of the results, are concerned. A broad use of this test is, at the moment, refrained by these difficulties.

MICRONUCLEUS TESTS

In most eukaryotes, chromosome and genome mutations result in the formation of micronuclei. These micronuclei are small intracytoplasmic masses of chromatin resembling small nuclei. They are formed from chromosome fragments or complete chromosomes which have not migrated to a spindle pole during anaphase. The formation of micronuclei stems either from chromosome fragmentation, or a malfunction of the mitotic apparatus. In the former case, micronuclei correspond to chromosome fragments which, having lost the centromere, have been unable to connect with the spindle fibers. In the latter case, they arise from complete chromosomes lagging at anaphase due to spindle abnormalities. Clastogenic compounds and spindle poisons both lead to an increase in the number of micronucleated cells.

The presence of many cells containing micronuclei is considered to be an indication of mutagenicity (Evans et al, 1959). <u>In vivo</u>, this test is mostly used on small terrestrial mammals (mice, rats, Syrian hamsters, Chinese hamsters, Guinea pigs, among others). Schmid (1976) gives a detailed procedure of the micronucleus test applied to bone marrow polychromatic erythrocytes from mammals.

For the <u>in vivo</u> study of genotoxicity of water, the use of aquatic organisms is prescribed. Among lower vertebrates, fish and amphibians, the red blood cells (RBCs) are nucleated and divide in the bloodstream; Furthermore, the blood smear technique is easily performed on them. Therefore, they represent a good potential material for the setting of the micronucleus test.

The Micronucleus Test in Fish

Some trials have been undertaken on <u>Umbra pigmaea</u> by Hooftman and de Raat (1982): they observed a very low rate of micronucleated RBCs after the treatment with the EMS. Metcalfe (1988) did not obtain any significant results on <u>Umbra limi</u> and <u>Ictalurus nebulosus</u> after the injection of EMS and B(a)P. In our laboratory a similar test using two other fish species (<u>Brachydanio rerio</u> and <u>Cyprinus carpio</u>) did not lead to statistically significant results (Jaylet and Deparis, unpublished results).

Therefore, at the moment, the fish micronucleus test does not seem capable of constituting a routine test for the study of genotoxicity of freshwater.

Amphibians Micronucleus Test

Krauter et al (1987) got an increase of the level of micronucleated RBCs by irradiation on <u>Rana catesbeiana</u>.

The most significant experiments concerning amphibians have been performed over the last few years, at the Centre de Biologie du Developpement of Toulouse. They allowed us to develop a micronucleus test on the three species: <u>Pleurodeles waltl</u> (Pleurodele), <u>Ambystoma mexicanum</u> (Axolotl) and <u>Xenopus laevis</u> (Xenopus) (Siboulet et al, 1984; Grinfeld et al, 1986; Jaylet et al, 1986a; Jaylet et al, 1986b; Jaylet et al, 1987; Fernandez and Jaylet, 1987; Fernandez et al, 1989; Zoll et al., 1988; Gauthier et al., 1989; Van Hummelen et al., 1989; Jaylet and Zoll, 1989, Jaylet et al, 1989a; Jaylet et al, 1989b). The tests are performed on the larval stages of these species. The stages of treatment correspond to fast growing larvae presenting an intense erythropoiesis together with a great number of divisions of the RBCs in the circulating blood (10% mitoses in the red peripheral blood cells on <u>Pleurodeles waltl</u> Deparis, 1973).

The larvae are maintained during a specific period of time in the water for which the genotoxicity is investigated (river water, effluents, drinking water or drinking water during the various stages of the treatment of raw water, water with divers xenobiotics added). After the intracardiac puncture and the blood smear are made, the rate of RBCs with micronuclei is evaluated and compared to the one of the larvae reared in reference medium. The details of the experimental procedures are described in the above-mentioned publications. The newt micronucleus test is the subject of an AFNOR descriptive (T90-325, 1987).

Studies with X-rays and Chemicals

a. The micronucleus test in pleurodele. Initially, the sensitivity and dose-response of the micronucleus test in Pleurodele was evaluated with X-ray irradiation, a well-known physical clastogenic agent. The measurements were made six days after X-ray irradiation. A dose of 6 rad (relatively weak) led to a significant effect. The dose-response is approximately linear up to 150 rad, after which the slope falls and the maximal effect is reached at 600 rad.

Results on 19 organic compounds have been published elsewhere (Fernandez et al., 1989). Aroclor 1254, butylated hydroxy-anisole, phenobarbital and 12-O-tetradecanoyl-phorbol-13-acetate produced negative results, while acridine orange, benzo(a)pyrene, e-caprolactam, cyclophosphamide, diethyl sulfate, epichlorhydrin, ethidium bromide, ethyl methane-sulfonate, ethyl dibromide (dibromoethane), N-ethyl-N'-nitro-N-nitrosoguanidine, N-ethyl-N-nitrosourea, hexa-methylphosphoramide, 3-methyl cholanthrene, pyrene and o-toluidine gave positive responses.

The cytogenetic effects of mercury compounds have been widely studied in plants, Drosophila and tissue culture cells, but to our knowledge they have not been evaluated in vertebrates _in vivo_. Pleurodele larvae were raised in water containing low concentrations of methyl mercuric chloride (CH_3HgCl) or mercuric chloride ($HgCl_2$) (Zoll et al., 1988). It should be noted that a low concentration of the two compounds (12 ppb) gave a positive result, and that at equivalent concentrations in the water both CH_3HgCl and $HgCl_2$ led to similar levels of micronucleated RBCs. The test gave positive results for concentrations below those often found in samples of contaminated water (Giraud and Guillet 1972). Both chromosome aberrations and abnormalities in cell division were observed in cells from animals treated with these two substances. Bioaccumulation of both compounds was also evaluated by determination of mercury levels in the larvae. After twelve days treatment, concentration factors (concentration in the organism/concentration in the water) of 1,200 and 600 were found for CH_3HgCl and $HgCl_2$ respectively. These findings can be compared to those for B(a)P reported by Grinfeld et al., (1986). The concentration of B(a)P in unfed Pleurodele larvae was 200-fold that in the water after 12 hours.

The marked bioaccumulation potential of newt larvae partially explains why it is not necessary to concentrate mutagenic micropollutants in samples of natural, or drinking water to detect genotoxic effects.

These results demonstrate the sensitivity and the reliability of the test for known genotoxic agents experimentally added to the rearing water. For example, a positive response is obtained with the B(a)P for a concentration corresponding to levels of pollution frequently observable in surface waters. On the other hand, even if the AFNOR descriptive recommends an optimal treatment lasting of twelve days, some preliminary works showed that a positive response could be registered as early as the first days of treatment.

b. The micronucleus test in the Axolotl. In order to investigate the generality of the micronucleus test in Pleurodele, larvae from another urodele, the Axolotl, were reared in water containing either of the compounds: B(a)P or EMS. The effects of the indirect mutagen (B(a)P), and the direct mutagen (EMS) were found to depend on both dose and exposure to the clastogen. For B(a)P, positive results were obtained after eight days treatment at a concentration of 0.025 ppm. After ten days treatment at a concentration of 0.1 ppm, numerous micronuclei were seen (> 250°/oo). Positive results were also obtained with EMS after eight days treatment at a concentration of 24 ppm. At 62 ppm, positive results were found after six days, while at 124 ppm positive results were found after only four days. The results with both these agents show that the Axolotl also holds promise as an _in vivo_ test system for the detection of low concentrations of clastogens in the aquatic environment. Although not of the same genus and of different geographical origin, the Axolotl is morphologically and biologically similar to the newt at both the embryonic and larval stages.

c. The micronucleus test in Xenopus. The third species used was Xenopus. It differs in a number of respects from the previous two species, especially with regard to its feeding behavior.

Three different variables: temperature, stage of larval development and frequency of renewal of the test substance were investigated using EMS as the clastogenic compound. In addition, a dose-response curve was established for B(a)P in order to determine the limits of sensitivity of the test (Van Hummelen et al., 1989).

With B(a)P, the lowest concentration (0.03 ppm) gave a negative response. From 0.06 ppm up to 0.50 ppm, an approximately linear increase in median value of cells with micronuclei was observed. This linear response indicates that the test is reliable, although the lower frequencies of cells with micronuclei at doses above 0.50 ppm, are probably accounted for by a lower rate of growth of the larvae exposed to high doses of B(a)P.

The effects of other organic products, different kinds of polluted water and mercuric salts are under investigation. With respect to mercury compounds, it should be noted that they are strongly accumulated by Xenopus tadpoles. For example, after twelve days the test is positive with only 2.5 ppb of methyl mercuric chloride.

In Situ Studies

Up to this point, the in situ studies are mostly worked out with the Pleurodele.

a. Detection of mutagenicity in drinking water. Groups of larvae were reared in tap water, while controls groups were reared in tap water which had been filtered over sand and active carbon to remove micropollutants. Seven separate tests carried out between October 1985 and May 1986 all gave positive results of varying degrees depending on the time of year. This test is therefore able to detect clastogens in normal drinking water (Jaylet et al, 1987). It could be used for quality control of drinking water during the various stages in the treatment of raw water without any requirement for prior extraction or concentration of micropollutants.

b. Study of the genotoxicity of drinking water during disinfection processes. The chlorine is frequently used in the water treatment process. Yet, the fact that residual chlorine, present in drinking water under free chlorine form (HOCL, OCL⁻), and chloramines can represent a genotoxic danger for the consumer has not been emphasized. However, the mutagen effects of hypochlorite (Wlodkowski and Rosenkranz, 1975) and of chloramine (Shih and Lederberg, 1976; Thomas et al, 1987) were shown on bacteria. The Pleurodele micronucleus test has been applied to the study of water samples which were previously submitted to the effect of chlorine and monocloramine. Larvae were placed in deionized water, free from organic substances and enriched with essential salts. To these samples sodium hypochlorite and monochloramine were added at various concentrations. The sodium hypochlorite at 0.125 ppm and 0.25 ppm and the monochloramine at 0.15 ppm both induced a significant increase of the level of micronucleated RBCs (Gauthier et al, 1989).

One of the alternatives to water disinfection by chlorine is to use ozone. It is well known that, in some experimental conditions, ozone presents a mutagen effect in vitro (Bourbigot et al, 1983; Kool and Hrubec, 1986). Pleurodele larvae were reared in the Seine water which had received different quantities of ozone. A weak ozonization induced a positive response of the test, but a strong one did not induce any genotoxic effect. Thus, weak ozone concentrations applied to a surface water induced the formation of mutagenic compounds which were destroyed with stronger concentrations (Jaylet et al, 1989a).

 c. The study of effluents genotoxicity. Some effluent samples from various petrochemical industries were added in various proportions to the rearing water of the Pleurodele larvae. A very significant rise of the rate of micronucleated RBCs was observed. The mutagenicity of the same samples were confirmed by the SCE-assay with Nothobranchius rachowi and the micronucleus test with Mytilus edulis.

The overall results stated above show that the amphibian micronucleus test is capable of revealing the genotoxic potentials of surface water without previous concentration.

CONCLUSIONS

As mentioned, the currently used in vivo tests for genotoxicity in freshwater are the test of chromosomal aberrations, the SCE test and the micronucleus tests. The choice between the different tests should be based not only on the specificity and sensitivity of the test, but also on the ability to reveal the various types of mutation (gene, chromosomal and genomic). In all these tests, the test system is a whole organism. Metabolic activation, detoxication, excretion and differential sensitivity of tissues are all taken into account. The ovarall effect of pollutants on the organism are evaluaed since interactions between substances, including possible synergistic or antagonistic activities are not disturbed.

The in vivo tests are thus more representative of natural conditions. In addition, the test organism is phylogenetically close to humans, and the results can be extrapolated more readily (less risk). However, these tests are more time consuming and costly than the in vitro tests. White and Champ (1983) have reviewed the factors influencing the results of such bioassays (age, sex, genotype, temperature, photoperiod, etc.). They have laid down three essential criteria for evaluation of bioassays: scientific value, validity, and extent of application.

Thus, a good test of mutagenesis should be rigorous, standardized and reproducible. A balance must be achieved between the number of controls carried out and the complexity of the procedure. Statistical methods need to be rigorous, although they also have limitations. They have to be adapted to each particular test. White and Champ (1983) recommended the use of a variety of species in order to obtain an overall picture. Practical mutagenesis tests are inevitably the result of a number of compromises, and the choice of test depends on the nature of the problem. The _in vitro_ tests are suitable as early warning systems, although they cannot supplant the _in vivo_ methods for more accurate evaluation of risks.

The notion of a threshold dose which is so useful and reassuring in toxicology may not be applicable in the case of mutagens. There may well be no net lower limit below which the total innocuousness of the mutagenic substance can be established. On the whole, the consequences of the actions of mutagens are similar on all living organisms. There are of course differences in sensitivity, and in some cases differences in mode of action.It is thus not always easy to extrapolate from animal to man. Notwithstanding this consideration, the most commonly used tests at present are those using bacteria, as they are both cheap and quick.

However, the risk from genotoxic agents is a hidden one. Although mutagenic micropollutants may not be lethal, they may alter the natural equilibrium in subtle ways or even sterilize fragile species. An enhancement in the rate of mutation may also accelerate the processes of evolution leading to a domination of certain species to the detriment of others, although here the precise risk is hard to estimate. In any event, extrapolation of the test results to the real world situation represents an exercise in judgement.

This publication is dedicated to the memory of Professor A. JAYLET.

REFERENCES

Afnor, 1987, Essais des eaux, Détection en milieu aquatique de la génotoxicité d'une substance vis-a-vis de larves de batraciens (Pleurodeles waltl et Ambystoma mexicanum). Essai des micronoyaux. Association Francaise de Normalisation (AFNOR). T 90:325.
Alink, G.M., Genotoxins in waters, 1982, _in_: "Mutagens in our Enrivornment", Alan Liss, R.L., Inc., New-York, 261.
Alink, G.M., Frederix-Wolters, E.M.H., Van Der Gaag, M.A., Van De Kerkhoff, J.F.J., and Poels, C.L.M., 1980, Induction of sister chromatid exchanges in fish exposed to Rhin water, _Mutat. Res._, 78:369.
Ames, B.N., Mac Cann, J., and Yamasaki, E., 1975, Methods for detecting carcinogens and mutagens with the salmonella mammalian microsome mutagenicity test, _Mutat. Res._, 31:347.
Barker, C.J., and Rackam, B.D., 1979, The induction of SCE in cultured fish cells (Ameca splendens) by carcinogenic mutagens, _Mutat. Res._, 68:381.
Bourbigot, M.M., Paquin, J.L., Pottenger, L.H., Blech, M.F., and Harttemann, P.H., 1983, Study of mutagenic activity in water in a progressive ozonizatio unit, _Aqua_, 3:99.
Deparis, P., 1973, Le sang circulant au cours de la croissance larvaire de P. waltlii Michah, _J. Physiol._, 66:423.

Dutka, B.J., Jova, A., and Brechin, I., 1981, Evaluation of
    four concentration/extraction procedures on water and
    effluents collected for use with the Salmonella
    typhimurium screening prosedure for mutagens, Bull.
    Environ. Contaminants Toxicol., 27:758.
Evans, H.J., Neary, G.J. and Willamson, F.S., 1959, The
    relative biological efficiency of single doses of fast
    neutrons and rays on Vicia faba roots and the effects
    of oxygen. II. Chromosome damage, The production of
    micronuclei, Int. J. Radiat. Biol., 1:216.
Fernandez, M., Gauthier, L. and Jaylet, A., 1989, Use of newt
    larvae for in vivo genotoxicity testing of water:
    results on 19 compounds evaluated by the micronucleus
    test, Mutagenesis, 4:17.
Fernandez, M. and Jaylet, A., 1987, An antioxidant protects
    against the clastogenic effets of benzo(a)pyrene in the
    newt in vivo, Mutagenesis, 2:293.
Gauthier, L., Levi, Y. and Jaylet, A., 1989, Evaluation of the
    clastogenicity of water treated with sodium hypo-
    chlorite or monochloramine using a micronucleus test in
    newt larvae (Pleurodeles waltl), Mutagenesis, 4:170.
Giraud, M. and Guillet, H., 1972, Teneur en mercure des
    milieux naturels. in: "La pollution par le mercure et
    ses derives", Rapport Ministere de l'Environnement,
    France.
Grinfeld, S., Jaylet, A., Siboulet, R., Deparis, P. and
    Chouroulinkov, I., 1986, Micronuclei in red blood cells
    of the newt Pleurodeles waltl after treatment with
    benzo(a)pyrene, dependence on dose, length of exposure
    post-treatment time and uptake of the drug, Environ.
    Mutagenesis, 8:41.
Hooftman, R.N., 1981, The induction of chromosome aberrations
    in Nothobranchius rachowi (Pisces: cyprinodontidae)
    after treatment with ethyl methane sulphonate or
    Benzo(a) pyrene, Mutat. Res., 91:347.
Hooftman, R.N., and De Raat, W.K., 1982, Induction of nuclear
    anomalies (micronuclei) in the peripheral blood
    erythrocytes of the eastern mudminnow Umbra pygmaea by
    ethyl methanesulphonate, Mutat. Res., 104:147.
Hooftman, R.N., and Vink, G.J., 1981, Cytogenetic effects on
    the eastern mudminnow Umbra pygmaea, exposed to ethyl
    methane sulfonate, benzo(a) pyrene, and river water,
    Ecotox. Environ. Safety, 5:261.
Jaylet, A., Deparis, P., Ferrier, V., Grinfeld, S. and
    Siboulet, R., 1986a, A new micronucleus test using
    peripheral blood erytrhocytes of the newt Pleurodeles
    waltl to detect mutagens in fresh-water, Mutat. Res.,
    164:245.
Jaylet, A., Deparis, P. and Gaschignard, D., 1986b, Induction
    of micronuclei in peripheral erythrocytes of axolotl
    larvae following in vivo exposure to mutagenic agents,
    Mutagenesis, 1:211.
Jaylet, A., Gauthier, L. and Fernandez, M., 1987, Detection of
    mutagenicity in drinking water using a micronucleus
    test in newt larvae (Plaeurodeles waltl), Mutagenesis,
    2:211.
Jaylet, A., Gauthier, L., Levi, Y., 1989a, Detection of
    genotoxicity in chlorinated or ozonated drinking water
    using an amphibian micronucleus test, in: "Genetic
    Toxicology of Complex Mixtures: Short-Term Bioassays in
    the Analysis of Complex Environmental Mixture, VI",

M.D. Waters, S. Nesnew, J. Lewtas, M.M. Mare and F.B.
     Daniel, Eds, Plenum Press, New-York, (in press).
Jaylet, A., Gauthier, L., and Zoll, C., 1989b, Micronucleus
     test using peripheral red blood cells of amphibian
     larvae for detection of genotoxic agents in freshwater
     pollution, in: "in situ evaluation of biological
     hazards of environmental pollutants", Sandhu, Ed.,
     Plenum publishing corporation.
Jaylet, A. and Zoll, C., 1989, Detection of pollution in
     freshwater using a mutagenicity test. CRC Critical
     Reviews in Aquatic Toxicology (accepted for publica-
     tion).
Kligerman, A.D., 1977, Umbra limi: An aquatic model for the
     study of sister chromatid differentiation and exchange
     in vivo, Ph. D. thesis, Cornell University, Ithaca, New
     York.
Kligerman, A.D., 1982, Fishes as biological detectors of the
     effects of genotoxic agents, in: "Mutagenicity, New
     Horizons in genetic Toxicology", Heddle, Ed., Academic
     Press, New York.
Kligerman, A.D., and Bloom, S.E., 1976, Sister chromatid
     differentiation and exchanges in adult mudminnows
     (Umbra limi) after in vivo exposure to 5-bromodeo-
     xyuridine, Chromosoma, 56:101.
Kligerman, R.N., Bloom, S.E., and Howell, W.M., 1975, Umbra
     limi: a model for study of chromosome aberrations in
     fishes, Mutat. Res., 31:225.
Kool, H.J., and Hrubec, J., 1986, The influence of an ozone,
     chlorine and chlorine dioxide treatment on mutagenic
     activity in (drinking water), Ozone Science and
     Engineering, 8:217.
Kool, H.J., Van Kreijl, C.F., Van Kranenen, H.J., and De
     Greef, E., 1981, The use of XAD-resins for the detec-
     tion of mutagenic activity in water, studies with
     surface water, Chemosphere, 10:85.
Krauter, P.W., Anderson, S.L., and Harrison, F.L., 1987,
     Radiation induced micronuclei in peripheral erythro-
     cytes of Rana catesbiana: An aquatic animal model for
     in vivo genotoxicity studies, Environ. Molecular
     Mutagenesis, 10:285.
Maruoka, S., and Yamanaka, S., 1982, Mutagenicity in Salmo-
     nella typhimurium tester strains of XAD-2-ether
     extract, recovered from Katsura River water in Kyoto
     City, and its fractions, Mutat. Res., 102:13.
Metcalfe, C.D.,1988, Induction of micronuclei and nuclear
     abnormalities in the erythrocytes of mudminnows (Umbra
     limi) and brown bullheads (Ictalurus nebulosus), Bull.
     Environ. Contam. Toxicol., 40:489.
Perry, P., and Evans, H.J., 1975, Cytological detection of
     mutagen-carcinogen exposure by sister chromatid
     exchange, Nature, 258:121.
Prein, A.E., Thie, G.M., Alink, G.M., Koeman, J.H., and Poels,
     C.L.M., 1978, Cytogenetic changes in fish exposed to
     water of the river Rhin, Sci. Total Environ., 9:287.
Quin, Y., Runrong, G., and Shiwei, Z., 1985, Effect of organic
     extract from drinking water on SCE frequency of CHO
     cells, Huanjing Kexue, 6:17.
Schmid, W., 1976, The micronucleus test for cytogenetic
     analysis. in: "Chemical Mutagens", Hollaender, A.,
     ed., Vol. 4, Plenum, New York.

Shih, K.L., and Lederberg, J., 1976, Effects of chloramine on *Bacillus subtilis* deoxiribonucleic acid, J. Bacteriol., 125:934.

Siboulet, R., Grinfeld, S., Deparis, P. and Jaylet, A., 1984, Micronuclei in red blood cells of the newt Pleurodeles waltl Michah: induction with X-rays and chemicals, Mutat. Res., 125:275.

Thomas, E.L., Jefferson, M.M., Bennett, J.J., and Learn, D.B., 1987, Mutagenic activity of chloramines, Mutat. Res., 188:35.

Van De Kerkhoff, J.F.J., and Van der Gaag, M.A., 1985, Some factors affecting optimal differential staining of sister-chromatids *in vivo* in the fish Nothobranchius rachowi, Mutat. Res., 143:39.

Van der Gaag, M.A., Gauthier, L., Noordsij, A., Levi, Y. and Wrisberg, M.N., 1989, Methods to mesure genotoxins in waste water: evaluation with *in vivo* and *in vitro* test, in: "Genetic Toxicology of Complex Mixtures: short-term bioassays in the analysis of complex environmental mixtures, VI", M.D. Waters et al., eds, Plenum Press, New York, in press.

Van der Gaag, M.A., and Van de Kerkhoff, 1985, Mutagenicity testing of water with fish: a step forward to reliable assay, Sci. Total Environ., 47:293.

Van der Hoeven, J.C.M., Bruggeman, I.M., Alink, G.M., and Koeman, J.H., 1982, The killifish *Nothobranchius rachowi* a new animal in genetic toxicology, Mutat. Res., 97:35.

Van Hummelen, P., Zoll, C., Paulussen, J., Kirsh-Volders, M. and Jaylet, A., 1989, The micronucleus test in xenopus: a new and simple *in vivo* technique for detection of mutagens in fresh water, Mutagenesis, 4:12.

White, H.H., and Champ, M.A., 1983, The great bioassay hoasc, and alternatives, in: "Hazardous and Industrial Solid Waste Testing", Second Symposium, ASTM STP 805, Conway, R.A., and Gulledge, W.P., 299.

Wlodkovski, T.J., and Rosenkranz, H.S., 1975, Mutagenicity of sodium hypochlorite for Salmonella typhimurium, Mutat. Res., 31:39.

Zoll, C., Saouter, E., Boudou, A., Ribeyre, F. and Jaylet, A., 1988, Genotoxicity and bioaccumulation of methyl mercury and mercuric chloride *in vivo* in the newt Pleurodeles waltl., Mutagenesis, 3:337.

CYTOGENETIC MONITORING OF INDUSTRIAL WORKERS EXPOSED TO CHEMICALS

N.P. Bochkov

Department of Genetic Monitoring
Research Center of Medical Genetics
Moskvorechie 1, 115 478 Moscow, USSR

The attention of hygienists and ecologists to genetic effects of environmental pollution in human population is not so serious as is needed. There are two main reasons for this situation according to me.

Firstly, it is because huge global ecological problems are arising and increasing. So the scientists, public health personnel and political leaders take into consideration mainly very serious disturbances in health and life of people and economics of their country.

Secondly, it is because the many awful disasters that have occurred during the last years (such as Seveso, Bhopal, Chernobyl, etc.). Both global pollution and big disasters distract scientific and public attention from the changes of human environment in broad sense, especially those which are caused by contacts with industrial factors, chemical ones in particular. Among the last there are undoubtedly many mutagens.

Environmental pollution today is not simply reality, but permanently increasing and now already a threatening danger. This conclusion is true for many countries, including the USSR. Soviet geneticists are disturbed about serious ecological situations in many of our towns. To be objective several examples from the published data will be given here (Izvestia, 4 October, 1989). During the first half of 1989 the soviet industry (only industry, i.e. without transport, and other kinds of polluters) threw away into the atmosphere 29 million tons of harmful chemicals. In spite of the hygienic reglementation for toxic substances in the atmospheric air, the allowable level is frequently exceeded in many towns. Several examples only are quoted in Table 1.

As one can see, these towns represent different regions of our big country, i.e. Middle Asia, Siberia, Ukrania, Caucasus.

It is quite understandable that on such scale of pollu-

Table 1. The Number of Times the Allowable Standards are Exceeded

Benzo(a)pyrene		Formaldehyde		Ammonia	
Towns	Increase	Towns	Increase	Towns	Increase
Frunze	14	Grozny	9	Rustavi	5
Novokuzneck	12	Dyushanbe	4	Nigny	
Magnitogorsk	11	Zaporozhye	2	Tagil	4
Donetsk	8				
Kemerovo	7				
Alma-Ata	7				

tion hygienists and physicians are concerned in the first place about the disturbance of health of living people.

Meanwhile, it is evident that it is important to consider the question of possible genetic consequences in workers from contacts with chemical substances and other environmental factors in industry because the big amount of population in the reproductive age. That is why the regular evaluation of different plants for mutagenicity may be considered as an important task for genetic toxicology and preventive medicine. Fortunately, as we know, there are several methods to carry out this task, mainly cytogenetic techniques on lymphocyte cultures. The analysis of chromosome aberrations, sister chromatid exchanges, micronucleus allow us to judge objectively the mutagenic activity of industrial pollutants.

The main goal of cytogenetic monitoring of industrial workers exposed to chemicals is the measurement of the safety of the working place, i.e., the measurement of a real exposure to industrial mutagens. People experience a combined action of factors at their working place. So the workers of rubber plants come into contact with several mutagenic chemicals (vulcanisation accelerators - altoks, tiuram, benzene, etc.) in combination with high temperature and noise. In the metallurgic industry there is a combination of chemicals and vibration. That is why it is necessary to introduce cytogenetic monitoring. This is an important task of occupational health service.

At the same time the cytogenetic monitoring data can be used in the future as a background for the evaluation of the mutagenic load for population and, consequently, for the prognosis of mutation process.

Though there are many publications about cytogenetic investigations of workers, it is necessary to stress that there are no systematic observations, e.g. the various types of industry, the exposed dose, or the duration of work. Definitely there is a strong necessity to summarize and systematize the data of cytogenetic monitoring and to outline the tasks for future investigations.

In this report we will summarize the data about the cytogenetic monitoring of workers of different industrial enterprises in the USSR.

Cytogenetic monitoring is carried out by specialists in institutes and laboratories of different profiles: hygienic, genetic, medical ones. It seems that results can be compared because the general methodical requirements in case of evaluation of frequency and types of chromosomal aberrations are similar to those which were elaborated by WHO consultation group in 1973.

The lymphocyte cultures were performed by micromethod. The cells were harvested during first mitosis. The metaphase analysis was standard. The following chromosome aberrations were recorded: single and double fragments, chromatid and chromosome exchanges. Gaps were not included. For the SCE analysis 5-bromodeoxyuridine was added to lymphocyte culture between 0-24 hours to the final concentration of 10 µg/ml, and the cells were fixed at the 72nd hour. This means that they passed two divisions in culture.

Analysis of the data published in the USSR shows that the number of individuals investigated in the exposed (professional) and control groups ranges from 5 to 100 persons and more. In each industrial enterprise the most exposed workers were investigated first. The number of analysed cells in the group varies from several hundreds to several thousands. The control groups, as a rule, were similar in the main characteristics to exposed groups. This means equal ratios between smokers and non-smokers, etc.

The results of cytogenetic examination of exposed and control groups were compared with summarized data on spontaneous level. During the last years we collected the results on the spontaneous level of aberrations and studied the differences in time and inter laboratories. It seems that variations between laboratories are not important.

After analysis of earlier investigations we decided to increase the number of metaphases analysed for each individual. (Bochkov et al., 1984). Thus in our and other laboratories the testing of mutagenicity of industrial factors is performed now by the analysis of 300 metaphases or more for each person, when the chromosomal aberrations are the genetic endpoints, and of 50 cells for the evaluation of SCE. In addition it is important that the sample contains no less than 10 individuals. Thus we analyse chromosome aberrations in 3000-4000 cells and sister chromatid exchanges in 500 cells for each exposed and control groups.

During the last 15 years in the USSR cytogenetic examinations of workers were carried out in many plants and factories. The principal goal of these examinations was the detection of mutagenic effects induced by industrial conditions. The list of industries, where the investigations were carried out, is shown in Table 2.

As it can be seen from the list of industries the workers of many enterprises (metallurgical, rubber, polymer and others) were examined. One should draw attention firstly to

Table 2. Types of Industries where Cytogenetic Monitoring of
          Workers was Carried Out

Types of industries	Authors
Metallurgical	Babajan et al., 1980; Sedova, 1989
Rubber	Alexandrov and Zurkov, 1983
Plastics	Fomenko et al., 1986; Katosova, 1973; Katosova and Pavlenko, 1985; Sazonova and Suskov,1982; Zurkov and Fitchidzhjan, 1977
Furniture	Chebotarev et al., 1985
Pesticides	Pilinskaya, 1982; Sharipov et al.,1989
Fertilizers	Sidenko et al., 1989

the wide range of different branches of industry and
consequently of chemicals. Secondly, a lot of workers are
engaged in these industries. Thirdly, we should take into
consideration that the examples are given from the towns of
different regions. So it is not local but wide mutagenic
exposure.

The frequency of chromosome aberrations observed in
workers from different enterprises is 2-3 times higher than in
control groups. For the general characteristics of cytogenetic
monitoring of workers let us take some examples from the
literature on each type of industry. Sedova (1989) summarized
the evaluation of mutagenic activity of industrial environment
among the workers from metallurgic enterprises. The results
of this investigation are given in Table 3. The workers
engaged in cast iron and steel smelting showed a significant
difference with the controls. These workers have come into
contact with manganese, chrome, nickel, lead and copper oxides
in the air. They work at high air temperature and are exposed
to intensive infra-red radiation.

Alexandrov and Zurkov (1983) have examined 82 workers
from three groups in the rubber industry. The workers of
these groups were exposed to a complex of harmful factors
(several vulcanization gases, benzene, dust of vulcanization
accelerators and others). Non-production staff was taken as
control group. It was shown that the manufacturing workers
have increased chromosome aberration frequency (2,63%) as
compared with controls (1,14%) and non-production staff
(1,34%). The authors suppose the mutagenic effect to be
connected with vulcanization accelerators.

The results of cytogenetic examinations of workers being
in contact with synthetic polymeric substances are represented
in the paper by Sazonova and Suskov (1982). The average
frequency of the aberrant cells as compared with control

Table 3. Chromosome Aberrations Frequency in Lymphocyte Culture of Metallurgic Industry Workers

Type of industry	Number of examined individuals	Number of metaphases	Number of aberrant metaphases (%)
Control	74	7691	1,15±0,09
Cast iron smelting	45	4500	2,35±0,13
Steel smelting	53	5300	2,69±0,15
Rolling shop	61	6100	1,02±0,08

(2,4±0,22% - 74 persons) is shown: 5,5±0,22 when contacting with epoxide tar (146 persons), 6,1±0,28 - with polyvinylchloride tar (52 persons), 5,0±0,4 - with phenolphormaldehyde tar (31 persons). Acentric single and double fragments prevail in all investigated groups. The levels of harmful chemicals in the air of working places were usually lower than existing allowable concentrations. The average age of the persons was about 39 years. The period of working was from 4 months to 30 years. In the control group the average age was 34 years.

Katasova, Pavlenko (1985) published the results of the cytogenetic investigation of the workers from several chemical industries. (Table 4). Significant increases in aberration cells frequency were observed in groups exposed to chloroprene (in all groups), vinyl chloride, dimethyl sulphate and lead. The relationship between the effect and working period was not revealed. The suggested safe concentrations according to Table 4 are: chloroprene: 0,5 mg/m^3, vinyl chloride: 0,1 mg/m^3, dimethyl sulphate: 0,1 mg/m^3, lead: 0,1 mg/m^3.

Chebotarev et al. (1985) examined the group of furniture workers, where formaldehyde is the main known harmful chemical. The control group included workers from the auxiliary shop of the same enterprise being in no contact with formaldehyde and other chemicals (Table 5). The increase frequency of chromosome aberrations (p = 0,025) and decreased unscheduled DNA synthesis (p = 0,049) is shown in peripheral blood lymphocytes of exposed workers. No difference in thiophosphamide induced chromosome aberrations was found between furniture workers and controls. There were no differences between groups in SCE frequencies.

Sharipov et al. (1989) studied the chromosome aberration frequency in 8 men and 31 women, who were in contact with the complex of pesticides in greenhouse farms. Ten women formed a control group. The workers of these farms were in contact with 18 pesticides, such as carbofos, sulphur, phosphamide, chlorophos. 4000 metaphases were studied in a professional group and 1000 in the controls. The aberrant metaphases frequency was 6,9±0,37% in the professional group and

Table 4. Cytogenetic Analyses of Lymphocytes from Workers Being in Contact with Chemicals in Industry

Chemicals	Concentration mg/m³	Number of investigated persons	Number of metaphases	Number of aberrant cells (%) M ± m	p
Chloroprene	45 −2,7	18	1666	4,77±0,57	< 0,001
Chloroprene	7,0−3,0	20	1748	3,49±0,51	< 0,001
Chloroprene	1,0−4,0	8	648	2,5 ±0,49	< 0,05
Control		9	572	0,65±0,56	
Vinyl chloride	111−1,8	37	3135	2,76±0,24	< 0,05
Control		12	1041	1,62±0,39	
Ethylene oxide	10,0−20,0	10	639	2,2 ±0,57	> 0,05
Control		8	529	1,19±0,29	
Dimethyl acetamide	5,0−10,0	20	1637	4,9 ±0,5	> 0,05
Control		16	1395	3,7 ±0,36	
Dimethyl sulphate	4,0−2,0	14	1300	2,59±0,24	< 0,05
Control		13	1235	1,57±0,21	
Lead	0,26−0,05	18	1780	3,5 ±0,51	< 0,05
Lead	0,003−0,115	12	1065	2,31±0,5	> 0,05
Control		45	4463	1,25±0,38	

Table 5. Chromosome Aberration Frequency in Lymphocytes of Workers being in Contact with Formaldehyde in Industry

Group characteristics	Spontaneous level		After thiophosphamide treatment (20 m kg/ml, 1h)	
	Number of individuals	Aberrant metaphases (%)	Number of individuals	Aberrant metaphases
Control group	14	1,64±0,39	19	50,05±1,54
Group being in contact with formaldehyde	37	2,76±0,26	32	50,18±1,67

2,1±0,45% in the control. The number of aberrations per 100 cells was 6,4 and 2,1 respectively.

As was shown by Sidenko et al (1989) the complex of factors of nitrogen fertilizers production induces a significant increase of chromosome aberrations. The aberrant cells from 17 women working in the ammonium production was 4,17±0,36% and 1,65±0,33% for the control group (11 persons). No effect on working period was demonstrated. Individual variability of cytogenetic effects of 2-3 times and more was demonstrated.

The data of cytogenetic monitoring of industrial workers allow to make the following conclusions:

1. In industrial situations of very different factories there are many factors, which increase the spontaneous level of chromosome aberrations in lymphocytes of peripheral blood to 2-3 times as large. Concerning the types of induced chromosome aberrations, the increases are observed for the single and double fragments. These were demonstrated in most investigations. Usually no differences are found in frequencies of exchange types of aberrations between control and exposed groups. Their increase cannot be statistically demonstrated in the sample size as published. At the same time the whole analysis of data shows that in numbers there is increase of exchange type of aberrations in all investigated groups. In any case independently from the types of induced aberrations we can speak about mutations from industrial chemicals. Sister chromatid exchanges as the method of cytogenetic monitoring was not so widely used by our specialists as the analysis of chromosome aberrations. There were no principal differences in conclusions between two methods, when investigation was performed in the same laboratory and on the same groups. But we consider that the evaluation of chromosome aberrations give us a more definite picture of mutagenic effects in real industry situation.

2. There are marked differences between individuals in all exposed groups. Such variations of cytogenetic damages may reflect many peculiarities at the same time: a difference in exposure to industrial pollutants of different workers, a difference in lifestyle (smoking, alcohol, food), a population polymorphism of xenobiotic biotransformation, a polymorphism of primary sensitivity of individuals on the DNA level, a polymorphism of the repair of DNA damages.

3. Until now the cytogenetic analysis was rarely performed in connection with measurement of industrial exposure. Those who carried out examination of workers on the plant know how difficult it is to measure the pollutant exposure, especially to a mixture. In most investigations no quantitative relation between cytogenetic damages and exposure was established. Apart from the absence of exact measurement of exposure, the small number of people investigated should be mentioned. That is why it is still difficult to understand definitely the connection between an exposure dose in the working place and the frequency of chromosome aberration.

4. No differences of chromosome aberration frequencies were discovered between men and women, and also between the subjects of different age.

5. An important question in the problem of cytogenetic monitoring of industrial workers is dependence of effects from the duration of work. A general conclusion in this direction is negative. For many other enterprises we have the same conclusions. The variations of working time at the same plant were from 1 to 25 and more years. These facts could be explained by mutation peculiarities on the organism level or by dynamic lymphocyte population in the blood, which circulate 1-2 years on the average in the peripheral blood. To clarify this question it is necessary to carry out prospective dynamic examinations of chromosome aberration frequency from the same individuals at least once per year. Such investigations allow us to answer a question about the individual sensitivity to a concrete mutagenic factor. It means that it would be possible to make an ecogenetic interpretation of individual chromosome aberration variations. During this examination, of course, it would be necessary to pay the strongest attention to possible variation of the exposures of occupational factors in the time.

6. The cytogenetic monitoring can be used for the population evaluation of industrial chemical mutagens, but not for individual prognosis. This is very important for the information in occupational health workers. The result of cytogenetic examination cannot lead to clinical conclusion on the examined person.

7. The cytogenetic monitoring of the workers exposed to industrial chemicals have shown that mutagenic effects occur in many people involved in industrial technology. It means that there is a real population danger from modern industry. Ecological improvement must be done not only in the atmosphere, water, soil, but also in the working places.

8. Though cytogenetic effects in somatic cells are evident, the data of increasing genetic endpoints in germ cells are absent or doubtful. The difference between the effects in somatic and germ cells needs special investigation.

9. It is not clear what is the clinical and prognostic significance of these alterations. At the same time it could be understood as the signal about unhealthy situations in the working place.

What would be a recommendation to occupational health and professional pathology? The connection between increased level of cytogenetic damages and clinical status of workers requires to be studied. It is quite desirable to include the prospective oncological examination of workers which have an increased level of chromosome aberrations.

10. Taking into account the importance of cytogenetic monitoring of workers it would be useful to organize the workshop on the minimal criteria of such examinations, especially professional data, and also the international data bank.

REFERENCES

Alexandrov, S.A., Zhurkov, V.S., 1983, Chromosome aberration in lymphocytes of rubber workers, Labour Hygiene and Occupat. Diseases (Russ.), 5:52.

Babajan, E.A., Bagramjan, S.B., Pogosjan, A.S., 1980, Influence of some chemicals of molybden production to chromosome of experimental animals and workers, Labour Hygiene and Occupat. Diseases (Russ.), 9:33.

Bochkov, N.P., Filippova, T.V., Jakovenko, K.N., 1984, The principles of cytogenetic monitoring for testing of occupational pollutants, Cytol. and Genetics (Russ.), 6:422.

Chebotarev, A.N., Titenko, N.V., Selezveva,T.G., 1985, Comparison of chromosome aberrations, sister chromatid exchanges and unscheduled DNA synthesis at evaluation of environmental mutagenicity, Cytol. and Genetics (Russ.), 19:109.

Fomenko, V.N., Katosova, L.D., Pavlenko, G.I., 1986, Cytogenetic analysis of peripheral blood of workers during the polymerisation of vynil chloride, Labour Hygiene and Occupat. Diseases (Russ.), 9:48.

Katosova, L.D., 1973, Cytogenetic analysis of peripheral blood of workers of chloroprene plant, Labour Hygiene and Occupat. Diseases (Russ.), 10:30.

Katosova, L.D., Pavlenko, G.I., 1985, Cytogenetic examination of the workers of chemical industry, Mutat. Res., 46:301.

Pilinskaya, M.A., 1982, Cytogenetic effect of pirimore (pesticide) in human lymphocyte culture in vivo and in vitro, Cytol. and Genetics (Russ.), 16:38.

Sazonova, L.A., Suskov, I.I., 1982, Cytogenetic effects of synthetic tars in the human, Genetica (Russ.), 7:1201.

Sedova, K.S., 1989, Summarized evaluation of mutagenic activity of metallurgic occupational conditions, Cytol. and Genetics (Russ.), 23:16.

Sharipov, I.K., Vishnevskaya, S.S., Mergembaeva, Kh. S., 1989, Chromosome aberrations in greenhouse workers in contact with pesticides, Cytol. and Genetics (Russ.), 23:60.

Sidenko, A.T., Sazonova, L.A., Vazhnik, L.A., 1989, The evaluation of effect of fertilizer production on chromosomes of workers, Labour Hygiene and Occupat. Diseases (Russ.), 1:19.

Zhurkov, V.S., Fitchidzhjan, B.S., 1977, Cytogenetic investigation of people exposed to chloroprene, Cytol. and Genetics, (Russ.), 3:210.

MONITORING CONGENITAL ANOMALIES IN POPULATIONS EXPOSED TO

ENVIRONMENTAL MUTAGENS

Radim J. Sram[1], Ivana Roznickova[2],
Vladimir Albrecht[1], Alena Berankova[1] and
Eva Machovska[3]

[1]Psychiatric Research Institute, 182 03 Prague
[2]District Hygiene Station, 415 68 Teplice
[3]District Hygiene Station, 466 00 Jablonec n.N.

INTRODUCTION

Congenital anomalies (CA) are believed to be one of the best mutation endpoints for the genetic monitoring in regions polluted by environmental mutagens (Crow, 1971). It is well known they are heterogenous by their etiology. Their fluctuation in time may correspond to the changing concentration of mutagens. The determination of CA should be based on medical-statistical data. But it may be possible only in the case of detailed identification of each CA in the proband and detailed analysis of other additional data (Czeizel and Sankaranarayanan, 1984; Bochkov and Sram, 1989).

CA may be understood as an endpoint, which is determined by the heritage of both parents, effect of mutagens to the gametes of both parents before the conception as well as on the foetus during the intrauteral development (which is simultaneously teratogenic). It means that pregnancy outcome is determined by the parents' genetic predisposition to diseases, by environmental pollution, occupational exposure and life style (smoking, alcohol, diet, drug abuse) before the conception; in the course of pregnancy also by the mother's infections, drugs and occupational exposure.

A part of CA may be then induced by mutagens in the environment and in the working place. In order to find out whether this is true we started to study the frequency of CA in the district of Usti n.L., which is one of the chemical industry centers in Czechoslovakia. The mutagenicity and carcinogenicity for man of some industrial chemicals has been already proved (haloethers - BCME, CMME). The analysis of pregnancy outcomes in the year 1972 indicated that CA frequency was higher in district Usti n.L. than in Prague. Simultaneously in parents with the occupational exposure in chemical industry, 20% increase of CA was observed in comparison with unexposed population (Sram et al., 1974). The analysis of CA in Usti n.L. was carried out to see if this

Mechanisms of Environmental Mutagenesis-Carcinogenesis, Edited by
A. Kappas, Plenum Press, New York, 1990

Table 1. Incidence of Congenital Anomalies in the
Period 1972 -1981 in District A

Year	N	CA (%)
1972	1995	3.2
1973	2206	3.3
1974	2630	4.0
1975	2347	5.9
1976	2213	7.2
1977	2118	5.8
1978	2079	7.0
1979	2056	8.2
1980	1647	8.4
1981	1526	7.6

result is only by random or specific for this particular
district. Incidence of CA in the period of 1972-1981 steadily
increased (Table 1). The increase of CA was expected to be
related to the new plant producing epichlorhydrin, increased
pollution from soft-coal power stations and open coast soft
mines, as well as the mutagenicity of drinking water.

This region is highly polluted by sulphur dioxide from
industrial sources due to geographic morphology and frequency
of inversion situations. This landscape is probably one of
the most polluted regions by sulphur dioxide in Europe. It is
believed that sulphur dioxide may affect the DNA repair and
induces chromosome aberrations in cow and ewe oocytes in vitro
(Jagiello et al., 1975). Therefore we put forward a hypothesis that a combined effect of sulphur dioxide and other
factors as nitrogen oxides, heavy metals, mutagens in working
place and drugs during pregnancy may create a specific risk
for this region. This became the reason why we analysed the
pregnancy outcomes and their relationship to various factors
in the three districts of Northern Bohemia with a specific
environmental pollution.

MATERIALS AND METHODS

The pregnancy outcomes were studied in: a) the district
Usti n.L. (district A), characterized by the chlorine chemical
industry, pollution from soft-coal power stations and open-
coast soft-coal mines; b) the district of Teplice (district B)
polluted by the soft-coal power stations and open-coast
soft-coal mines; c) the district of Jablonec n.N. (district
C), called control district, which is partially polluted by
the soft-coal power stations from Poland (Fig. 1).

Information about pregnancy outcomes was obtained from
medical records at the maternity hospitals. Diagnosis of CA
according to the International Classification of Diseases
(WHO, 1977) was verified at the child's age of 1 year. The
study lasted from 1982-1986. Altogether 18 060 pregnancy
outcomes were examined.

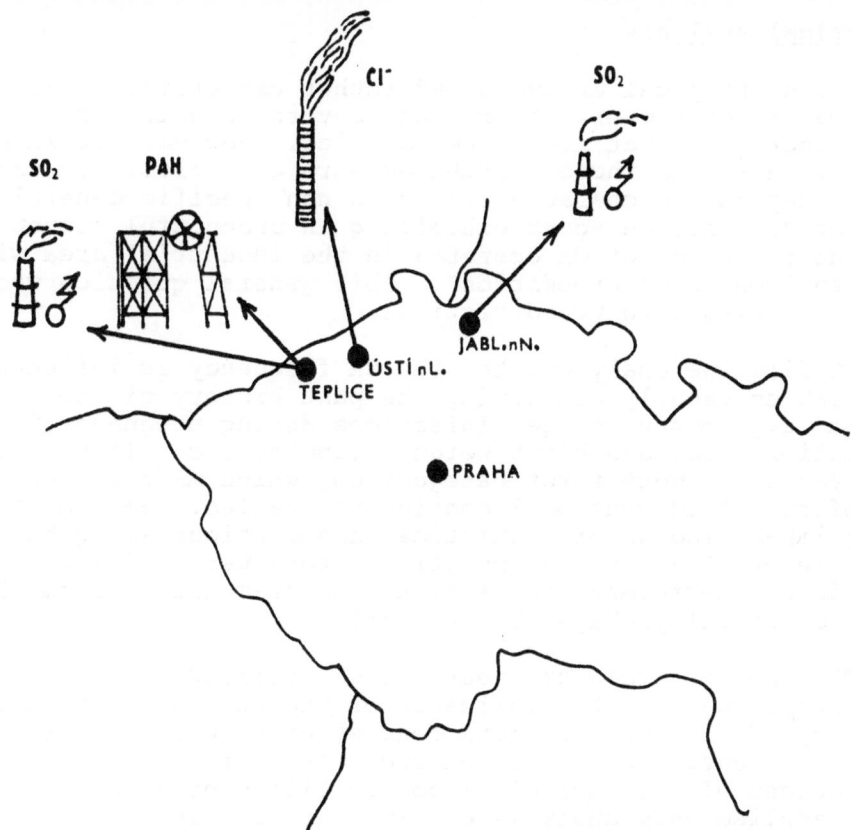

Fig. 1. Air pollution sources in Northern Bohemia.

The following items were registered from medical records:

a) Child: name, date and place of birth, sex, vitality (alive, mature, CA), number of fetuses, week of delivery, weight in g, length in cm; number of children and number of dead children born to this mother, delivery (in time, induced with drugs, surgical), course of delivery, injury during delivery, transfusion; new-born child's (fetus) death, clinical cause of death.

b) Mother: name, age, marital status, education, occupation, menstrual cycle, pregnancy (normal, at risk), number of previous pregnancies, spontaneous abortions, artificial abortions; during this pregnancy: X-ray, drugs, infections, occupational contact with chemicals.

c) Father: name, age, occupational contact with chemicals.

As markers of environmental pollution, the data on sulphur dioxide and nitrogen oxides were obtained from monitoring stations.

Statistical Analysis

The statistical analysis had rather exploratory than confirmatory character. There were several a priori formulated hypotheses that had to be verified. However, it should be noted that even these hypotheses were of preliminary nature only. They can be characterized by a non specific general question "whether an agent exhibiting an undoubtful connection with the frequency of CA operates in the industrial area with heavy environmental exposition". This general question was, in fact, analysed to two directions.

At first we analyzed, how the CA frequency is influenced by variables usually confirming the past history of the proband, e.g. mother's age, infections during pregnancy, drugs consumption, etc. and birth weight, length, etc. If necessary these variables were first categorized, which made it possible to conform multidimensional contingency tables. Several tens of log-linear models of statistical associations among the incidence of CA and the anamnestic factors were examined. The analysis was performed via using the program BMDP4F from the BMDP statistical package (Dixon, 1985).

The second circle of questions was related to the detection of what is the influence of the sulphur dioxide on frequency of CA. The SO_2 data consist of time series of daily average concentrations obtained from monitoring stations. Among dozens of methodological possibilities of data processing we applied only analysis of time series. The first problem that was to be solved consists in the determination of the time interval into which we convert the original data for purposes of further statistical analysis. It concentrated on the problem how the month's average of SO_2 influences the month's frequency of CA. We used the spectral analysis to obtain more interpretable insight into the time dynamics of these two time series.

RESULTS

Total number of pregnancies in the district A was 7644 cases with 9.8% CA, in the district B 7190 cases with 8.2% CA, in the district C 3226 cases with 6.4% CA. The frequency of CA is presented in Table 2.

The highest level of CA was reached in the district A in the year 1982. Both mining districts A and B differ from district C. Similarly, in the same districts, there were born more children with weight lower than 2500 g at delivery (Table 3).

If the frequency of CA was analysed according to groups of anomalies in each district, there were no significant changes between the examined period. Only in the district A, there was a tendency for the decrease of CA of the cardiovascular system (Table 4).

When the spectrum CA in each district was compared with the data from USA and Hungary, in the districts A and B may be observed the tendency for the higher frequency of CA of

Table 2. Frequency of Congenital Anomalies (%) of Total Births

District	Year				
	1982	1983	1984	1985	1986
A	11.1	9.7	9.9	9.1	8.7
B	8.5	7.9	7.8	7.8	8.7
C	6.7	6.0	6.5	-	-

Table 3. The Frequency of Children with Weight Lower than 2500 g at Delivery

District	Year				
	1982	1983	1984	1985	1986
A	8.1	8.4	7.7	7.5	8.7
B	8.3	8.3	9.2	7.9	6.5
C	5.5	6.5	4.3	-	-

Table 4. The Spectrum of Congenital Anomalies in District A. (% among total births)

Anomaly of	1982	1983	1984	1985	1986
Brain and back bone	0.4	0.4	0.8	0.4	0.1
Eye	0.1	0.3	0.8	0.3	0.3
Ear, face and neck	0.6	0.9	0.5	0.2	0.6
Cardiovascular	1.8	1.0	1.3	0.9	0.8
Respiratory	0.2	0.2	0.1	0.2	0
Digestive	0.1	0.3	0.3	0.1	0.1
Urinary and genital	1.9	1.4	2.0	1.9	1.5
Skeletal	4.3	4.2	3.6	4.0	3.6
Integument	1.9	2.2	1.6	1.7	2.2
Others	1.2	0.7	0.7	0.4	0.8

cardiovascular system, urinary system and genital organs and of integument (Table 5).

According to the anamnestic characteristics of mothers, the districts A and B differ from the district C in the ratio of single mothers, risk pregnancies, artificial abortions, whereas A from B and C by the consumption of drugs during pregnancy; infections were more frequent in the district B (Table 6).

Table 5. The Spectrum of Congenital Anomalies in Northern
          Bohemia, USA (Myrianthopoulos and Chung, 1974) and
          Hungary (Czeizel and Sankaranarayanan, 1984)
          (% among total births)

Anomaly of	District A	B	C	United States	Hungary
Brain and back bone	0.4	0.4	0.3	0.5	0.3
Eye	0.3	0.3	0.2	0.24	0.03
Ear, face and neck	0.5	0.5	0.5	0.4	0.2
Cardiovascular	1.2	1.0	0.6	0.86	0.8
Respiratory	0.1	0.1	0.2	0.14	0.03
Digestive	0.2	0.3	0.2	0.62	0.28
Urinary and genital	1.8	1.9	1.1	1.15	0.94
Skeletal	4.0	2.3	3.1	4.38	3.14
Integument	1.9	1.8	0.7	1.0	0.08
Others	0.8	0.5	0.5	0.13	0.21

Table 6. Anamnestic Characteristics of Mothers

Characteristics	A	District B (%)	C
Single	14.1	14.0	10.7
Pregnancy at risk	45.6	45.0	30.1
Abortions			
Spontaneous	13.2	13.7	13.5
Artificial	20.1	20.5	15.2
During this pregnancy			
Drugs	34.2	46.3	44.0
Infections	18.5	21.5	16.6

From factors affecting CA in all districts significant
p < 0.05) were the following characteristics: sex, maternity,
week of delivery, weight, delivery in time, drugs during
pregnancy. CA were affected in the districts A and B also by:
course of delivery (surgical and with complications), death of
previous child, vitality and perinatal mortality. Surprisingly, infections did not increase CA in all districts, except
trichomoniasis in the district A. From drugs, only the effect
of analgesics and psychotropics was analysed for the time
being. Significant increase of CA was related to the therapy
of chlorpromazine in districts A and B, diazepam and all
psychotropic drugs (if all of them were put together) in
district B. No effect was observed in district C.

Effect of sulphur dioxide to the CA, if related to the
month of conception, was compared in the district A (Table 7).

Special attention was focused on the occupational exposure to mutagens. Several high risk groups were observed. Due to the small samples, these results are not statistically significant, but may be understood as the indication of a possible genotoxicity. If the occupational exposure of mothers (Table 8) is analysed in smaller groups, differences between chemical plants in districts A and B were observed (the increase of CA related to non-exposed group was +39% and +9%, N = 94, N = 102). Painters on glass may also be evaluated as a high risk group, (+85%, N = 49) as well as workers in ceramic industry (+93%, N = 47).

Several high risk groups were observed also with regard to the occupational exposure of fathers (Table 9). Significant increase of CA was observed among printers (p < 0.001) and there was indication of a possible genotoxicity among exposed groups in soft-coal gas production, chemical and machine industry and mines. If this aspect is analysed according to each district, an increase of CA was observed, compared to non-exposed groups, in glass workers (district B +31%, N = 610; district C +32%, N = 135) and in district B, in miners (+20%, N = 914), machine industry (+29%, N = 388) and chemical industry (+12%, N = 77).

DISCUSSION

The frequencies of CA in all three districts seem to be stable in the period 1982-86, with a higher level in mining districts. A particularly high level frequency of CA was observed in the district A in the year 1982. That year was a very specific one from the point of view of meteorology, as during winter period there were several weeks of continuous smog situation (concentrations of pollutants were above the maximum allowable concentrations during more than six weeks in January and February). This is the reason why a possible effect of SO_2 as a marker of the total air pollution was studied (Table 7).

There were still two other factors affecting CA in the year 1982. The harvest period in 1981 was under heavy rains. Also grain accepted to stores was very wet. Under these conditions, the storage system with a stable temperature stimulated the growth of fungi producing mycotoxins. These grains were proved to be mutagenic in Ames test and teratogenic using chicken embryo test. At that time, the district A used as one source for making drinking water also the water from Labe river with a high content of organic compounds. Due to the chlorination, such samples of drinking water were mutagenic in Ames test (Cerna et al., 1987). This source of water was replaced later on the grounds of these results.

Sulphur dioxide was used as a marker of air pollution. Its concentration was related to the number of newborn children in each month and to the frequency of CA in the district B. The spectral analysis confirmed that there is a strong circa-annual periodicity in the SO_2 data (Fig. 2). This periodicity can be modelled by a simple sinusoidal wave, i.e. by the function

A sin (f.t) + B cos (f.t)

Table 7. Effect of SO_2 to CA, Related to the Month of
          Conception in District A (Year 1982)

Month	SO_2 ($\mu g/m^3$)	CA (%)
1	492	13.7
2	344	12.3
3	168	14.6
4	114	12.2
5	78	13.1
6	48	9.2
7	56	9.9
8	69	5.7
9	99	5.8
10	136	8.0
11	165	11.4
12	224	13.6

Table 8. Congenital Anomalies and the Occupational Exposure
          Mothers (Δ Related to Non-exposed Group)

Group	N	CA(%)	Δ (%)
Non-exposed	13 069	8.8	–
Printers	16	25.0	+184
Hair-dressing	61	11.5	+31
Gas production	63	11.1	+26
Agriculture	353	11.0	+25
Chemical industry	279	10.4	+18
Hospitals	504	9.3	+11

Table 9. Congenital Anomalies and the Occupational Exposure
          Fathers (Δ Related to Non-exposed Group)

Group	N	CA(%)	Δ (%)
Non-exposed	12 111	8.4	–
Printers	18	33.3***	+296
Gas production	91	14.3	+70
Chemical industry	349	11.7	+39
Machine industry	504	10.3	+23
Miners	1 111	9.5	+13
Medical doctors	190	7.4	-12

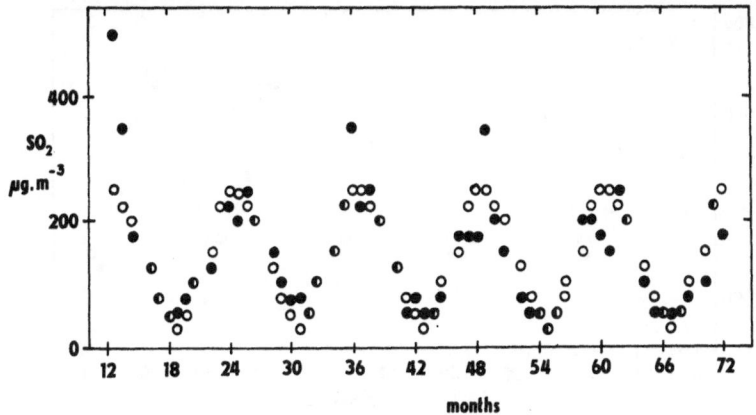

Fig. 2.   SO$_2$ circa-annual periodicity. (●) observed
data; (o) theoretical wave.

where A and B are unknown amplitudes, parameter f is adjusted
to obtain the annual period.   The amplitudes A and B were
fitted by a simple linear regression using the program BMDP1R
from the BMDP package.   This annual component explains 74% of
the variance of SO$_2$.

We tried to detect whether the SO$_2$ annual component
induces a driving effect in the month's frequency of CA.   For
this purpose we used the coherence function indicating some
but not too much convincing connection between SO$_2$ an CA.   We
also tried to explain the dynamics of CA in terms of models
with intervention variables consisting of the instant values
of SO$_2$ and their time lagged twins.   In these models the CA
rate was adjusted to the date of conception.   None of these
models explained more than 10% of variance of CA.   This is
probably due to the fact that CA do not exhibit any circa-
annual periodicity.

We may conclude that the population is exposed to such
concentrations of SO$_2$, which are harmful to embryos in their
early stage.   For this reason we turned our attention to the
analysis of conceptions of children born in the district B.
The birth frequency in Bohemia has a massive annual periodicity explaining more than 63% of its variance (Fig. 3). The
corresponding data from the district B have not such a strong
annual periodicity.   It explains only 9% of the variance of
the month's birth dynamic (Fig. 4).

This prompted us to focus attention on the phase relationship between SO$_2$ and frequency of conceptions in the
district B.   No doubt that whenever SO$_2$ is above 200 µg/m^3
then the conception frequency wave-curve is in its declining
part (Fig. 5).   Simultaneously, the annual components of the
birth frequency in the district B attains maximum a month
before the maximum in Bohemia.   This finding could contribute
to the explanation why the annual periodicity in district B is
more than 7 times smaller than the amplitude of the annual
periodicity in the whole Bohemia.

We may speculate, that the increase of SO$_2$ suppressed the

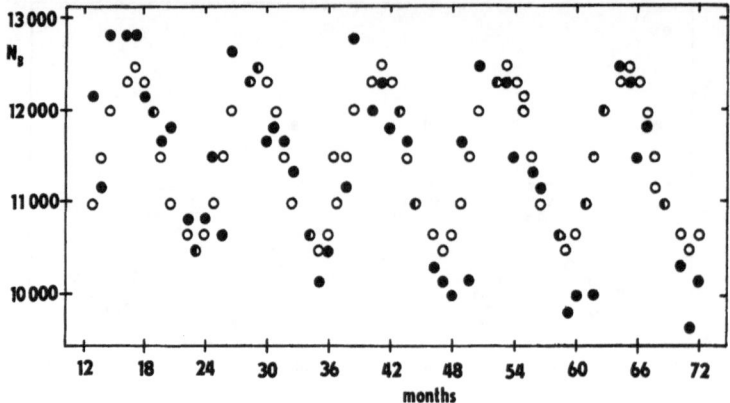

Fig. 3. Birth frequency in Bohemia. (●) observed data;
(o) theoretical wave; (N_B, number of children
born per month).

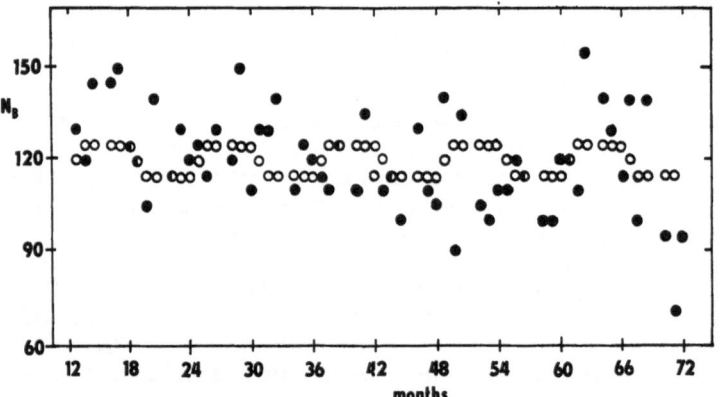

Fig. 4. Birth frequency in Teplice. (●) observed data;
(o) theoretical wave; (N_B, number of children
born per month).

number of successfully fertilized oocytes. We may hypothes-
ize, that other mutagens inducing mutations in gamets were
potentiated by SO_2 due to the suppression of DNA repair
mechanism. Other reasons may be the induction of chromosome
aberrations by SO_2 and other pollutants. Oocytes carrying
such changes are not implanted or they are aborted during very
early stage of pregnancy. We tried to relate these results to
the number of days when the level of SO_2 was higher than
maximum allowable concentration (MAC) that amounts to 150
$\mu g/m^3$ during 24 h in Czechoslovakia. If the level of SO_2 was
2 times higher during 10 days in month than MAC, a tendency to
the increase of CA was observed.

It should be stressed that observed changes were not only
the results of SO_2 exposure. They were combined with NOx, but
its annual periodic component was not so intensive as that of
SO_2. The soft coal from North Bohemia also contains heavy

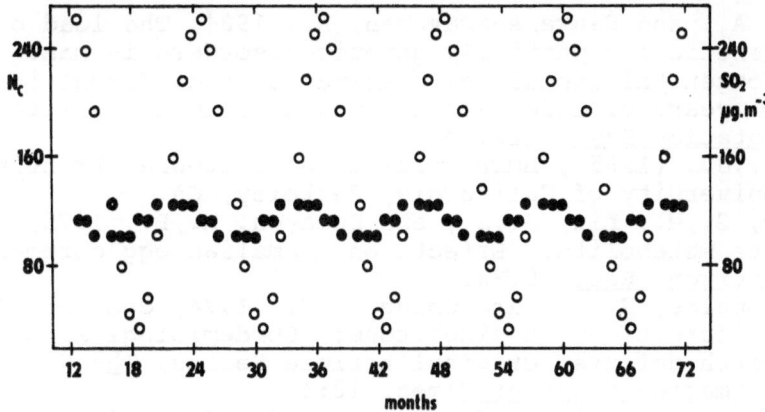

Fig. 5. Annual periodicity of conceptions (●) and SO₂
(o) at Teplice (Nc - number of conceptions per
month).

metals, which may suppress DNA repair, too. Other surprising
results were the increase of CA due to the occupational
exposure of mothers and fathers. Especially, the effect of
father's exposure indicates that factors acting on gametes in
polluted regions may be very important. Occupational exposure
in gas production and mines is mostly the result of exposure
to PAH, whereas in machine industry it may be the result of
emulsions used for machine tools containing PCB and mineral
oils.

     The results of occupational exposure as well as the
effect of air pollution indicate that this region may be quite
specific in all Europe. Injury observed in the districts A
and B may be the results of combined action of several
factors. If these factors would act individually, there would
be perhaps no harm done to developing fetuses. Our results
indicate the significance of the environmental exposure to
high concentrations of SO₂ and parents occupational exposure.
The high level of CA seems to postulate the presence of
several mutagens in the environment of Northern Bohemia. These
results should stimulate the study of these specific conditions more thoroughly.

REFERENCES

Bochkov, N.P., and Sram, R.J., 1989, Modern trends and
     realities of genetic monitoring, Biol. Znet. bl.,
     108:341.
Cerna, M., Hajek, V., Stejskalova, E., Dobias, L., and Zudova,
     Z., 1987, Chemical contaminants in water and their
     possible late effects. II Practical experience with
     examination of the mutagenic activity of drinking
     water, (in Czech), Cs. Hyg. 32:355.
Crow, J.,F., 1971, Human population monitoring, in: "Chemical
     Mutagens. Principles and Methods for their Detection",
     Vol. 2, A. Hollaender, ed., Plenum Press, New York.

Czeizel, A., and Sankaranarayaman, K., 1984, The load of genetic and partially genetic disorders in man. I. Congenital anomalies: estimates of detriment in terms of years of life last and years of impaired life, Mutation Res., 128:73.

Dixon, W.,J., (1985), BMDP Statistical Software, Printing, University of California, Berkeley, CA.

Jagiello, G.,M., Liu, J.,S., and Ducayeu, M.,B., 1975, SO_2 and its metabolite: Effects on mammalian egg chromosomes, Environ. Res., 9:84.

Myrianthopulos, N..G., and Chung, C.S., 1974, Congenital malformations in singletons: Epidemiological survey, birth defects, original Article Series, The National Formation March of Dimes, 10:1.

Sram, R.,J., Kopacilova, G., Bartova, J., Roth, Z., and Seemanova, E., 1974, Delivery records analysis as an approach to determine possible effect of environmental pollution to the pollution genetic load (in Czech), Cs. Hyg., 19:426.

World Health Organization, 1977, International Classification of Diseases. Manual of the International Statistical Classification of Diseases, Injuries and Causes of Death, Vol. 1 and 2, WHO, Geneva.

EUROPEAN COMMUNITY RESEARCH ON GENETIC EFFECTS OF

ENVIRONMENTAL CHEMICALS AND ON BIOMONITORING OF HUMAN EXPOSURE

A.I. Sors[1] and E. Marafante[2]

[1]Environment Research Programme
Commission of the European Communities
Rue de la Loi 200, 1049-Brussels, Belgium

[2]Environment Institute
Joint Research Centre, Ispra Establishment
Ispra (Varese), Italy

INTRODUCTION

This paper* is divided into two distinct but interrelated parts. The first provides an overview of environmental research by the European Community concerning the genetic effects of environmental chemicals. The second part focusses on one particular coordinated project in this field; biomonitoring of human populations exposed to genotoxic environmental chemicals. As well as being one of the focal points of this EEMS meeting, this coordinated project is the latest addition to the EC R & D Programme and is a good example of collaborative European research.

EC ENVIRONMENTAL POLICY

The European Community was among the first to respond to the 1972 Stockholm Conference on the Human Environment. The First EC Action Programme on Environment commenced in 1973 and spelled out in some detail a number of specific actions to be taken. While these were mainly responding to acute, environmental problems, the current (4th) Action Programme is more concerned with long term, broad scale environmental problems and with the adoption of a preventive approach to environment policy.

With the signing of the Single European Act in 1987, the Community was given significant new responsibilities for environmental protection. It is probably fair to say that environmental policy formulation is now predominantly carried out at Community level.

---

*The views expressed in this paper are those of the author's and are not necessarily those of the Commission of the European Communities.

Mechanisms of Environmental Mutagenesis-Carcinogenesis, Edited by
A. Kappas, Plenum Press, New York, 1990

The Single European Act stipulates that: "in preparing its action relating to the environment, the Community shall take account of available scientific and technical data".

This is the basis for EC environmental R & D. The objective are as follows:

- to provide scientific and technical data which support the Community Environment Policy

- to address longer-term environmental problems thus preparing the way for the development of preventive and anticipatory policies

- to serve as an instrument for enchancing further, at Community level, the coordination of research activities in the environmental field.

The current R & D Programme, called STEP (Science and Technology for Environmental Protection) is for the period 1989-1992 and contains 9 research areas, shown in Table 1.

Of these, Area 1, and in particular Area 2, contain work relating to genotoxic effects of environmental chemicals.

Research Area 1: Environment and Human Health

The overall aim of this Research Area is to provide a scientific basis for the continuing development of preventive environmental health policies.

In the past, work in this area focussed mainly on individual, high priority pollutants such as lead, cadmium, asbestos, etc. However, with EC regulations on these substances now in force, emphasis has shifted towards new problems based upon a preventive research approach to policy and supporting research.

The rationale for the approach adopted in this Research Area is as follows:

Table 1. Research Areas of the STEP Program

---

1. Environment and human health
2. Assessment of risks associated with chemicals
3. Atmospheric processes and air quality
4. Water quality
5. Soil and groundwater protection
6. Ecosystem research
7. Protection and conservation of the european cultural heritage
8. Technologies for environmental protection
9. Major technological hazards and fire safety

---

i) For the general population within the EC, exposure to single pollutants are unlikely to be of major health significance

ii) However, combination(s) of environmental factors are likely to cause some health impairment

iii) Therefore, environmental health protection, and supporting research, should be directed towards systems which can identify exposure and early effects in the target organs and/or indicate, at an early stage, possible health impairment of exposed population(s)

The Research themes within this Area are shown in Table 2.

Research Area 2: Assessment of Risks Associated with Chemicals

Research Area 1 deals primarily with pollutants present in the environment to which exposure has already occurred. However, clearly the best preventive approach is to assess risk prior to the release of the substance into the environment (for new chemicals) or, at least, before they cause discernible effects (existing chemicals).

The control of toxic chemicals is of high priority in EC environmental policy. Among the most important policy initiatives at the present time are:

- Further development of Directive 79/831/EEC (6th Amendment) for classification and labelling of new chemicals

- Collection of information and evaluation of risks of existing chemicals

- Reduction of, and alternatives to, the use of laboratory animals in chemicals testing.

Table 2. Research Themes of Research Area 1

---

1. Biomonitoring of human populations exposed to genotoxic environmental chemicals

2. Early indicators of adverse health effects from exposure to environmental pollutants, nephrotoxic effects, neurotoxic effects, immunotoxic effects

3. Development of environmental epidemiology within the European Community (by coordination only)

4. Indoor air quality and its impact on men (includes air pollution epidemiology) (Implemented by COST 613)

This is the regulatory background to the current EC R & D on environmental chemicals; in particular assessment of their genetic effects.

Collaborative EC research in this area started about 15 years ago. This has now become a significant coordinated effort, with over 20 laboratories in almost all EC Member States. The new (STEP) Programme will also be open to participation by Finland, Norway and Sweden.

The overall R & D content of Research Area 2 are shown in Table 3.

At the present time the work on Genetic effects of chemicals contains three closely defined projects. The objectives, present and future content and current participation are indicated in Tables 4, 5 and 6.

There are a number of EC initiatives to find alternatives to the use of animals in toxicity testing. One of these is within the current EC Environment R & D Programme; although the project is of modest size (see Table 7), it is intended to increase activities within this topic in STEP.

The four coordinated projects mentioned above plus biomonitoring, as well as being directed towards immediate environmental policy considerations, are generating important data and understanding on the mechanism of environmental mutagenesis (Fig. 1).

The overall scientific output of this Programme to date was examined by a bibliographic survey, which was commissioned by an external Evaluation Panel. This survey found that in all areas covered by the projects on genetic effects of environmental chemicals, the number of published papers in quality peer-review Journals, as a proportion of total European and World outputs, was far in excess of its relative size in terms of financial and manpower resources.

Table 3. Research Themes of Research Area 2

1. Development and validation of protocols for the assessment of health risks, with emphasis on genetic effects of environmental chemicals

2. Alternatives to the use of animals in test protocols

3. Assessment procedures for the abiotic degradation of chemicals

4. Test systems for the assessment of ecological effects of chemicals

5. Refinement and application of quantitative structure/activity relationships (QSAR's)

Table 4. Molecular Dosimetry of Alkylating Agents

- To study use of DNA adduct formation as a parameter
  (exposure indication for comparative studies)

- Correlation of DNA damage at molecular level and
  biological endpoints

- Current: series of ethylating agents used as model
  compounds

- Future: - extend to methylating agents
          - apply techniques in biomonitoring
          - extend to "environmental"
            genotoxins(?)

- Participation: 10 laboratories in 7 EC countries

Table 5. Tests for Genomic Mutations (Aneuploidy)

- To develop and validate test systems for detecting
  chemicals capable of causing aneuploidy; to study
  models of action

  Current: A number of test chemicals (up to 16)
  are used in various test systems both in vivo and
  in vitro

  Future: - further study of mechanisms of action
          - development of convenient in vitro
            assay?

  Participation: 8 laboratories in 5 Member States

Table 6. Developing Tests for Non-Genotoxic
          Carcinogens

- To develop and validate methods (short-term
  assays) for identification of non-genotoxic
  carcinogens

  Current: - various in-vivo and in-vitro models
             used to cover range of effects
           - focus on cell-transformation assays

  Future: - validation of protocols
            for cell transformation
          - elucidate mechanisms of action

  Participation: 8 laboratories in 5 Member States

Table 7. Alternatives to the Use of Vertebrate Animals in Toxicity Testing

- To develop and validate metabolic activation, systems in mutagenicity testing, not derived from animals

<u>Current</u>: - use of human red cells, plant extracts, cell lines
- common use of reference compounds
- end points studied in bacteria and in mammalian cells in culture

<u>Future</u>: - continue validation using various cell strains
- continue enzymatic characterization
- develop suitable (long-term) cell cultures from various tissues for toxicity testing (e.g., neurotoxicity tests with nerve fibres)

<u>Participation</u>: 5 laboratories in 5 EC Member States

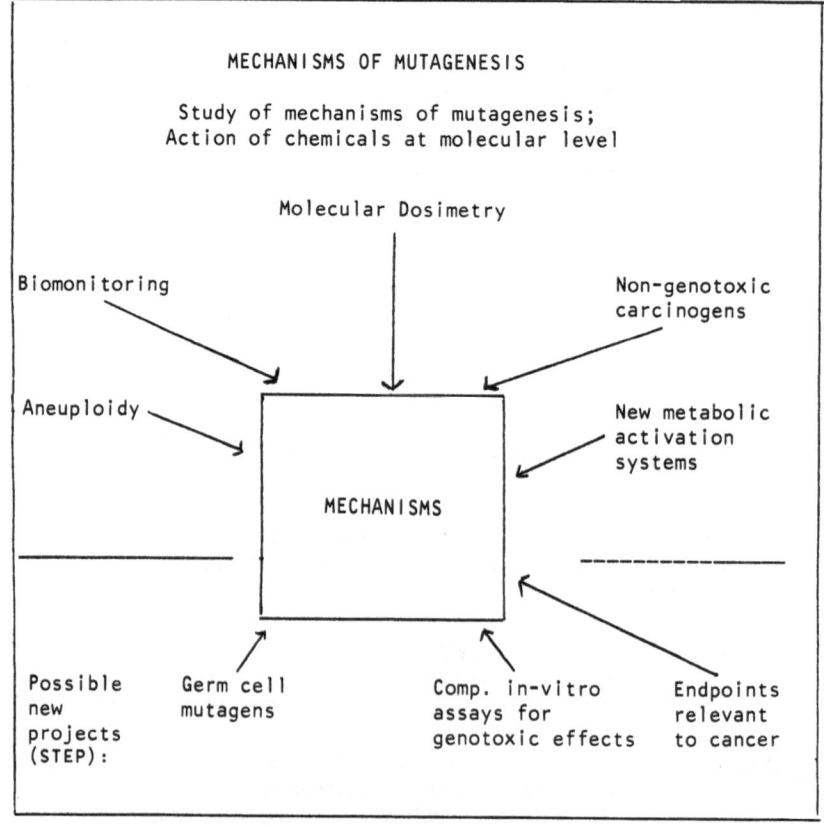

Fig. 1. Study of Mechanisms of Mutagenesis

Part of the reason for this high scientific quality is the balance between scientific interest and policy relevance. The former ensures participation by the leading institutes within the Community, while the latter assures the Commission and its advisory committees that the resources provided are efficiently utilized.

COORDINATED PROJECT: BIOMONITORING OF POPULATIONS EXPOSED TO GENOTOXIC CHEMICALS

In 1988 the CEC launched a research project on the Biomonitoring of Human Populations Exposed to Genotoxic Environmental Chemicals. This project was intended to assist the development of population monitoring systems designed to quantify the exposure to potential mutagenic chemicals in the environment and to detect possible early effects.

The rationale is that during the last decade great progress has been made in methodologies for assessing exposure to and biological effects of genotoxic chemicals. New methods for the identification of DNA damage _in vivo_ offer the possibility to quantify individual exposure. When related to selected genetic end-points, such as the indication of micronuclei formation, sister chromatid exchange, chromosomal aberrations and point mutations, these methods could provide consistent indications on the exposure to genotoxic chemicals and on their biological effects.

In the present phase, the Biomonitoring Project consists of the development, comparison and intercalibration of advanced techniques for the quantification of molecular (target) dose and its comparison to external exposure. The approach is based on the formation of DNA and protein adducts, in relation to selected genetic end-points.

Several laboratories cooperate closely on various aspects of methodological development, for (i) the identification and measurement of DNA adducts by 32P-Postlabelling method, immunological assays and advanced Mass Spectrometry technique; (ii) the detection of hemoglobin and plasma protein adducts by Gas Chromatography-Mass Spectrometry; (iii) the detection of chromosomal aberrations, SCE and micronuclei in peripheral lymphocytes; (iv) the quantitative determination of HPRT-mutant frequency in lymphocytes and mutations in hemoglobin gene. Participating laboratories in the project are listed in Table 8.

In Table 9 the methods presently considered in relation to their specificity in exposure studies are reviewed. The detection of microscopically visible damage in the genetic material of human cells (cytogenetic assays), as well as the detection of genetic mutations in somatic cells at the level of the individual, require a careful and detailed evaluation of age, sex, smoking and medical history and all other life-style factors which may influence the results. On the ohter hand, the estimation of direct DNA damatge by different methods (chemical, biochemical and immunological), based on the characterization and quantification of DNA adducts formed by reactive chemicals or metabolites in accessible targets (i.e., DNA in lymphocytes) may provide direct evidence on the identity and quality of the exposure (Molecular Dosimetry).

Table 8. Participating Laboratories

Danish Cancer Society	DK
University of Patras	GR
MRC, Carshalton and Sussex University	UK
University of York	UK
Universidad Autonoma de Barcelona	E
Vrije Universiteit Brussel	B
State University of Leiden	NL
University of Wurzburg	D
University of Essen	D
University College of Swansea	UK
Institute of Occupational Health	SF

Table 9. Markers for Genotoxic Chemicals

<u>Cytogenetic Assays</u>: Scoring of chromosomes (SCE, CA, MN)
Relatively low sensitivity, aspecific for inducing agents

<u>Genetic assays</u>: Gene expression (HPRT), Hemoglobin mutations
Low sensitivity, aspecific for inducing agents

<u>Immunoassay</u>: Antibodies for DNA-adducts, (USERIA, ELISA)
High sensitivity, very specific (single adduct)

<u>Biochemical Methods</u>: DNA-adducts by 32-P-post-labelling
High sensitivity (specific) semi-quantitative

<u>Chemical Methods</u>: DNA/protein adducts by mass spectroscopy (GC-MS; µHPLC-MS/MS)
High specificity (high sensitivity) reference methods

Well defined population cohorts and well characterized chemicals to which the exposure occurs have been identified in order to obtain a correct assessment of external and internal exposure (Table 10).

Blood and urine are used as accessible biological materials. Individuals representative of different situations were selected, according to the following criteria:

i) Positive controls for calibration purposes: cancer patients exposed to chemotherapeutic agents

Table 10. Populations - Chemicals

Positive Controls for Calibration Purpose

Chemotherapeutics: Patients under chemiotherapy
(melphelan, cis-platinum, cyclophosphamide,
bleomycine)

External/Internal Exposure to Active Metabolites

Occupational exposure: Workers exposed to styrene

Environmental/Occupational Exposure

Low to high level risk assessment: ethylene oxide
rural/urban populations, workers

Complementary

Hydrazine (occupational exposure)
aromatic amines (occupational exposure)
1,3-Butadiene (occupational/environmental)

Table 11. Full Content of the Coordinated Project

Chemicals	Population and Targets	Methods
Arom. Amines	Workers in chemical industry	Cytogenetic: - Micronuclei - SCE
Hydrazine	Occupational exposure	- Chromosomal aberrations
Styrene	Occupational exposure	
Chemioth. agents - Melphaan	Cancer patients - Blood	Point mutations: - HB (PCR) - HB mutations
- Bleomycin - Cyclophosphamide	- Urine	Protein adducts - GC-MS - HPLC
1,3-Butadiene	Occupational exposure	DNA adducts - 32P-post labelling
Ethylene oxide	Urban/rural population	- Immunoassays - Mass spectro-metry

ii) Workers exposed to occupational chemicals for which a good external/internal exposure relationship could be envisaged

iii) Rural versus urban populations for which individual agents of exposure could be defined

Table 11 shows the overall content of the project. Samples from individuals are analysed by the participating institutes for each of the assays reported. Although not indicated in this Figure, samples from appropriate reference groups are also considered. Most efforts in the present phase are devoted to the methodological development, including intercalibration activities between the participating laboratories. The latter relate mainly to blood samples spiked in vitro with the chemicals studied as well as on in vitro adducted DNA.

It is expected that the integrated approach followed in the present project will give consistent indications for the establishment of methodologies for biological monitoring systems. However, a clear understanding of future possibilities for a specific methodology requires also a reliable evaluation on the development of related quality assurance programmes, as well as the availability of appropriate reference materials. Efforts in these directions have to be undertaken before extending similar studies to wider ranges of population and exposure situations.

CONTRIBUTORS

Veena Afzal, Laboratory of Radiobiology and Environmental
    Health, University of California, San Francisco,
    CA 94143-0750 USA

Vladimir Albrecht, Psychiatric Research Institute,
    182 03 Prague, Czechoslovakia

Raghbir S. Athwal, Department of Microbiology and Molecular
    Genetics, UMDNJ-New Jersey Medical School, 185 South Orange
    Avenue, Newark, NJ 07103, USA

Romualdo Benigni, Laboratory of Toxicology and Ecotoxicology,
    Istituto Superiore di Sanita, Rome, Italy

Alena Berankova, Psychiatric Research Institute, 182 03
    Prague, Czechoslovakia

Nikolai P. Bochkov, Department of Genetic Monitoring, Research
    Center of Medical Genetics, Moskvorechie 1, 115478 Moscow,
    USSR

Bryn A. Bridges, MRC Cell Mutation Unit, University of Sussex,
    Falmer, Brighton BN1 9RR, UK

Ivan Chouroulinkov, Institut de Recherches Scientifiques sur
    le Cancer, Centre National de la Recherche Scientifique,
    B.B. No. 8, 94802, Villejuif Cédex, France.

John W. Drake, Laboratory of Genetics, National Institute of
    Environmental Health Sciences, Research Triangle Park,
    NC 27709, USA

Vincent Ferrier, Centre de Biologie du Développment, Univer-
    sité Paul-Sabatier, 118 Route de Narbonne, Toulouse Cédex,
    France

Judith H. Ford, Genetics Department, The Queen Elisabeth
    Hospital, Woodville, 5011 South Australia, Australia

Laury Gauthier, Centre de Biologie du Développment, Université
    Paul Sabatier, 118 Route de Narbonne, Toulouse Cédex,
    France

Ramadevi Gudi, Department of Microbiology and Molecular
    Genetics, UMDNJ-New Jersey Medical School, 185 South Orange
    Avenue, Newark, NJ 07103, USA

Erich Hecker, Institute of Biochemistry, German Cancer
    Research Center, Im Neuenheimer Feld 280,
    D-6900 Heidelberg 1, FRG

Etta Käfer, McGill University, Biology Department, Montréal,
    Canada H3A 1B1

Andreas Kappas, Institute of Biology, National Research Center
    "Democritus", 153 10 Athens, Greece

Ryuichi Kato, Department of Pharmacology, School of Medicine,
    Keio University, Shinjuku-ku, Tokyo, Japan 160

Gursurinder P. Kaur, Department of Microbiology and Molecular
    Genetics, UMDNJ - New Jersey Medical School, 185 South
    Orange Avenue, Newark, NJ 07103, USA

Claude Lasne, Institut de Recherches Scientifiques sur le
    Cancer, Centre National de la Recherche Scientifique, B.P.
    No. 8, 94802, Villejuif Cédex, France.

Eva Machovska, District Hygiene Station, 46600 Jablonec n.N.,
    Czechoslovakia

Erminio Marafante, Environment Institute, Joint Research
    Center, Ispra Establishment, Ispra (Varese) Italy

A.T. Natarajan, Department of Radiation Genetics and Chemical
    Mutagenesis, Sylvius Laboratory, Medical Faculty, State
    University of Leiden, Wassenaarseweg 72, AL 2333 Leiden,
    The Netherlands

Torsten Nilsson-Tillgren, Institute of Genetics, University of
    Copenhagen, Copenhagen, Denmark

Gregorio Olivieri, Dipartimento di Genetica e Biologia
    Molecolare, Università degli Studi di Roma, Rome, Italy

Luz Orfila, Institut de Recherches Scientifiques sur le
    Cancer, Centre National de la Recherche Scientifique, B.P.
    No. 8, 94802, Villejuif Cédex, France.

Neidhard Paweletz, Institute of Cell and Tumour Biology, German
    Cancer Research Center, Im Neuenheimer Feld 280, D-6900
    Heidelberg 1, FRG

Claes Ramel, Institute of Genetic and Cellular Toxicology,
    Wallenberglabolatory, University of Stockholm, S-106 91
    Stockholm, Sweden

Michael A. Resnick, Yeast Genetics/Molecular Biology Group,
    National Institute of Environmental Health Sciences,
    Research Triangle Park, NC 27709, USA

Friedrick Rippmann, Institute of Biochemistry, German Cancer
    Research Center, Im Neuenheimer Feld 280, D-6900 Heidelberg
    1, FRG

Ivana Roznickova, District Hygiene Station, 415 68 Teplice,
    Czechoslovakia

Shahbeg S. Sandhu, Genetic Toxicology Division, US Environmen-
    tal Protection Agency, Research Triangle Park, NC 27711,
    USA

Barbara Sedgwick, Imperial Cancer Research Fund, Clare Hall
    Laboratories, South Mimms, Potters Bar, Herts EN6 3LD, UK

Constantine E. Sekeris, Institute of Biological Research and
    Biotechnology, National Hellenic Research Foundation,
    116 35 Athens, Greece

Frank Solomon, Department of Biology and Center for Cancer
    Research, Massachusetts Institute of Technology, Cambridge,
    MA 02139, USA

Andrew I. Sors, Environment Research Programme, Commission of
    the European Communities, Rue de la Loi 200, 1049 Brussels,
    Belgium

Radim J. Sram, Psychiatric Research Institute, 182 03 Prague,
    Czechoslovakia

A.D. Tates, Department of Radiation Genetics and Chemical
    Mutagenesis, Sylvius Laboratory, Medical Faculty, Univer-
    sity of Leiden, Wassenaarseweg 72, AL 2333 Leiden,
    The Netherlands

Baldev K. Vig, Department of Biology, University of Nevada,
    Reno, NV 89557, USA

Brant Weinstein, Department of Biology and Center for Cancer
    Research, Massachusetts Institute of Technology, Cambridge,
    MA 02139, USA

Sheldon Wolff, Laboratory of Radiobiology and Environmental
    Health, University of California, San Francisco,
    CA 94143-0750, USA

Yasushi Yamazoe, Department of Pharmacology, School of
    Medicine, Keio University, Shinjuku-ku, Tokyo, Japan 160

Catherine Zoll, Centre de Biologie du Developpment, Université
    Paul-Sabatier, Route de Narbonne 118, Toulouse Cédex,
    France

INDEX